Resistance Is Fertile

Wilhelm Peekhaus

Resistance Is Fertile
Canadian Struggles on the
BioCommons

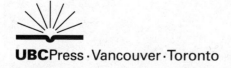

UBCPress · Vancouver · Toronto

21 20 19 18 17 16 15 14 13   5 4 3 2 1

Printed in Canada on FSC-certified ancient-forest-free paper
(100% post-consumer recycled) that is processed chlorine- and acid-free.

Library and Archives Canada Cataloguing in Publication

Peekhaus, Wilhelm C.
    Resistance is fertile : Canadian struggles on the biocommons / Wilhelm Peekhaus.

Includes bibliographical references and index.
Issued also in electronic formats.
ISBN 978-0-7748-2310-4 (bound); ISBN 978-0-7748-2311-1 (pbk.)

    1. Agricultural biotechnology – Canada. 2. Agricultural biotechnology –
Government policy – Canada. 3. Agricultural biotechnology – Economic aspects –
Canada. I. Title.

S494.5.B563P43 2013   630.971   C2012-908135-3

Canadä

UBC Press gratefully acknowledges the financial support for our publishing program of the Government of Canada (through the Canada Book Fund), the Canada Council for the Arts, and the British Columbia Arts Council.

This book has been published with the help of a grant from the Canadian Federation for the Humanities and Social Sciences, through the Awards to Scholarly Publications Program, using funds provided by the Social Sciences and Humanities Research Council of Canada.

UBC Press
The University of British Columbia
2029 West Mall
Vancouver, BC V6T 1Z2
www.ubcpress.ca

Contents

Acknowledgments

I would like to take this opportunity to thank all of the interview respondents who participated in this research, all of whom gave so generously of their time and their insights about the struggles on the terrestrial and knowledge commons in respect of biotechnology in this country. The dedication and passion they exhibit toward these issues were a great source of optimism and inspiration.

Abbreviations

APHIS	Animal and Plant Health Inspection Service
BACC	Biotechnology Assistant Deputy Minister Coordinating Committee
BMCC	Biotechnology Ministerial Co-ordinating Committee
Bt	*Bacillus thuringiensis*
CBAC	Canadian Biotechnology Advisory Committee
CBAN	Canadian Biotechnology Action Network
CBD	United Nations *Convention on Biological Diversity*
CBI	Council for Biotechnology Information
CBS	Canadian Biotechnology Strategy
CFIA	Canadian Food Inspection Agency
CGIAR	Consultative Group on International Agricultural Research
COP	Conference of the Parties to the United Nations Convention on Biological Diversity
DNA	Deoxyribonucleic acid
EFSA	European Food Safety Authority
ETC Group	Action Group on Erosion, Technology, and Concentration
FAO	Food and Agriculture Organization of the United Nations
FBCN	Food Biotechnology Communications Network
FDA	US Food and Drug Administration
GURT	Genetic Use Restriction Technology
HUGO	Human Genome Organization
ISAAA	International Service for the Acquisition of Agri-Biotech Applications
MOSST	Ministry of State for Science and Technology

NBAC	National Biotechnology Advisory Committee
NBS	National Biotechnology Strategy
NFU	National Farmers Union
NGOs	Non-governmental organizations
rBGH	Recombinant Bovine Growth Hormone
TRIPS	*Agreement on Trade Related Aspects of Intellectual Property Rights*
WTO	World Trade Organization

Resistance Is Fertile

Introduction

When it comes to biotechnology, the mainstream media, governments, and self-styled life science companies articulate one uniform message: biotechnology promises to yield a wealth of benefits that will improve health, lifestyles, diet, environment, and economy.[1] It is claimed that this science and its attendant technological applications will help medical researchers identify disease genes and develop prophylactic measures or new treatments for ailments. Similarly, human vaccines produced with bacteria through recombinant deoxyribonucleic acid (DNA) procedures are purported not only to reduce costs, thus theoretically making them available to a wider range of the population, but also to be safer than vaccines manufactured using animal organs (for example, the pancreas of cattle or swine were previously used to produce insulin for the treatment of diabetes). Biotech "pharming" (pharmaceutical farming) will apparently allow scientists to genetically engineer animals to induce their bodies to produce proteins and medicines that offer human therapeutic value. Environmentally focused biotechnologies are billed as the next technological panacea that will help reverse the environmental degradation wrought on the world by the last two centuries of industrial development. New applications in the realm of agricultural biotechnology are supposed to increase crop yields and thus help the world feed itself while simultaneously reducing the volume of harmful chemicals required for agricultural production. These are just some of the benefits that biotechnology proponents in the corporate world and government foretell for this science and its technological applications.

As Robert Bud (1993, 3-4) points out, such rhetoric has helped pave the way for intensive investment in biotechnology: "In the early 1980s when prophesies of genetic engineering were articulated and employed to win resources, the seers foretold a world in which wealth would relate to the ability to manipulate the new science. After numerous reports extolled its importance and power, every major industrial power invested heavily."

Indeed, by the beginning of the current millennium, a number of developed and developing countries had articulated leadership in biotechnology as a national economic goal.[2] Canada is among the growing number of countries around the globe that have embarked on a path of government-sanctioned commodification of biotechnology. In fact, Canadian governments have been actively attempting to position this country as a world leader in bio-technology for almost three decades. According to the International Service for the Acquisition of Agri-biotech Applications, as of 2011, Canada ranked fifth in the world behind the United States, Brazil, Argentina, and India in terms of arable land cultivated with genetically engineered crops – some 10.4 million hectares (James 2011). However, as we will see in Chapter 2, the veracity of the statistics published by this organization is subject to debate.

Canadian government biotechnology policy, which has been informed by a national strategy since 1983, is typically articulated in the following terms:

> Biotechnology is a powerful "enabling technology" with applications in many industrial sectors and holding much promise for the future. It has great potential to add to industrial efficiency, output and jobs, enhance the productivity and competitiveness of Canada's important natural resource sectors, safeguard the environment and enhance our quality of life through improved pharmaceuticals, diagnostic medicine and food production. Many people see biotechnology as the next important "change maker" after the convergence of information, computer and telecommunications technolo-gies, which have transformed our lives. All Canadians – producers and consumers across the country, including people in smaller communities and rural areas – will benefit from the new transformation. (Industry Canada 1998, 2)

This passage, and, indeed, the full policy document from which it derives, sums up succinctly what the proponents of biotechnology contend this sci-ence and its attendant technological developments portend for humanity. Aside from hundreds of millions of dollars in direct federal funding of bio-technology research, such pronouncements are supported concretely through the allocation of additional federal funds to finance biotechnology com-mercialization projects such as: the Industrial Research Assistance Program, which invested $60 million between 1998 and 2006; Technology Partner-ships Canada, which provided $293 million between 2001 and 2006; the Canadian Institutes of Health Research, which disbursed $13.8 million be-tween 2001 and 2006; the Scientific Research and Experimental Development Tax Incentive Program, which furnished over $3 billion in tax assistance to Canadian businesses in 2006; and the Business Development Bank of Canada, which has committed $154 million to life science projects, with plans to increase this amount to $191 million over the fiscal 2006-10 planning period (Canadian Biotechnology Advisory Committee 2006b; Treasury Board of Canada Secretariat 2006).[3]

Specific to agricultural biotechnology, the Government of Canada claims that biotechnology helps "develop better diagnoses and treatments of human, animal and crop diseases, breed new crops that are more stress tolerant, nutritious and higher yielding, and reduce the need for pesticides and fertilizers in food production" (Government of Canada 2004, 16). In an attempt to realize some of these purported benefits, the federal government developed the Canadian Crop Genome Initiative, which was a major project led by Agriculture and Agri-Food Canada, the federal department responsible for agricultural matters, including food and feed crops developed using biotechnology. This project was tasked with developing corn, soybean, canola, and wheat varieties that were resistant to disease, insect attack, cold, and drought and that yielded improved crops in terms of both quantity and quality (ibid.). As of mid-2012, Agriculture and Agri-Food Canada has approved sixty-three genetically engineered seed events for unconfined release into the environment, and Health Canada has approved eighty-four foods for human consumption (although the actual marketed crop varieties that have been genetically engineered in this country are limited to canola, soybean, corn, and sugar beets).[4] Perhaps more indicative of the degree to which genetically engineered organisms have become commonplace in Canada is the estimate by groups such as Greenpeace and the Council of Canadians that 70 percent of processed foods sold in this country contain at least one genetically engineered ingredient. Biotechnology companies have also been attempting to develop virus-resistant transgenic seeds that are designed to protect the resulting crops from particular blights and viruses. Synthetic chemicals and genetic engineering techniques are also being employed to speed up the maturation process of meat animals while simultaneously reducing their required feed levels. These very brief examples only touch the surface of the range of agricultural products and applications that researchers believe are yet to be developed through genetic engineering.[5]

However, despite two decades of research and almost fifteen years of active marketing in a number of countries around the world, genetically engineered crops still fail to deliver consistently higher yields, enhanced stress tolerance, or improved sustainability through a reduced need for tilling and chemical fertilizer and pesticide applications. Moreover, both industry and government continue to downplay the mounting evidence about the safety and environmental risks that are connected to genetically engineered organisms, and public debate in regard to biotechnology has been severely circumscribed, to the point of being almost non-existent. Instead, what passes for information and discussion about this technoscience is often limited to one-dimensional and celebratory accounts.

It is against such a backdrop that this book seeks to develop a critical analysis of agricultural biotechnology in Canada, including the Canadian Biotechnology Strategy (CBS), a policy that poses considerable ramifications for the developmental trajectory of biotechnology in this country. Biotechnology in its Canadian context is a large, profitable domain, and one that proffers basic questions about the organization of our social life. The following examination will define and answer some of these queries by adapting a Marxist theoretical framework and relating it to documentary evidence and interview data. The research reported in this book was driven by the following three overarching questions: how does the federal government's CBS facilitate the commodification of science in general and of biotechnology in particular; what counterstruggles have emerged in Canada that attempt to re-appropriate or foreclose the products and processes developed by biotechnological capital; and what are the information and knowledge issues encompassed by such struggles? Marxism is uniquely suited to this explanatory task because of its conceptual strength in engaging social domination and conflict in a historical context of episodic change in society's relations of production and exchange. However, before elaborating on the appositeness of a Marxist critique of contemporary Canadian biotechnology, I want to rehearse briefly some of the relevant extant literature that establishes both the historical context in which this technoscience has been developed and some of the ways this development has been conceptualized.

Situating and Conceptualizing the Historical Context of Biotechnology Development

Basic biological processes such as fermentation and the use of yeast have been exploited for millennia. Moreover, in the mid-nineteenth century, the Austrian Augustinian monk Gregor Johann Mendel experimented with peas to determine the way in which genetic traits are passed on to a plant's progeny.[6] Similarly, the roots of gene mapping lie in studies from the 1910s that examined the common fruit fly, *Drosophila*, although large-scale mapping would not be technically possible until the advent of powerful computer technologies in the 1980s.[7] It has been more than half a century since James Watson and Francis Crick, in their now famous letter to the journal *Nature* in 1953, described DNA as a molecule composed of two twisting, paired strands that run in the opposite direction of one another, giving rise to its double helix shape.[8] Their work furnished the conceptual advances that fostered the expansion of the biological sciences, as researchers went on to determine that DNA facilitates the synthesis of proteins. Messenger ribonucleic acid was subsequently discovered in 1960. By 1967, scientists had succeeded in isolating the enzyme (DNA ligase) that is responsible for joining DNA chains. In 1970, restriction enzymes (endonuclease) that cut DNA strands at specific sites were isolated. Within three years, scientists were using these newly discovered enzymes to introduce foreign DNA into bacteria by cutting and splicing gene fragments in a process known as "recombinant DNA" since the genetic information contained within the DNA is recombined *in vitro*.

Despite these starting points in biological research, it was not until the late 1980s that biotechnology as we know it today began in earnest. To a large extent, the growth of biotechnology as a science and industrial sector was enabled by the expanding computing capacity of increasingly sophisticated information and communication technologies developed in this same period. Steven Best and Douglas Kellner (2001) employ the term "biocybernetic era" to capture the synthesis between genetic engineering and computers. Thomas Mitchell (2003, 483) speaks of "biocybernetic reproduction." According to Manuel Castells (2000, 63), prominent sociologist and theorist of the "network society," "technological convergence increasingly extends the growing interdependence between the biological and microelectronics revolutions, both materially and methodologically. Thus, decisive advances in biological research, such as the identification of human genes or segments of human DNA, can only proceed because of massive computing power." Although this contention understates some of the fundamental advances made in the fields of biology, cybernetics, and information theory, the importance of advanced computer applications to the massive growth in biotechnological research and application development remains clear.

Seizing on the vital connections between biology and information, some individuals go so far as to assert that biology is being transformed into an information science and that genetic engineering is a type of information technology (Voigt 2008; Yoxen 1981; Zweiger 2001). As one prominent molecular biologist has argued, "the decisive, energizing perception of biotechnology since the Second World War, the key to its strength and vigour, has been one that treats organisms as information-processing machines ... biology has become a kind of flatland in which the only activity is the processing and transmission of genetic information" (Yoxen 1983, 18-19). As a number of observers have elaborated, the computer metaphor has come to be employed widely in discourses on biotechnology, through which life is represented as being reducible to a genetic code that can be read, edited, and copied (Boyle 1996; Castells 1989; de Landa 1991; Levidow and Tait 1995; Rifkin 1998).

The problematic notion of DNA as code actually goes back to Watson and Crick's early work on the structure of DNA as well as to the research conducted by Marshall Nirenberg and Heinrich Matthaei, which determined that messenger ribonucleic acid transcribes genetic information from DNA by directing the assembly of amino acids into complex proteins. Colloquially referred to as "breaking the genetic code," this discovery represented molecular biology's Rosetta Stone (Crick 1962a, 1962b, 1963; Matthaei et al. 1962; Nirenberg 1963). Prominent scientific historian Lily Kay (2000) offers a detailed history of the intersections of molecular biology and cybernetics and information theory, including the rise to prominence of the information metaphor in biology. Critical of the unproblematized manner in which information metaphors were easily imported from information theory into biology, she characterizes her work as a study of the "epistemic rupture from purely material and energetic to an informational view of nature and society" (328). However, as she demonstrates in her well-documented historical account, the importation of informational metaphors into molecular biology was based on a utilitarian calculus; it was not "because they worked in the narrow epistemic sense (they did not) but because they positioned molecular biology within post-war discourse and culture, within the transition to a postmodern information-based society" (609). Kathleen McAfee (2003, 203) is similarly critical of the trope of the "genetic code," claiming that its attendant conception of easily adding, deleting, or otherwise manipulating genetic material belies the empirical evidence and experience of most molecular scientists, who are unable to transfer genetic material between organisms with any exacting degree of precision.

Framing DNA as information is a reductionist discourse that disregards the adeptness of living organisms. In ways still not well understood by science, organisms adjust to additions or deletions of genetic material by substituting alternate pathways for those altered in some way, thus preventing – or otherwise reacting to – the organic change that the genetic engineering was designed to induce. More problematic, at least from the perspective offered in this volume, is that the informational metaphors that have infiltrated biology discourses have quickly become literalized in a way that promotes an informational approach to the study of biotechnology that lends itself to corporate capture (Fox Keller 2000; Haraway 1997; Kay 2000). As a result of the technological advances in biological information processing software and hardware, the metaphor of "life as information" can now be given material reality that lends itself to commodification, particularly through the expanding contemporary intellectual property regime (Bowker 2000; Sunder Rajan 2006). In Chapter 4, we will return to a more in-depth examination of how capital is employing the notion of "gene equals information" to facilitate accumulation.

In their trenchant assessment of the development of biotechnology, Dorothy Nelkin and Susan Lindee (2004, xiii) write that "[i]n the 50 years since the famous Watson and Crick paper, genetics has become an important corporate enterprise, and much of the promotion of genes and DNA reflects this commercial nexus." In the 1980s, a number of commentators who were already critical of the increasing commodification of biotechnology began to identify the inability of regulatory and other scientific assessment mechanisms to accommodate social interests such as ecological impacts, human health effects, ethical considerations, distributive justice, social need, economic productivity, and market demand (Krimsky 1991; Yoxen 1981). Instead, the development of new biotechnological products was, and continues to be, driven largely by market incentives to create new commodities and improve production methods. The result, according to Best and Kellner (2004, 198), is that "all natural reality – from microorganisms and plants to animals and human beings – is subject to genetic reconstruction in a commodified 'Second Genesis.'"

Biotechnology also offers a solution, at least a temporary one, to the finite territory problem that plagues capital. As members of the Critical Art Ensemble (2002, 30) emphasize, "[t]he molecular invasion of the body is the new frontier where untold resources and profits may be appropriated." Nature, according to one geography scholar, "is consequently now undergoing an 'involution' much as space did in the first few years of the twentieth century when planetary expansion was effectively at an end ... when productions of space no longer pushed the borders of the unknown so much as re-worked its internal subdivisions. Faced with the loss of extensive nature, capital re-grouped to plumb an everyday more intensive nature" (Katz 1998, 47). These developments in molecular biology have prompted others to consider the biology-information nexus from a political economic perspective, situating biotechnology firmly within the domain of "information capitalism," in which information assumes a prominent role as both a factor of production across multiple economic sectors and a commodity in its own right (Dyer-Witheford 1999; Heller 2001; Schiller 2007).

Within this broader political economic context, the conceptual redefinition of nature as information, communication, and control supports and reinforces its increasing commercial exploitation. Facilitating this process, chemistry and biology have been applied as instruments to link organic processes to their technological and commercial exploitation (Guattari 1992). Two decades ago, Jack Kloppenburg (1988) left room for the possibility that biotechnological material would increasingly derive its value from its informational content rather than from its material form. Given the ability of emerging technology to transform fundamental genetic and biochemical properties, other writers of the time postulated that the tradition of valuing biotechnological materials as material resources alone would have to accommodate a new assessment that also reflects their informational resource aspects (Doyle 1985; Mooney 1983).

In more recent work, Eugene Thacker (2005) analyzes the impact of broader globalization processes on the development of biotechnological knowledge and practice (focusing predominantly on biomedical and genetic research). The bulk of his work, which seeks to address how ontological questions in regard to biotechnology are insinuated in broader social, economic, and cultural questions, is devoted to an investigation of the interstices between biology and informatics, including the way advances in the latter have facilitated innovation in the former. He is, therefore, also interested in interrogating the often contradictory dichotomies found in biotechnology between the natural and the artificial, the biological and the technological. Building on the work of Castells in regard to the informatization of biotechnology, Thacker examines the way genetic materials are rendered into digital forms that capture the informational content of the embodied artifacts and, in turn, are manipulated and recombined to produce novel biological materiality.[9]

Biotechnology and its accompanying information technologies now render it possible to not only derive biochemical information from organic material but also use such information independently of the original biological sample. That is, these new biotechnologies interact synergistically with the information embodied in the organic specimens to produce additional types of information that become sources of productivity and value. Increasingly, it is precisely the derived biotechnological information rather than the material form that is of interest to researchers. As Bronwyn Parry (2004, 59) reminds us, "[a]lthough this may not seem of any great significance, changing the way in which the information is embodied or presented proves to have profound effects on the dynamics of trade and exchange" as well as on production.

In our current conjuncture, we are witnessing an expansion of capitalist social relations into more fields of social labour and a consequent growing application of market exchange to a mounting range of commodified objects, including information. Martin Kenney (1986, 4) made explicit the political economic implications of the connection between biology and information over two decades ago, at the outset of what some have coined the "biotechnology revolution." He stated that "[b]iotechnology is an information-intensive technology and will very easily fit into a restructured economy based on information. Indeed, biotechnology will provide one of the new economy's crucial underpinnings." However, as Sheldon Krimsky (1991, 17) pointed out at around the same time, "[w]e do not yet know whether the revolution in applied genetics will establish higher standards for civilization as a whole, will respect diverse forms of life and habitats, will liberate us from disease or enslave us to a genetic determinism, whether its achievements will be shared equitably, or whether its significance will be mixed with a favorable outcome to narrow interest groups." Close to two decades later, this current book seeks to pick up on some of these considerations and sketch a picture of how they are being worked out in Canada.

Situating the Present Work

While such a brief overview of some of the existing literature clearly overlooks much of the depth of the arguments and theories developed by these leading scholars, it is sufficient to demonstrate that the nexus between the informational and biological inherent in contemporary biotechnology has been a topic of discussion since at least the early 1980s. Moreover, this subject has been debated from a variety of perspectives, ranging from the ontological, to the metaphorical, to the political and the economic. Within this broader context, the present work seeks to situate empirically and theoretically some of the ways in which agricultural biotechnology has been appropriated by capital as an element of broader accumulation strategies, including the way that the increasing commodification of both biotechnological information and materiality have been insinuated in the trajectory of contemporary capitalist social relations. Where this book differs from its predecessors, particularly those that have looked specifically at Canada, is in how it emphasizes the major instances of social resistance that are being mobilized against particular aspects of biotechnology in this country. As it turns out, these aspects tend to revolve predominantly around issues implicated in agricultural biotechnology.

While biotechnology has yielded some groundbreaking discoveries in a variety of areas, a plethora of issues that go beyond the science and that have wide-ranging implications for contemporary society follow in the wake of this new technoscience. A more balanced account of the development of biotechnology must challenge the assumption, either implicit or explicit, that scientific progress derives from the capacity of scientists to surmount technical barriers, public apprehension, or the obstacles placed by bureaucratic and political elites. Rather than blindly adopting an uncritical assessment of innovation, we must understand scientific research and advancement in terms of the social and political processes that constitute its context of development. That is, science and technology need to be conceptualized as social and cultural practices constituted within and by the dominant power structures and values of the societies in which they are a part.

Members of the Critical Art Ensemble (2002, 43) poignantly sum up the dangers of uncritically equating technological development in the service of capital with progress:

> Of course there is no real gain, only relative gain. Class structure replicates itself in the technology ... decades of technoculture have taught us only that the greater the intensity of technology, the greater the workload. Much the

same is true of efficiency. Improved efficiency only means more profit and speed for capital, while the implied promise of individual benefit never seems to materialize. Taken together, a working definition of progress emerges that means nothing more than the expansion of capital, but presents itself as advancement of the common good.

We thus need to engage in political economic analyses that problematize the role of capital and the state in the developmental trajectory of science and, particularly, of biotechnology. As Karl Marx (1993, 704) pointed out long ago in the *Grundrisse*, scientific innovation migrates with relative ease into the realm of capital: "Invention then becomes a business, and the application of science to direct production itself becomes a prospect which determines and solicits it."[10] Harry Braverman (1974, 156) develops this idea in his own work when contrasting the Industrial Revolution of the nineteenth century to the scientific-technical developments of the twentieth century: "Science as generalized social property incidental to production and science as capitalist property at the very center of production."

While certainly not subscribing to a Marxist perspective, the Canadian government nonetheless has long realized the exchange value potential of biotechnology. For example, federal policy documents tend to attribute a variety of economic benefits to biotechnology: "In the last twenty years, biotechnology has become an increasingly important engine for economic growth and social development. It is now widely held that the transformative power of biotechnology will change forever the way we do things and interact with each other and the natural world, and that it will concomitantly change the culture of societies" (Canadian Biotechnology Advisory Committee 2006b, 3). Rejecting the underlying teleological tone of this contention, this book strives to contribute to the continuing academic, policy, and popular debates surrounding the role of biotechnology as a critical site within the present conjuncture of capitalist development.

Indeed, biotechnology might be considered to present a microcosm of the dominant characteristics of contemporary capitalist social relations: high capital intensity as reflected in the relatively small number of firms and workers involved in this sector; extensive control and command over the labour force; the involvement of both the state and capital in research and funding; the international focus of the large multinational biotechnology firms; and the intense use of information and communication technologies. As this book endeavours to argue, in order to inject some modicum of democratic control over the trajectory of biotechnology, we need to move beyond conceptualizing this technoscience as merely a particular combinatory set of scientific and technological knowledge and innovation and admit a discussion of the broader implications that it portends for contemporary social relations.

Introduction to a Marxian Analysis of Biotechnology

In an effort to conceptualize the empirical evidence presented in the following pages, I will develop and draw upon a theoretical framework informed by Marx's foundational notion of "primitive accumulation" and its relationship to enclosures and commons. Together, these concepts provide suitable registers for apprehending both the commodification of biotechnology and its consequent opposition by those social subjects that typically fall outside the more orthodox Marxist pre-occupation with the factory and skilled labour. At its most elemental level, Marx's (1992, 899) conceptualization of primitive accumulation conceives of it as a practice that separates producers from the means of production: "Thus were the agricultural folk first forcibly expropriated from the soil, driven from their homes, turned into vagabonds, and then whipped, branded and tortured by grotesquely terroristic laws into accepting the discipline necessary for the system of wage-labour." This process in England began as early as the fifteenth century and reached its zenith between the late seventeenth and early nineteenth centuries when members of the aristocratic class appropriated previously common lands, to which traditional usufruct rights of access and use for all were attached, and transformed them into deeded, private property.

In contradistinction to traditional exegetical accounts of Marx's primitive accumulation, recent scholars posit a basic ontological connection between primitive accumulation and expanded reproduction, believing that, for Marx, accumulation in general is a form of intensified primitive accumulation. Primitive accumulation should thus be understood as a continuous process that endures beyond the traditionally cited historical examples of land enclosure. Moreover, primitive accumulation, according to these contemporary theorists, should be conceived of as the extra-economic prerequisite to capitalist production that not only endures in contemporary society but also has been extended across the globe (Bonefeld 2001, 2002; De Angelis 2001, 2007; Glassman 2006; Harvey 2003). For example, structural adjustment programs foisted on Africa in the 1990s by the International Monetary Fund as a response to the debt crises in the 1980s resulted in massive shifts of land ownership that robbed local populations of their traditional means of subsistence. Through international trade agreements, Western intellectual property regimes were forced on developing nations in a way that commodified indigenous knowledge. Capital was also engaging in concerted efforts to privatize and commodify a variety of common resources, including such elemental ones as water, genetic material, and information.

Taking primitive accumulation and the consequent organization of capital around private property and wage labour as its conceptual starting points, the analysis in this book rejects the traditional dichotomies of state and market and political and economic. Instead, it conceives of these concepts as different expressions or forms of the conceptually prior separation between the actual producers and their means, processes, and products of production that give rise to antagonistic social relations. Articulated in broad strokes, the following chapters will rehearse the various practices of corporate and government enclosures of both terrestrial and informational commons that can be conceptualized as contemporary instances of primitive accumulation. Put more explicitly, by engaging the concept of primitive accumulation, we will see how the exploitation of biotechnology renders organic existence as a new source of profit in service of capitalist accumulation imperatives.

Yet in keeping with the goal of interrogating counterstruggles, we will also see that capital's drive to engulf expanding realms of social and biological existence within the logic of its social relations ignites multiple revolts by resistant political subjects. Juxtaposing the concept of the commons with the enclosing practices of primitive accumulation will help bring into focus a critical thread in this book that interweaves diverse points of resistance among social subjects struggling against various aspects of agricultural biotechnology beyond the immediate point of production. Indeed, a basic contention of the present work is that any analysis of contemporary capitalist society should proceed from the underlying assumptions that inherent to that society is an antagonistic relationship between class subjects and that attempts at domination provoke resistance aimed at emancipation – a resistance that will be interrogated throughout this book.

In the same way that Marx's methodological orientation is based on the perspective of the working class rooted in its own historical activity within the capital-labour relation,[11] contemporary efforts at understanding and situating the current conjuncture of capitalist social relations can be advanced through research into the genealogy of social and political opposition movements. Remaining consonant with the theoretical framework informing this work, social movements that resist the extension of capitalist relations of exploitation into the sphere of social reproduction might be construed as being just as much a part of contemporary class struggles as are the workers at the direct sites of production (Bieler and Morton 2006). Following Harry Cleaver (1992), the exposition offered in the subsequent chapters seeks to interrogate the "nature of the totality/globality that capital has sought to impose, the diversity of self-activity which has resisted that totality and the evolution of each in terms of the other." An extended analysis of the opposition in Canada that resists the enclosure of the biological and knowledge commons (or what I will sometimes group together throughout this volume under the rubric "BioCommons") will allow us to engage with the second query driving this research – namely, what counterstruggles have emerged in Canada that seek to challenge and foreclose the products and processes developed by biotechnological capital? I employ the term "commons" to refer less to a specific institutional form and more to its inherent social practices that structure the way resources – be they material or immaterial – are accessed, used, and managed by a group of people beyond the logic of the capitalist market. The commons thus involves the production of meanings and values through active engagement among subjects who struggle to maintain or regain social control over social wealth through opposition to capitalist and state practices of enclosure.

By interrogating the various forms of struggles that have emerged in defiance of the progressive enclosure of biotechnology, this monograph aspires to go beyond past investigations of biotechnology that tend to give short shrift to, or completely ignore, issues of resistance.[12] My elaboration of these counterstruggles will place particular emphasis on the implications they have for both the biological and the knowledge commons. This focus will permit me to respond to the third question informing the current work – what are the information and knowledge issues encompassed by such struggles? As mentioned earlier, the increased importance of information as both a commodity that possesses exchange value and as an input into a variety of production processes across economic sectors has significantly intensified the enclosure of information and knowledge commons. Biotechnology provides a particularly apropos example of the scope and implications of such enclosure.

At this point, I would like to clarify some terminological use throughout this volume. Taking a cue from Massimo De Angelis (2007), who himself follows Marx's usage of the term, I will avoid the *ism* of "capitalism," discussing instead "capital" and "capitalist social relations." Indeed, Marx never referred to capitalism, instead preferring to talk about the capitalist mode of production. By adopting this perspective, Marx was in a position to develop a critique of capital as an all-encompassing social relation or, what a more contemporary theorist refers to, as a system of social metabolic control: "Capital is not simply a material entity. We must think of capital as a historically determinate way of controlling social metabolic reproduction. That is the fundamental meaning of capital" (Mészáros 2008, 75). Such a conceptualization permits us to apprehend capital, or a capitalist mode of production, as one mode of organizing livelihoods that co-exists with, and is related to, others. In this way, we can conceive of the social field as a space open to strategic contestation among different forces. Similarly, I would like to clarify my use of the term "capital." My intent is not to hypostatize capital as a monolithic entity that develops and executes an internally consistent program of accumulation. Instead, I employ "capital" as shorthand for the aggregation of individual capitalists that, in general, represent a class in the broader system of capitalist social relations.[13] Finally, this work consciously employs the term "genetically engineered" in direct opposition to the industry preferred appellation "genetically modified," which is meant to convey a sense of naturalness and line of continuity from conventional (natural) agricultural, to breeding practices, to modern biotechnology.

A Brief Note on Methodology and Scope

The methodology that guided the research presented in this book represents a synthetic approach that reflects a combination of two strains of Marxist thought – a classical political economic analysis of capital and the state and an autonomist bottom-up approach that commences with struggle. While I certainly privilege struggle, as demonstrated by my choice of interview informants, the study began with a documentary analysis designed to sketch the broad contours of the biotechnology industry and biotechnological development in this country. The documentary analysis focuses on the CBS, and, aside from the actual documents that articulate this government policy, position briefs, publications, and other relevant material produced by federal bodies such as the Canadian Biotechnology Advisory Committee and the Biotechnology Assistant Deputy Minister Coordinating Committee of the Government of Canada were also examined. Having established this macro context, interviews were then conducted with key informants involved in the resistance against various aspects of biotechnology, which, as already mentioned, revolve predominantly around agricultural and food issues. After compilation, the results were offered back to the interview participants for further review and comment. Given the oppositional nature of such struggles, it was unlikely that documentary analysis alone would have revealed their true scope. Indeed, a lack of resources often circumscribes the documentary material that resistance movements can make available. Interviewing ensured that important evidence was not omitted from the study. These struggles, which themselves were approached as interruptions in the circuit of capital, were then analyzed in terms of their development, content, direction, and means of circulation along the entire circuit of capital. Finally, the analysis moved upward to relate all of these aspects of struggle to the broader capitalist initiative in terms of general social planning, investment, and technological innovation, as established through the documentary analysis.

Although it is perhaps already clear, the material presented in this book focuses predominantly on the Canadian domestic context. In part, this decision speaks to what I perceive to be a gap in the literature on biotechnology and, particularly, the resistance to biotechnology in Canada. Nonetheless, I think that a number of the issues and lessons illustrated in this work might find a certain degree of resonance in other countries grappling with agricultural biotechnology questions. Certainly, some cursory international comparisons are drawn where applicable and, of course, all of the biotechnology companies examined are multinational corporations, but, in general, this work does not engage with the international context. This is due, in part, to the richness of the material in the Canadian context. It is, similarly, a result of the empirical fact that many groups opposing agricultural biotechnology in this country focus on the domestic environment. And, finally, it comes from my belief that the scope and breadth of the issues surrounding biotechnology at the international level warrant their own separate, future research project.

Moving Forward

Before proceeding to the substantive content of this book, I would first like to offer a roadmap to help guide the reader in following the argumentation as it is developed in the pages that follow. In an attempt to answer our first research question about the prescriptive role of the CBS in facilitating the commodification of biotechnology in Canada, and in a manner that sets the context for responding to the other two, Chapter 1 is dedicated to outlining federal government policies and other pronouncements that serve to harness the vitality of the biotechnology sector as a motor for scientific innovation and economic growth. In addition to articulating the major elements of the CBS, I will introduce the players mobilizing against various aspects of agricultural biotechnology in Canada. Chapter 2 contemplates the capitalist appropriation of seeds and agriculture. This chapter also examines some of the resistance being organized against a particularly insidious example of corporate control of seeds – Terminator technology. This chapter also will begin to outline the theoretical constructs highlighted earlier that I propose can most usefully help assess the empirical findings with respect to the corporate capture of agricultural biotechnology. In part, the Terminator discussion provides a segue into Chapter 3, which rehearses the major past and present battles fought against specific genetically engineered technologies. Here, in the context of resistance, an elaboration of the concept of the "commons" will complete our theoretical framework. Chapter 4 investigates and elaborates the ways through which the intellectual property regime might be conceived of as a contemporary form of primitive accumulation that facilitates the enclosure of biological information and resources. In addition to presenting three of the major judicial cases waged to date in Canadian courts over genetically engineered organisms, this chapter will offer findings that demonstrate ways in which contemporary patent practices in regard to biotechnology actually offend against many traditional justifications invoked in support of the intellectual property system. Chapter 5 analyzes the biotechnology regulatory regime as a mechanism that facilitates enclosure of this science and its attendant technological applications. Deaf to pleas by both civil society and some scientific actors to expand the terms of reference of our current system of regulation, Canadian policy-makers remain steadfastly committed to an increasingly deficient linear model of scientific assessment that stubbornly refuses to admit broader social and political economic concerns into its deliberations. Chapter 6 describes some of the strategies and tactics employed by government and business to construct a deliberately circumscribed discourse around biotechnology, in what can be understood as a purposive enclosure of the knowledge commons.[14] The concluding chapter offers an overview of the major empirical findings that emerge from this research project.

1
Canadian Biotechnology Policy and Its Critics

The intent of this chapter is to illustrate, in broad strokes, the major federal policies over the last three decades in regard to biotechnology, placing particular emphasis on the current Canadian Biotechnology Strategy (CBS) developed in 1998. Rather than engage in a detailed historical analysis of federal biotechnology policy since its inception in 1980, which has already been done by several capable authors, the current account will establish the prevalent policy context, which will then lead into a later interrogation of the ways in which the BioCommons are being enclosed and how such acts of primitive accumulation are being resisted.[1] The second part of the chapter will introduce the dominant organizations and their main foci in mobilizing against various aspects of agricultural biotechnology in Canada.

Canadian Biotechnology Policy

The Canadian federal government, similar to governments in most other developed countries, has long considered science and technology to be integral to Canada's economic development and prosperity, and biotechnology has been no exception in this regard. Indeed, in the early 1980s, the Science Council of Canada advised the federal government of the need to achieve what it called "technological sovereignty," which places an emphasis on the importance of integrating technology into modern industrial processes (as we will see in the following chapter, this is a major theme worked out long ago by Marx). Attaining technological sovereignty, according to the Science Council of Canada, demanded economic restructuring based on expanded co-operation between the public and private sectors and specialization in a limited number of critical technologies as a way to strengthen domestic technological innovation and capacity. The three technologies tapped by the Science Council of Canada were engineering, information and computer technologies, and biotechnology.

It was also in the early 1980s that the now defunct federal Ministry of State for Science and Technology (MOSST) (1980) developed the policy document titled *Biotechnology in Canada*, which articulated the need for Canada to develop a federal framework designed to promote and develop biotechnology. Citing Canadian comparative weaknesses in expertise and funding for the biotechnology industry vis-à-vis the United States and Europe, this report suggested that Canada should commit to those areas of biotechnology that could best be exploited given the domestic context. Following this report, MOSST created a private sector task force that was charged with advising the minister on the viability of biotechnology policies, including the means for expanding research and development. The action plan delivered by the task force recommended a ten-year development project designed to stimulate a climate of research innovation and investment that could provide the basis for a varied and vibrant domestic biotechnology industrial sector. The task force similarly suggested that the federal government take the lead role in co-ordinating the plan. Some of the ways the task force's report envisioned buy-in from industry included: the provision of financial and tax incentives from government; the expansion of a scientific base sufficient to the developmental needs of biotechnology; an increase in collaboration with international actors; and attention to the Canadian regulatory environment to ensure Canada was not placed at a strategic disadvantage relative to other international players (Task Force on Biotechnology 1981).

The work of MOSST and its task force quickly bore fruit. Assuming that biotechnology would come to play an increasingly significant role in the economic growth of industrialized nations, the Government of Canada adopted the National Biotechnology Strategy (NBS) in 1983. The government of the day allocated $22 million over two years to develop and implement the strategy and an additional $100 million to fund national biotechnology research centres. The 1983 NBS focused predominantly on developing Canadian research capacity and a hospitable investment climate. This strategy, which fell under the jurisdiction of Industry Canada (a departmental placement that itself is quite telling of the economic emphasis accorded biotechnology), articulated the following four objectives:

1 To strategically focus biotechnology research and development on areas of importance to Canada;
2 To ensure a sufficient supply of well-trained personnel to work in the Canadian biotechnology sector;
3 To encourage cross-disciplinary and cross-sectoral research and communication of research results;
4 To develop an investment environment that would attract corporate investment in Canadian biotechnology. (Industry Canada 1998)

Indeed, developing commercial capacity by expanding the Canadian science base, facilitating technology transfer and development programs, strengthening intellectual property laws to entice commercial investment, and augmenting collaboration both within and without Canada remained key themes articulated by the National Biotechnology Advisory Committee (NBAC) throughout its lifetime. The NBAC was established under the auspices of the 1983 NBS to advise the minister of industry on the economic and industrial aspects of biotechnology. Given the broader political economic climate of the day, in which neo-liberalism was already establishing a strong foothold within developed countries around the world, including Canada, this commercial emphasis should not appear that surprising. If anything, as we will see later in this chapter, it has provided a line of continuity in Canadian government policy in regard to biotechnology up to this day. As other Canadian biotechnology policy researchers point out, "'innovation' provided the context for investing in new technologies such as biotechnology, and the National Biotechnology Strategy was one component of a broader policy designed to promote new technologies. Genetic engineering promised enormous economic, social, scientific and industrial potential at a time when Canada was struggling at both the national and international level" (Abergel and Barrett 2002, 141). However, as will be elaborated in the Chapter 5 discussion of Canada's regulatory regime, the characterization of biotechnology as innovative collides with and contradicts the contention made by industry and government that, for regulatory purposes at least, genetically engineered organisms are "substantially similar" to conventionally bred crops and animals.

Up to this point, the regulatory situation with respect to biotechnology was confused, and no federal department or agency had developed, let alone implemented, regulations specific to genetically engineered organisms. In 1988, MOSST published a policy document titled *Biotechnology Regulations: A User's Guide* that stipulated that seeds and plant varieties developed through genetic engineering would fall under the provisions of the *Seeds Act,* as administered by Agriculture Canada, which was the name of the department at this time.[2] A more formalized regulatory system was introduced in early 1993 with the Federal Regulatory Framework for Biotechnology. Its stated goal was to "minimize environmental risks while fostering competitiveness through timely introduction of biotechnology products to the marketplace" (Canadian Food Inspection Agency 1993, para. 5). In particular, the framework was developed to:

1 Maintain Canada's high standards for the protection of the health of workers, the general public and the environment;
2 Use existing legislation and regulatory institutions to clarify responsibilities and avoid duplication;

3   Continue to develop clear guidelines for evaluating products of biotech-
    nology which are in harmony with national priorities and international
    standards;
4   Provide for a sound scientific database on which to assess risk and evalu-
    ate products;
5   Ensure both the development and enforcement of Canadian biotechnol-
    ogy regulations are open and include consultation; and,
6   Contribute to the prosperity and well-being of Canadians by *fostering a
    favourable climate for investment, development and adoption of sustainable
    Canadian biotechnology products and processes.* (Ibid., para. 3 [emphasis
    added])

As subsequent chapters will elucidate in greater detail, the second, fourth,
fifth, and sixth elements of this regulatory framework remain contentious
among the critics of agricultural biotechnology in this country. Particularly
problematic is the fact that the regulatory review is triggered by the novelty
of the traits expressed by a plant or the novel attributes of a food or food
ingredient (the new regulatory category "plant with novel traits" was established) rather than by the specific process by which the plant or food
is created.[3] As we will also see, this product-based approach to regulation is
becoming increasingly untenable as new evidence emerges about the imprecision of genetic engineering techniques (thus calling into question the
fourth mandate of the regulatory framework). Given this lack of process
distinction, all agricultural commodities and food products, whether developed using conventional breeding methods, mutagenesis,[4] or genetic
engineering, are regulated by the same acts. Indeed, part of the rationale
for this decision was that the Canadian Food Inspection Agency (CFIA) had
a number of already existing product-based pieces of legislation that would
allow the new regulatory regime to respond to the second directive of the
framework – that is, build on existing legislation rather than develop and
implement a *sui generis* regulatory regime.[5] As one bureaucrat involved in
drafting the new regulations made clear, "a critical principle in the framework is that no new institutions or legislation respecting the regulation of
biotechnology products will be developed" (as cited in Kuyek 2002, 35).
    The justification offered for why Canada has failed to enact legislation
specific to biotechnology and genetic engineering is that the *Canadian Environmental Protection Act* ostensibly functions as a safety net to catch all
those products that might not be covered by other legislative devices.[6] Such
rhetoric notwithstanding, the Chrétien government of the day made the
conscious decision to maintain the regulatory status quo by leaving the
regulation of biotechnology firmly in the hands of Agriculture and Agri-
Food Canada and Health Canada. The government completely ignored the
recommendation issued by the House of Commons Standing Committee

on Environment and Sustainable Development to move regulatory oversight for biotechnology to Environment Canada, which would have been given increased power through a proposed section to the *Canadian Environmental Protection Act*. Under the proposed consolidated regulations that would have fallen within the jurisdiction of Environment Canada, genetically engineered organisms would have been regulated as being potentially "toxic substances" and would have had to undergo environmental assessment by the department. Given the possibility of more stringent regulation, as well as the usurpation of a major Agriculture Canada policy mandate, it is perhaps not surprising that this option never made it beyond policy debates (Abergel and Barrett 2002; Kuyek 2002).

The 1993 regulatory framework was revamped in 2000 when ministers from the Treasury Board approved the Canadian Regulatory System for Biotechnology in order

> to enhance Canada's regulatory capacity and to ensure that Canadians have an efficient, credible and well-respected biotechnology regulatory system that safeguards health and the environment as a priority and, thereby, permits safe and effective products to enter the market. The strategic objectives of the CRSB [Canadian Regulatory System for Biotechnology] are to meet technical capacity and human resource needs; improve public awareness of and confidence in the regulatory system; increase the efficiency, effectiveness and timeliness of the regulatory system; and generate knowledge to support the regulatory system. (Treasury Board of Canada Secretariat 2006, paras. 8-9)

The increasing importance of biotechnology to Canada's overall science and technology framework was re-affirmed in the 1997 Speech from the Throne, in which biotechnology was singled out as a "key sector." This same throne speech articulated the creation of a team of seven ministers – led by John Manley, minister of industry – to oversee the renewal of Canadian biotechnology policy. As a result, the Canadian Biotechnology Strategy Task Force was struck in 1997 to review and further develop the 1983 NBS, which had lagged in responding to the extreme pace of development in biotechnology that was occurring at the time. Early in 1998, this task force engaged in consultations with provincial officials, industry, non-governmental organizations, scientists, academics, and other relevant stakeholders about the vision, goals, and principles of a renewed national biotechnology strategy as well as its potential impacts on the biotechnology industry and research and development.

The task force also solicited input from Canadians about how their interests could best be reflected in policy development, although the entire exercise has been criticized as being "a contrived public consultation process" (Kuyek

2002, 41). According to the federal government, over 5,000 individuals and organizations participated in the deliberations that led to the latest iteration of federal biotechnology policy, the CBS, which was published in 1998 to address a broader range of issues in regard to biotechnology (Government of Canada 2004). However, another publication issued by Industry Canada admits that the figure of 5,000 includes "web site 'hits'" (Industry Canada 1998, 3). Yet neither a definition of what was considered a website hit nor an exact number of such hits is provided. The possible inflation of public consultation numbers through such means lends legitimacy to Kuyek's critical assessment of the process.

The CBS was conceptualized by the federal government as a "blueprint to enhance the quality of life of Canadians in terms of health, safety, the environment and social and economic development by positioning Canada as a responsible world leader in biotechnology" (Government of Canada 2004, 3). The CBS drew on the expertise of a number of federal departments and agencies that were involved in regulatory activities, research and development, technology transfer, and investment and trade related to biotechnological products and services. According to the federal government, the CBS represented a policy framework that incorporates social, ethical, health, economic, environmental, and regulatory issues into decisions about biotechnology in a manner that "fully reflects and responds to Canadians' demands that biotechnology be developed in ways that do not pose a danger to humans, the environment or animals" (ibid.). However, as we will see in subsequent chapters, this contention was widely contested by various groups that mobilized against particular aspects of agricultural biotechnology.

The CBS is based on the following nine pillars:

1 Ensure that Canadians have access to, confidence in and benefit from safe and effective biotechnology-based products and services;
2 Ensure an effective scientific base and make strategic investments in research and development to support biotechnology innovation, the regulatory framework and economic development;
3 Position Canada as an ethically and socially responsible world leader in the development, commercialization, sale and use of biotechnology products and services;
4 Be sensitive to the need for developing countries to build indigenous capacity to assess and manage the risks of biotechnology;
5 Improve public awareness and understanding of biotechnology through open, transparent communications and dialogue;
6 Solicit broadly-based advice to the government on biotechnology;
7 Promote awareness of, and maintain excellence in, Canada's regulatory system, based on the Federal Regulatory Framework for Biotechnology

(1993), to ensure the country's continued high standards for protecting health, safety and the environment;

8 Support the development of a Canadian biotechnology human resources strategy to ensure an adequate supply of highly qualified personnel;

9 Work with the provinces, territories, business, academia, and consumer and other interest groups to develop and implement action plans addressing stewardship issues (for example, health, safety, environment, and social and ethical matters), sectoral opportunities and horizontal challenges (for instance, research and development, regulations, human resources, investment, innovation, technology transfer and market access). (Canadian Biotechnology Advisory Committee 2006b, 24)

Despite the broader range of issues consolidated under the new and expanded Canadian biotechnology policy, federal documents made it explicitly clear that commercial imperatives would drive the CBS: "Biotechnology is a powerful 'enabling technology' with applications in many industrial sectors and holding much promise for the future. It has great potential to add to industrial efficiency, output and jobs, enhance the productivity and competitiveness of Canada's important natural resource sectors, safeguard the environment and enhance our quality of life through improved pharmaceuticals, diagnostic medicine and food production" (Industry Canada 1998, 2).

During the 1998 consultations conducted by the task force developing the CBS, a number of stakeholders articulated the need for an independent advisory body that would operate at arm's length from the government to advise on crucial policy matters. Based on recommendations by the NBAC and the Standing Committee on Environment and Sustainable Development, and in an attempt to respond to the fifth element of the new CBS about "improv[ing] public awareness and understanding of biotechnology through open, transparent communications and dialogue," the new Canadian Biotechnology Advisory Committee (CBAC) was created in 1999 to replace the NBAC. The CBAC, which was disbanded on 17 May 2007, was composed of volunteer members from the areas of science, business, nutrition, law, philosophy, ethics, and environmental and public advocacy.[7] The CBAC consisted of a chair and twelve to twenty members with varying subject areas and degrees of expertise. Membership was ostensibly based on individual attributes of a particular member and not on the basis of representing a particular interest, although this claim is widely contested by observers who are critical of biotechnology in this country. The nomination process was public, but a Biotechnology Deputy Ministers Selection Panel reviewed the nominations according to specific criteria such as expertise, knowledge, and experience. This selection panel issued recommendations about the appointments, with the final decisions being reserved for the ministerial members of the Biotechnology Ministerial Co-ordinating Committee (BMCC). The chair and committee members were usually appointed for a three-year term, although extensions were permissible at the discretion of the BMCC ministers.

Rather than report directly to Parliament, as was recommended by the Standing Committee on Environment and Sustainable Development, the CBAC received direction from, and provided advice to, the BMCC, which itself was supported by the Biotechnology Assistant Deputy Minister Co-ordinating Committee (BACC). From an operational perspective, the BACC is the management committee of the CBS. The BACC has a permanent representative from each department that receives CBS funding and that possesses significant biotechnology expertise and experience (Agriculture and Agri-Food Canada, Environment Canada, Fisheries and Oceans Canada, Health Canada, Industry Canada, International Trade Canada, Justice Canada, and Natural Resources Canada) as well as one member from the National Research Council, one member from the CFIA, and one member representing the tri-councils (the Canadian Institutes of Health Research, the Social Sciences and Humanities Research Council of Canada, and the Natural Sciences and Engineering Research Council). The BACC is responsible for setting priorities and providing horizontal co-ordination of CBS initiatives (Canadian Biotechnology Advisory Committee 2006b).

The CBAC issued advisory memoranda to the federal and provincial governments of Canada in order to apprise policy-makers of issues related to biotechnology that would require immediate or medium-term attention as well as the effects of government policy on biotechnological developments in Canada (Canadian Biotechnology Advisory Committee, 2004a). The scope of the policy advice provided by the CBAC ranged from ethical to social, regulatory, economic, scientific, environmental, and health issues associated with biotechnology (Government of Canada 2004). However, as will be discussed in subsequent chapters, the opposing nature of some of these roles, particularly those between regulation and promotion, resulted in conflicts of interest, leading some commentators to assert that the CBS and the CBAC are biased heavily in favour of the biotechnology industry. According to Devlin Kuyek (2002, 4), Canadian "biotechnology policy has been the private domain of a small number of corporate executives, the offices of the Prime Minister and the Privy Council, a selection of senior government bureaucrats ... university presidents and board members of governmental/industry promotion and granting agencies."

The CBAC originally identified five special research projects: the regulation of genetically engineered food; intellectual property concerns related to biotechnology; issues around novel uses of biotechnology, such as stem cells; the integration of ethical and social issues into biotechnology policy; and the consequences for privacy that emerge around biotechnology (Government of Canada 2004). Although the CBS and the CBAC documents articulate the importance of social and ethical dimensions in informing biotechnology policy development, the overall tone of both the federal biotechnology strategy and the various reports published by the CBAC leave the reader with the unmistakeable conclusion that the commercialization of this science and its resulting technological applications is the driving motivation.

In one of its reports, the CBAC recommended that the Government of Canada update the CBS (the CBAC first called upon the federal government to renew and revise the CBS in December 2004) in a manner that "integrates the economic, environmental, ethical, legal, regulatory, scientific and social considerations pertaining to biotechnology and its implications for Canadian society and its long-term interests." Failure to do so would apparently hinder "Canada's ability to access, apply and harness the power of biotechnology to best serve the public's social and economic interests" (Canadian Biotechnology Advisory Committee 2006b, 3). The report went on to urge Canadian governments to create "a supportive business environment for biotechnology firms by addressing the factors that influence innovative capacity; namely, university systems, university-industry technology transfers, intellectual property laws, the pool of scientists and engineers, and availability of venture capital funding" (5). Even with respect to environmental concerns (encompassed by the ninth pillar of the CBS), the CBAC, following recommendations made by its internal Expert Working Party on Biotechnology and Sustainable Development, urged the federal government to develop environmental initiatives in a manner that would employ biotechnological innovations for economic development (Expert Working Party on Biotechnology and Sustainable Development 2006).

The environmental catastrophe wrought on the earth over the previous two centuries under the dominance of capitalist social relations and the corresponding corporate hindrance of concerted and effective efforts to respond to this situation have seriously undermined the proposition that capitalist-inspired economic development and environmentalism can coexist in any equal manner. What quickly becomes apparent from the policy documents emanating from the CBAC is that this advisory body parroted that same tension inherent in the CBS between the roles of promotion and regulation of biotechnology. It is thus perhaps not surprising that the CBAC was dismissed by a number of critics almost from the outset of its existence for privileging its promotion mandate (discussed more fully in Chapter 6).

Indeed, an emphasis on innovation and economic expansion has long informed Canadian biotechnology policy: "Biotechnology is one of the world's fastest-growing technologies. It offers significant economic benefits, particularly in exports and job creation" (Industry Canada 1998, 1). The federal government has concretized its commitment to biotechnology policy through substantial levels of public funding (almost $7.8 billion between 1997 and 2009), which is designed to develop and sustain a domestic biotechnology industry (Statistics Canada 1998, 2001, 2002, 2003, 2004, 2005, 2010).[8] Of course, the CBS focus on exploiting biotechnology for commercial purposes finds its precursor in Canada's broader science and technology policy, which has long stressed the link between science and technology research and economic growth and the perceived corresponding need to foster partnerships between businesses, academic institutions, and government (Industry Canada 1996). This, no doubt, explains why universities comprise by far the largest source of biotechnology company spinoffs. Some of the major elements of Canadian science and technology strategies have included federal government support for private sector research and development through tax relief, the development of industry-led consortia involved in pre-competitive research, and targeted assistance for specific firms engaged in high-risk commercial projects. Another articulated policy goal has been the purported need to re-tool the intellectual property regime in ways that would render it more amenable to private sector commercialization of the innovations developed through federally supported research (Government of Canada 1996).

Some of these same elements and emphases on jobs, growth, and industrial efficiency are reflected in Canada's *Growing Forward Framework Agreement*, which was struck between the federal, provincial, and territorial agriculture ministers in July 2008. This new framework legislation, which replaced the 2002 *Federal-Provincial-Territorial Framework Agreement on Agricultural and Agri-Food Policy for the Twenty-First Century*, provided $1.3 billion in funding for agricultural investment from all three levels of government over five years (2008 to 2012) on a 60:40 basis between the Government of Canada and the provincial and territorial governments. The *Growing Forward Framework Agreement* commits all three levels of government to achieving three strategic outcomes: a competitive and innovative agricultural sector; a sector that contributes to society's priorities; and a sector that proactively manages risk. The first goal of ensuring a competitive agricultural sector is to be achieved by offering assistance in commercializing innovation and developing what the agreement refers to as a bioeconomy.[9] Ensuring innovation also apparently hinges on improving Canada's regulatory performance, which, in this case, means providing "science and other support to help the industry generate approvals for health claims" and offering "support for industry-led marketing strategies" (Agriculture and Agri-Food Canada 2008, paras. 8, 9).

The active engagement called for in this agreement – to provide industry with science and other resources in support of applications for regulatory approval – reflects the tension inherent in Canadian biotechnology policy between the government's dual functions of regulator and promoter of biotechnological applications and processes. A government commitment to support industry marketing strategies indicates just how far this tension is being pushed in favour of commercialization imperatives. Indeed, the executive summary of the framework agreement goes on to claim that society's priorities (which are not articulated) can be met by supporting and recognizing food-safety systems (although these remain undefined) that producers will only need to adopt "where the market demands it" (Agriculture and Agri-Food Canada 2008, para. 12).

Yet, as we will see in subsequent chapters, beyond the disquietude about the appropriation and capitalist exploitation of science, a number of the wider social, political, economic, ethical, and environmental implications that attach to biotechnology are engendering opposition in this country precisely because of their omission, or at best superficial treatment, in policy discussions and development. However, before delving into specific instances and broader themes of resistance against the appropriation of biotechnology in Canada, we need to first introduce the major organizations leading this mobilization.

The Players Mobilizing against Agricultural Biotechnology

Although the research driving this work engaged in documentary analysis of the corporate and government/regulator positions adopted with respect to biotechnology, the selection of interview informants was consciously restricted to those subjects actively engaged in struggle against biotechnology. In part, this decision speaks to a perhaps iconoclastic challenge to the traditional norms of research "balance" and "neutrality." Instead, following the call by Walter Benjamin (1998) for cultural producers to remain cognizant of the class interests their work serves, I believe that capital already enjoys an unbalanced advantage in terms of the resources it can assemble to deliver its message in respect of biotechnology. While a significant amount of the present work engages with analyses of capital and government actions in respect of agricultural biotechnology, these themes are included precisely because they were articulated by interview informants as part of the substantive issues with which they are grappling. Thus, in helping to give a voice to those subjects opposing various aspects of agricultural biotechnology in Canada, the capital/government nexus necessarily must be interrogated.

Similarly, the evidence marshalled in this chapter attests to the methodological importance I attribute to the interrogation and analysis of points of struggle as moments that are revelatory of the power mechanisms at work in the broader context of capitalist society. The accent on resistance provides an analytical lens through which to view the social forces that determine a particular phenomenon, making manifest the contradictions and tensions that dynamically structure a given conjuncture of social relations. As Antonio Negri (2005, xiii) asserts, "struggles are the great teachers when it comes to our knowledge of social development, and are the engines of revolutionary theory." And John Holloway (1992, 159) reminds us that

> the more intense the social antagonisms, the less securely established will be the fetishised self-presentation of social relations. It is not theoretical reflection but anger born from the experience of oppression that provides the cutting edge to pierce the mystifications of capitalist society. The role of theory is not to lead the way but to follow, to focus on the contradictory nature of experience, to give more coherence to the vaguely perceived interconnections, to broadcast the lessons of struggle.

As the account advanced in the following sections should attest, there is a vibrant landscape of active resistance that is animated by a passionate and well-informed group of activists committed to responding to, and checking, the capitalist-dominated developmental trajectory of agricultural biotechnology in this country.

Canadian Biotechnology Action Network (CBAN)

A prominent umbrella group in Canada that is mobilizing resistance against biotechnology is CBAN, which is composed of various environmental, social justice, and consumer groups that previously had worked collaboratively on the campaign to stop the regulatory approval of Monsanto's bovine growth hormone for dairy cows. Building on this success, which was based on sharing information and co-ordinating common actions on issues raised by genetic engineering, biodiversity, sustainable farming, and corporate control in agriculture, these groups were also successful in pressuring Monsanto into abandoning its plans to introduce genetically engineered wheat into the Canadian marketplace (as we will elaborate more fully in Chapter 3). According to Lucy Sharratt, an interview respondent and co-ordinator of CBAN, this informal collaboration began in 1999 and was made more permanent in 2006 when a number of the involved participants concluded that any concerted effort against genetic engineering in this country would depend upon the insertion of new momentum and resources into the struggles against biotechnology.

Thus was born CBAN, which is designed to function as a hub to assist information exchange, support grassroots action, and co-ordinate activities at the national and international levels.[10] While there may be grassroots members involved in CBAN mobilization and work who would tend to define themselves as anticapitalist, such a position is not an explicit articulation of the organization. Some of the groups aligned with CBAN have brought forth what might be perceived as an anti-corporate stance, but in terms of self-identification, CBAN perceives itself to be pro-democracy and pro-justice rather than anti-anything, according to Sharratt. While CBAN has taken the lead in Canada on the Ban Terminator campaign (genetically engineered seed sterilization technologies), Sharratt makes it clear that the motivation behind its creation was to establish a movement that goes beyond a single campaign with limited project definition and, instead, to adopt an orientation based on long-term struggle around the multiple issues raised by biotechnology. Ensuring a broader, enduring perspective for CBAN is something that Brewster Kneen, another interviewee and prominent food system and anti-biotechnology activist, believes is critical to biotechnology opposition in this country. As he explains, part of the broader mandate must include discussion and analysis of the corporate control of seeds and agriculture in a very explicit manner so as to frame the issues in ways that speak to the general public. According to him, resistance to biotechnology, including the work of CBAN, must become bolder both in terms of the way it looks at the system and the way it communicates its message to people. Kneen believes deeply in the need to elicit broader societal discussion about the current conjuncture of neo-liberal capitalist society within which biotechnology is being developed.

CBAN seeks to develop itself in a manner consistent with strong social justice and international solidarity imperatives that position farmers and agriculture at the centre of the discourse on genetic engineering. According to Sharratt, when CBAN talks about genetic engineering, its members want to open a dialogue about farmers and farming in a way that will help to increase the political voice of farmers in Canada, foster an urban understanding of farmers, and expand connections between non-governmental organizations (NGOs) and farmers. That having been said, CBAN also recognizes that biotechnology poses environmental, health, and moral concerns, although the latter are not something CBAN seeks to actively define. From CBAN's perspective, debates about the developmental trajectory of biotechnology encompass issues of democracy that are not being adequately, if at all, addressed by the Canadian government. It therefore seeks to highlight what it perceives to be a major deficiency in Canada's regulatory system, namely, that it offers no room for consideration of social and economic factors in its decision-making processes.

One of the goals of the network is, therefore, to democratize the decision-making processes around genetic engineering. CBAN and its member groups advocate the need for debates in Parliament and the development of regulations that work for people. These regulations must include consideration of economic and social factors. A corollary to this goal, according to Sharratt, is the concern about the power that corporations hold and are able to wield in political and economic fora. In terms of social justice for farmers and agriculture, CBAN emphasizes three main concerns. One is the farm income crisis that stems from the high cost of farm inputs and relatively low prices for farm outputs. Aside from Canada's low-priced food policy, there is a plethora of additional policies that obstruct farmers from obtaining sufficient money for their product. Indeed, as discussed in Chapter 2, the latest Canadian statistics demonstrate that agricultural input suppliers now account for almost the entire amount of revenue produced by farmers in this country, leaving the actual producers with an average of $1.45 per acre per year over the past twenty-four years (National Farmers Union 2009). The second and related concern is that genetic engineering increases farmers' dependence on corporate inputs such as seeds and herbicides, which not only exacerbates the high cost of such inputs but also facilitates increasing corporate control of agriculture and the food supply. The third Terminator-specific concern is that this particular application of genetic engineering is explicitly designed to create dependence on the part of farmers such that corporations could exclusively control seed as a farm input. For CBAN and its members, Terminator technology represents another, albeit in this instance particularly insidious, biological tool designed to facilitate increased corporate control over seed markets.

One of the major dynamics of CBAN is that it functions as a coalition designed to take up campaigns in a very active manner. In terms of strategies, CBAN traditionally has sought to bring together research and information on genetic engineering that is sometimes rather difficult to locate. The diversity of its membership base means that CBAN must be able to provide a variety of information to its members that they can then employ when mobilizing their respective networks. As Sharratt points out, CBAN recognizes a pressing need for organizations to do research that they can then make readily available to the public. For example, the Canadian government does not provide any comprehensive list of genetically engineered crops that is, in any way, easily accessible to the public. Seeking to fill this public information gap, CBAN has compiled and placed such a list on its website, and it also includes details about which corporations own the intellectual property rights to genetically engineered seeds.

Similarly, CBAN perceives a real need to correct misinformation and provide new information and analysis that goes beyond the often one-sided and celebratory informational content disseminated by the biotechnology proponents in industry and the government. To this end, CBAN organizes public lectures and other types of events that engage prominent speakers dedicated to informing the public about the range of scientific, social, economic, and political issues that attend biotechnology. For example, CBAN has organized a number of screenings across the country of the film *The World According to Monsanto*. Groups interested in showing the film locally can contact CBAN for help in putting together a showing, including advice and information sheets for engaging the audience about the subject matter of the film. In April and May 2008, CBAN, in collaboration with a number of social justice, environmental, health, food, and farming groups, organized a cross-Canada speaking tour about the global impacts of biofuels on food, farmers, and human rights. Speakers from Colombia, Mexico, Mali, Argentina, Paraguay, the Philippines, the United States, the United Kingdom, and Canada participated in these free public events, which were held in Charlottetown, Halifax, Montreal, Ottawa, Winnipeg, and Saskatoon.

In terms of other information events, CBAN, in collaboration with Canadian Organic Growers, the Ecological Farmers Association of Ontario, and the Ontario chapter of the National Farmers Union, held a public panel discussion at the University of Toronto in November 2008 titled "Genetic Engineering in Your Food: Things You Need to Know." These same groups hosted a booth together at the Royal Agricultural Winter Fair to share information on organics, ecological agriculture, and genetic engineering with the public of Ontario. In December 2008, CBAN held a fundraiser in Ottawa to coincide with the launch of the book by Shiv Chopra about his experience at Health Canada where he and three colleagues were harassed and ultimately dismissed because of their unwillingness to acquiesce to the internal and external pressures to approve products they had determined posed potential health risks to animals and humans.

Through its information collection and research activities, CBAN seeks to ensure that it remains abreast of current and upcoming issues. For example, should it be discovered that a genetically engineered crop variety, such as wheat or alfalfa, is slated to receive imminent regulatory approval, CBAN will investigate and consider the possibilities and political opportunities in order to develop a potential strategy that can be brought forward to the grassroots organizations and NGOs in order to mobilize them in different ways. As we will see in the following discussion, such activity is occurring once again in respect of genetically engineered wheat.

As it happens, a type of product-by-product protest has characterized the history of such mobilization, as was the case with the opposition organized against genetically engineered wheat, genetically engineered alfalfa, and recombinant bovine growth hormone. However, Sharratt contends that it is not the intention of CBAN to fight the government product by product. Rather, the aim is to maintain a rigorous and concerted critique of regulation in a very detailed way over a number of years in order to achieve systematic results. Yet in the interim, CBAN also recognizes the urgent obligation to address the corporate push to develop and commercially release new genetically engineered organisms, including Monsanto's genetically engineered sugar beets and its new stacked variety of genetically engineered corn (which is elaborated more fully in Chapter 5).

Money, according to Sharratt, is the major obstacle currently impeding the work and information-dissemination activities of CBAN. In fact, it was not until October 2007 that CBAN, thanks to a small grant, was able to launch its own website. Prior to this, the CBAN website was only a holding page, forcing the organization to piggyback on the Ban Terminator website. Similarly, the high cost of graphic design and printing services limits the amount of informational resources CBAN is able to develop and circulate. At an even more fundamental level, lack of financial resources severely circumscribes the amount of research CBAN is able to conduct. For example, Sharratt has been attempting for a number of years to secure funding in order to research the situation in Canada in regard to genetically engineered trees. Although field trials are currently being executed, there is a complete dearth of publicly available information on the issue.[11] Similarly, dairy companies and farmers are concerned about the impact genetically engineered alfalfa could have on their business, but they have as yet been either unwilling or unable to fund such research. As Sharratt passionately points out, CBAN has the expertise; it just needs the money.

Another difficulty related to information dissemination arises from the complicated task of communicating to a broad public about a fairly complex subject that can easily come across as a technical issue. So a major challenge for CBAN, according to Sharratt, is to develop the right informational materials that not only speak to people but also serve to catalyze them into action. CBAN has therefore decided that it will only ask people to act on those issues that hold out the promise of yielding results. As Sharratt points out, "there is an abundance of things people can do, but if such actions are not going to have a result or if they are not supported by other works that other people are doing and it goes nowhere, then ultimately no one, at least not those opposed to particular aspects of biotechnology, will be served" (Interview). Contingent upon funding, CBAN is planning to develop a fact sheet on genetic engineering that is included within a fact sheet on farming for urban consumers. One strategy for reaching urban consumers is to disseminate the information via farmers markets and events in urban communities. Similarly, outreach to farmers about genetic engineering will be pursued via rural fairs and farm shows. In the interim, CBAN has produced a variety of fact sheets, pamphlets, and guides for consumers and farmers alike, all of which are made available on its website. In fact, CBAN's website has developed into a valuable information source that, in addition to the above, provides a wealth of resources about agricultural biotechnology and food sovereignty and ecological justice issues more broadly.

Greenpeace

Greenpeace is a global organization that maintains a presence in roughly sixty countries with approximately forty-five offices throughout these different countries. According to Eric Darier, interviewee and the agriculture co-ordinator of Greenpeace Canada, its main strategy is to follow large multinationals around the world to make sure they do not engage in detrimental practices in certain (usually less-developed) nations after having been blocked from engaging in those practices in other (usually better-developed) nations by civil society and organized groups. For example, Greenpeace prioritizes Asia, China, and India because of the rapid economic growth occurring in these countries and because multinationals are moving there, especially large chemical companies.

This global perspective applies equally to issues of genetic engineering. Greenpeace recognizes that the success of its efforts against genetic engineering will be influenced heavily by its ability to co-ordinate an international movement against biotechnology and genetic engineering. According to Darier, the key to Greenpeace opposition efforts is exerting pressure on the companies and governments promoting biotechnology in precisely those places where the biggest difference can be achieved. Similar to the sentiments raised by Sharratt in respect of CBAN's opposition efforts, Darier points out that Greenpeace is cognizant of the fact that it is unable to do everything, so, instead, it chooses to be very strategic about what it does and where it is doing it.

Given that Greenpeace is mainly an environmental group, its perspective focuses on the preservation of biodiversity, which encompasses the issue of the environmental release of genetically engineered organisms. In order to protect biodiversity, this group is a strong advocate of the precautionary principle.[12] Greenpeace was, in fact, instrumental in pushing quite hard for the international community to adopt the UN Cartagena Protocol on Biosafety to the Convention on Biological Diversity, which finally happened in Montreal on 29 January 2000.[13] According to Darier, Greenpeace realized quickly that, faced by large multinationals such as Monsanto and governments heavily influenced by these same companies, it was vital for the protection of biodiversity to achieve a multinational agreement. Greenpeace therefore lobbied forcefully for the development of some form of international agreement that would afford a degree of civil society control over genetic engineering and the biotechnology industry.

Its environmental focus notwithstanding, Greenpeace is also conscious of, and receptive to, various other points of entry for opposition to biotechnology. For example, corporate concentration and the resulting control of the world's food supply are major concerns, and ones that, as Darier points out, are being facilitated in the biotechnology industry through the merger and acquisition activities among seed and chemical companies. Other concerns relate to ethical reservations about biotechnology, particularly with respect to the patenting of life forms. Still others emerge from a Third World solidarity perspective, which also includes the fight against the commodification of life forms. In fact, the Greenpeace office in Germany conducts a significant amount of work in fighting and trying to revoke life patents. However, as Darier laments, Greenpeace does not possess the resources to pursue similar strategies in Canada. With limited resources available in this country, Greenpeace focuses on mandatory labelling for genetically engineered foods.

Darier posits the existence of an informal division of labour among various opposition movements. In those areas that go beyond the immediate focus of Greenpeace, but to which Greenpeace is nonetheless sympathetic, it engages in partnerships with those groups that are most focused on that particular issue, allowing them to take the lead on it. Similarly, Greenpeace strives to build linkages with other groups that have an interest in the issues upon which it concentrates. For example, in the case of labelling genetically engineered food, Greenpeace works with consumer groups since labelling can also be approached from a consumer perspective. This type of strategy is premised on informal and formal arrangements that exploit the synergies between the different groups involved, such that each partner focuses on its particular area of expertise based on what is feasible given the level of available resources. In terms of general strategies, Greenpeace follows what might be called a "weakest link in the chain" battle plan, preferring to exert pressure on the most vulnerable corporation in that particular industry sector in order to compel others to follow suit. For example, mandatory labelling in Europe, although finally adopted by the European Commission, followed on the heels of massive consumer pressure on some of the largest European retailers to withdraw genetically engineered ingredients from their products and shelves.[14]

In terms of disseminating information, Greenpeace makes use of multiple channels. The first level consists of front-line staff, who, through the Greenpeace canvassing program, contact individuals at the door, on the street, and over the phone. Canvassers receive a briefing paper that articulates both the issues and the campaigns in which Greenpeace is actively involved. According to Josh Brandon, interview respondent and genetic engineering campaigner for the Greenpeace Vancouver office, Greenpeace conducts about 1,400 conversations a week with Canadians through this canvassing program. Similarly, Greenpeace has a number of volunteers who collect signatures for different things and do tabling for various events so that, in general, Greenpeace conducts a substantial amount of one-on-one grassroots conversations. As Brandon points out, Greenpeace engages in public fora and talks at school groups and so on and also has very extensive e-mail lists as well as a monthly magazine, which is sent to its 80,000 members country-wide in both electronic and paper format. Some of the obstacles Greenpeace faces in disseminating information, beyond the gatekeepers in corporate media, include internal capacity on particular issues and financial constraints. That is, events and campaigns have to be co-ordinated nationally, which means that not all issues can receive the same attention.

National Farmers Union (NFU)

The NFU is a direct membership, voluntary organization structured according to districts that are organized on a more-or-less provincial basis, with the exception of Quebec, which has its own distinct provincial farmers' organization. With a focus on promoting and preserving family farming as a basic food production unit in Canada, the NFU's mandate also includes an emphasis on agricultural and social policies that facilitate agriculture and community around family farming. According to Terry Boehm, interviewee and NFU vice president, his organization has conducted a number of successful campaigns around seeds, genetically engineered wheat, and attempts to curb restrictive plant breeders' rights legislation, among others. At the outset of all of its campaigns, the NFU generally puts out a press release, although, as Boehm points out, it does have problems getting them picked up by the mainstream media given the high degree of media concentration. Relationships with other like-minded groups, environmentalists, and social justice activists who are interested in food and food policy tend to be more essential to the NFU in disseminating its messages. Aside from these networks, the NFU also relies on public meetings, its own membership discussions, co-operation with receptive parliamentarians and parties, and the provision of witness testimony at parliamentary committee meetings to advance its policy positions.

The NFU also engages in some government consultation processes. For example, it participated in the national forum on seeds in an attempt to ensure a voice for its questions, ideas, and position. Unsure of how successful such consultative exercises are, Boehm is quick to add that there is always a fine line between participation and co-optation. Similarly, participating in such processes requires the organization to devote a substantial amount of its time and resources to a relatively defined debate, which impinges on what can be effected on other fronts. For the most part, Boehm believes that it is the activities beyond Parliament that tend to be more successful.

As Boehm outlines, the Internet is an important communication medium for the NFU in disseminating its information. In particular, the NFU relies on listservs and its website, although the organization is finding it increasingly difficult to keep the latter current given the speed with which things seem to be happening and the number of different people working on various issues. The NFU also engages in a variety of ongoing activities designed to help inform the broader public about some of the issues involved in biotechnology. It sends out opposite-the-editorial pieces to newspapers, and it authors papers and briefs that articulate the implications of biotechnology from farmers' perspectives. These papers and briefs are distributed to members and others in order to help people understand the concerns around biotechnology, to inform people about what actions are being undertaken, and to provide analysis of current and proposed legislation in terms of how it does or might affect farmers. Overall, the NFU places a great deal of emphasis on providing rigorous analysis and disseminating the results of such analysis in order to help people understand and navigate the multiple concerns encompassed by biotechnology. Boehm maintains that the research and analysis the NFU produces are well respected both nationally and internationally, as made evident by the multiple requests it receives from other groups to use its material.

ETC Group (Action Group on Erosion, Technology, and Concentration)
The ETC Group (which was formerly the Rural Advancement Foundation International) approaches biotechnology issues from social and economic perspectives, although as the ETC executive director and co-founder, Pat Mooney, makes clear, there is a variety of cultural, ethical, environmental, and health concerns that also attach to biotechnology. So while the ETC Group does not deny that these too are aspects of the issue, it focuses on who owns and controls this technology, what the implications of such ownership patterns are for the types of products being developed, and the social ramifications of those products once they are introduced into society. At a fundamental level, the driving goal of the ETC Group is to ensure that those people and groups who are most affected by biotechnology are informed adequately so that they are enabled to formulate their own approaches to this science and its technologies. Beyond disseminating the necessary information to the people who need to know about it, a second goal of the ETC Group is to inject information about the issues that surround biotechnology into broader fora for debate, such as the United Nations. Since as early as 1987, the ETC Group has therefore lobbied tirelessly at the UN Food and Agriculture Organization (FAO) to push the biotechnology issue onto the international agenda. The ETC Group has been successful in convincing the FAO to agree that there should be a code of conduct in biotechnology, although the issue has not moved further than an agenda discussion point. Nonetheless, Mooney points out that there has been an ongoing debate within the FAO because of the efforts of the ETC Group.

Mooney believes that both sides have focused the biotechnology debate too narrowly on health and environmental issues, something that continues to frustrate him and his colleagues. One of the failures of organized opposition to biotechnology, in Mooney's opinion, has been that many groups who instinctively, and for political reasons, are appropriately concerned about these technologies have struck the path of least resistance by concentrating on health and environmental risks. And while these are real and serious concerns, many of the groups that have chosen that level at which to mobilize resistance have made, according to Mooney, a tactical error. Independent of the ETC Group, prominent food system and anti-biotechnology activist Brewster Kneen raises similar sentiments, asserting that, while safety issues are important, the debate must go beyond these to consider a broader range of principled points associated with biotechnology and corporate control over agriculture and the food system. In Kneen's own words, "I wish activists would spend a little more time on their homework and learn the science better and think more about the broader context of this stuff and where it's coming from, its implications and so on." Unfortunately, as Kneen also readily admits, he has not been as successful as he would like in winning this argument.

As might be expected, the Internet is an important tool for the ETC Group to disseminate information. However, unlike other opposition groups, the ETC Group has existed for three decades. Over the course of these years, it has worked successfully with a plethora of other organizations, which has helped it establish a great deal of familiarity and credibility in activist, corporate, and government circles. Moreover, according to Mooney, the quality of its investigations and analysis has ensured that people, or at least its partners, tend to trust what the ETC Group says. So when it distributes informational content, it receives accurate and good responses from media and from its partners, who, in turn, circulate the information to other interested people.

Having introduced the main groups in Canada involved in resistance against agricultural biotechnology, the following chapters turn attention to the specific issues and campaigns around which these groups are mobilizing. In addition to opposition directed against specific genetically engineered products, these groups are engaging with broader structural and systemic concerns that range from opposition to corporate control of seeds and agriculture through to critiques of Canada's intellectual property and regulatory regimes, efforts to redress the genetic contamination of the terrestrial commons, and endeavours to expand the terms of the biotechnology debate in this country.

2
Enclosure and Resistance on the BioCommons

> Few enclosures of the commons threaten to usher in as many profound changes as the commodification and privatizing of genetic structures ... There are many *complicated dimensions to biotechnology*, but one worrisome trend is the conversion of our shared genetic heritage into a privately owned inventory managed for commercial gain. The genetic structures of life, which have "belonged" to everyone from time immemorial, are being propertized in order to move them from the public commons to private markets, with all the shifts in power, accountability, and moral norms that that conversion entails.
>
> – David Bollier, *Silent Theft: The Private Plunder of Our Common Wealth* [emphasis added]

This chapter and those that follow seek to elucidate the "complicated dimensions to biotechnology" that David Bollier (2002) articulates by sketching out the contested landscape of this science and its technological applications. Drawing on documentary analysis of government and corporate reports and other texts in respect of biotechnology, as well as interviews conducted with key informants from organized movements opposed to aspects of agricultural biotechnology in Canada, the discussion will illustrate the parallels and connections between enclosures of biological commons and knowledge/ informational commons. For ease of presentation, I have organized the empirical part of this research thematically into chapters, although there is some overlap in places where a strict separation would unduly hinder the flow of the narrative. By interweaving the evidence derived from the documentary analysis and the interviews, my intent is to demonstrate how activists are responding to, and mobilizing against, contemporary corporate and state strategies designed to bring agricultural biotechnology firmly within the purview of corporate control.

As we saw in the previous chapter, Canadian biotechnology policy is informed by the underlying presumption that this science and its attendant applications promise to deliver a wealth of social, environmental, and economic benefits. Situated against this general policy backdrop, we need to interrogate the various mechanisms employed by capital to obtain and maintain corporate control over seeds and agriculture. Taken together, this chapter and the previous one establish a number of the broader structural constraints within and against which opponents of agricultural biotechnology must engage in this country. This chapter will also introduce the first example of some of the sustained resistance being organized against a specific genetic engineering technology – Terminator technology. Following this presentation, I will elaborate on the theoretical framework that informs the analysis of the empirical evidence presented throughout the book.

Corporate Biotechnological Control of Seeds and Agriculture

The scientific breakthrough that figured prominently in the subsequent commodification of biotechnology was the technique of recombinant deoxyribonucleic acid (DNA) that Herbert Boyer, from the University of California at San Francisco, and Stanley Cohen, from Stanford University, invented in 1973. Three years later, Boyer and Robert Swanson, a venture capitalist, founded one of the first biotechnology companies, Genentech (a shorthand version of "genetic engineering technology"), which continues to be a dominant biotechnology firm operating today, although the Swiss firm Roche now largely owns it. Following the hype surrounding the announcement in 1978 that Genentech had managed to clone human insulin using recombinant technology and bacteria, the biotechnology industry quickly expanded.[1] Wall Street picked up on these developments in August 1979 when Nelson Schneider, a financial analyst at E.F. Hutton who specialized in the pharmaceutical sector, wrote a paper directed at institutional investors titled "DNA: The Genetic Revolution." At a presentation one month later, at which Schneider was scheduled to present his ideas about the investment opportunities he perceived in this newly emerging technology, over 500 individuals were in attendance, which dwarfed the thirty participants that were expected. For analysts at E.F. Hutton and other investment banking firms, biotechnology appeared poised to become an enabling technology along the lines of information and communication technologies (Bud 1993).[2] Aside from a plethora of start-up biotechnology companies, a number of the major pharmaceutical and chemical multinationals began developing in-house research programs using recombinant DNA, signed research contracts with some of the start ups, and even began acquiring equity stakes in a number of them.[3]

Wall Street's prognostications notwithstanding, it is important to bear in mind the difficulties in assessing the magnitude of agricultural biotechnology on both a national and global scale. Although Statistics Canada collects annual data about the production of principal field crops in this country, it only disaggregates statistics according to conventional and genetically engineered crops for corn and soybeans and only in the provinces of Ontario and Quebec. When I queried a statistician at Statistics Canada about this limited collection of information specific to genetically engineered crops, I was told that the department follows the data requests provided by Agriculture and Agri-Food Canada and the Canadian Food Inspection Agency (CFIA). My requests to these latter two organizations asking for a rationale for the partial data collection remain unanswered as of writing. The same Statistics Canada employee also told me that the decision to not track the amount of genetically engineered canola was made in consultation with the Canola Council of Canada.[4]

Internationally, the situation appears somewhat similar, with very few countries differentiating between conventional and genetically engineered crops in their annual agricultural statistics. As a result, tracking this type of information has been left largely to one organization, the International Service for the Acquisition of Agri-Biotech Applications (ISAAA), which publishes its annual report, *Global Status of Commercialized Biotech/Genetically Modified Crops,* outlining the global hectarage of genetically engineered seeds according to crop type and number of farmers. In fact, the ISAAA report has become so ubiquitous that even the Organisation for Economic Co-operation and Development and the UN Food and Agriculture Organization tend to rely on the ISAAA's data in their own publications. As a staunch agricultural biotechnology promoter that counts among its donors a number of major seed companies, including Monsanto and Bayer CropScience, as well as industry lobbyist CropLife International, the ISAAA has been criticized by various non-governmental organizations (for example, the United Kingdom-based GMWatch, LobbyWatch, and GM Freeze and the African Centre for Biodiversity) for continually overstating the impact of genetically engineered crops across the globe.

Part of this critique revolves around doubts about the veracity of the data put forward by the ISAAA. For example, in 2005, the amount of land claimed to be cultivated with genetically engineered maize in the Philippines and genetically engineered cotton in South Africa was challenged as being too high (*Monsanto and ISAAA's Hype Exposed* 2006). It is important to point out that neither of these countries tracked these types of statistics. Even in the case of countries where official statistics are available, there are data reliability problems. Lobby Watch, for example, maintains that the ISAAA consistently inflates figures for the United States by between 2 and 9 percent. In 2006, the ISAAA's annual report claimed that Iran was growing genetically engineered rice, a claim that was challenged by the International Rice Research Institute (James 2006). All references to Iran were quietly removed in the

2007 report (James 2007). Independent analysis of genetically engineered maize crops in Europe revealed that the 2007 ISAAA figures were inflated by a factor of four. In 2008, the ISAAA's report placed incredible emphasis on the commercialization of 700 hectares of genetically engineered maize in Egypt as being of strategic importance for the African continent (James 2008). Yet, according to a US Department of Agriculture Foreign Agricultural Service Gain Report from 2009, progress had been stalled on commercial planting approvals in Egypt and the country had not produced any commercial biotechnology crops (Mansour 2009).

Based upon my own calculations employing Statistics Canada data, the 2011 ISAAA report overestimates the amount of land sown with genetically engineered crops in this country by over one million hectares. According to the ISAAA, 10.4 million hectares of land were planted with genetically engineered crops, but my numbers show just under 9.2 million (James 2011).[5] In fact, the ISAAA has continually embellished the amount of arable land cultivated with genetically engineered crops in this country since at least 2006 (the first year Statistics Canada began differentiating between conventional and genetically engineered crops and, thus, the first year any real estimate can be calculated – although, as already mentioned, only for corn and soybeans and only in the provinces of Quebec and Ontario) – by over 500,000 hectares in 2006 and 2007, by about 200,000 in 2008, by almost 500,000 in 2009, and by over 600,000 in 2010.[6] I inquired with the ISAAA about their data sources for Canada and was informed in an initial e-mail response that figures were obtained from the Canola Council of Canada, the Government of Canada, industry, and personal communications between Clive James (chair of the ISAAA and author of the annual reports) and Canadian scientists.[7] When I followed up and asked for specific contact details, I was informed that none could be provided and that James had obtained his source from the Canola Council of Canada through a personal communication. When I subsequently spoke with a vice president from the Canola Council of Canada, she was somewhat perplexed and made clear that, although canola is approximately 93 percent genetically engineered in Canada as of 2009, the council is not in a position to speak about any other crop.

Commenting on the 2008 ISAAA annual report, Pete Riley, director of the United Kingdom's GM Freeze, points out that "once again ISAAA have massaged the data on GM crops to make a struggling industry look impressive. The vast majority of hundreds of millions of farmers in the world use conventional seeds to produce food and this is where we need to concentrate research rather than on a technology that has delivered very little" (GM Freeze 2009, 1). Helen Holder, GMO campaign co-ordinator at Friends of the Earth Europe concurs, adding that "a tiny proportion of farmers in the world are growing crops on a very small fraction of the world's agricultural land. The biotech industry is inflating the figures in an attempt to convince the media and politicians that GM crops are a success. The latest industry figures for Europe are highly unreliable which indicates that this is probably the case for other regions as well" (as cited in GM Freeze 2009, 1).

Perhaps the main messages to be derived from the criticisms of the ISAAA reports are not only that their data should be treated with some caution but also that their overall claims should be subject to more critical assessment. For, even based on the statistics offered by the ISAAA, less than 3 percent of global arable land is devoted to the cultivation of genetically engineered crops. Of the mere twenty-five countries worldwide that grow genetically engineered crops, the top six account for almost 95 percent of the global genetically engineered hectares: the United States, Brazil, Argentina, India, Canada, and China. Despite the hype that is promoted by industry about the multiple applications for genetic engineering among a variety of crop varieties, soybeans, maize, cotton, and canola continue to account for over 99 percent of all global crop land sown with genetically engineered seeds. Put another way, no new genetically engineered crop varieties have been adopted on any significant scale since genetically engineered crops first began to be grown commercially. Probable inflated figures notwithstanding, it is certain that the major players in the agricultural biotechnology sector are expanding and consolidating their control.

In fact, according to Jeffrey Smith (2003, 1), major American corporations are crafting strategic plans to control the world's food supply:

> This was made clear at a biotech industry conference in January 1999, where a representative from Arthur Anderson Consulting Group explained how his company had helped Monsanto create that plan. First, they asked Monsanto what their ideal future looked like in fifteen to twenty years. Monsanto executives described a world with 100 percent of all commercial seeds genetically modified and patented. Anderson Consulting then worked backward from that goal, and developed the strategy and tactics to achieve it. They presented Monsanto with the steps and procedures needed to obtain a place of industry dominance in a world in which natural seeds were virtually extinct.

Although Monsanto has not yet achieved its goal of completely saturating markets with patent-protected genetically engineered seeds, it and a handful of other multinational corporations have captured around two-thirds of the global proprietary seed market. Table 2.1 outlines the world's top seed companies according to seed revenue, which together have accounted for around 55 percent of the commercial seed market since 2007 (worth US$26.7 billion, US$33 billion, and US$36 billion in 2007, 2008, and 2009, respectively. These figures are based on information generously provided by Context Network). This relatively stable level of control over commercial seed markets exercised by the top seed companies over the past few years represents a marked increase over the last decade. For example, in 1996 these large seed companies accounted for only 37 percent of the world seed market. The degree of control by these multinational corporations over the proprietary seed market (that is, seed varieties subject to intellectual property restrictions) is even more pronounced – in 2009, these companies controlled 61 percent (US$19.7 billion) of a US$32 billion market. Monsanto alone controls 23 percent of the global proprietary market, and the top three companies (Monsanto, Dupont, and Syngenta) together control 45 percent of this market. Perhaps most indicative of the level of control is the estimate that 87 percent of the total land area across the globe planted with genetically engineered seeds is sown with Monsanto products, either directly or through licences to other companies (ETC Group 2008).

Through a concerted campaign based on aggressive corporate mergers and acquisitions, as well as partnerships with some of the other large players in the agricultural biotechnology sector, Monsanto has required less than a decade to emerge as the largest player in this industry. With such market clout has come a corresponding ability to dictate onerous terms on both growers and seed distributors. In addition to its Bt line of insect-resistant crops that the company markets under the name "YieldGard,"[8] Monsanto vigorously markets in Canada canola,[9] soybean,[10] and corn[11] seeds that have been genetically engineered to resist glyphosate, the active ingredient in Roundup herbicide formulations. Marketed as "Roundup Ready," these seeds produce crops capable of surviving post-emergent applications of Roundup, a broad-spectrum, non-selective systemic herbicide that is typically sprayed and absorbed through the leaves of a plant. Its mode of action is to inhibit an enzyme involved in the synthesis of the amino acids tyrosine, tryptophan, and phenylalanine, which are critical to plant survival.[12] Since it is a non-selective herbicide, glyphosate kills a broad spectrum of plants/weeds unless they have been genetically engineered to be resistant.

The use of glyphosate had made it easier for farmers to prepare fields for planting and to control weed growth, although evolving weed resistance is rapidly becoming a major problem. Farmers who wish to plant Roundup Ready canola must contract with Monsanto to pay for the seed as well as a technology fee, which, prior to 2009, amounted to $15 per acre. Beginning in the 2009 growing season, Monsanto Canada altered its technology fee collection practices for Roundup Ready canola by implementing a "bag licence fee." Under this new model, growers no longer pay per acre. Instead, the technology fee is incorporated into the price of a bag of seed – the current price is $7.40 per kilogram of seed.[13] Sales of genetically engineered corn

Table 2.1

World's top seed companies by seed revenue, 2007-09 (US$)

Rank and company	2007		2008		2009	
	Seed revenue ($ millions)[a]	% of global proprietary seed market ($22 billion)	Seed revenue ($ millions)[b]	% of global proprietary seed market ($28.4 billion)	Seed revenue ($ millions)[b]	% of global proprietary seed market ($32.2 billion)
1 Monsanto (US)	4,964	23	6,369	22	7,297	23
2 Dupont/Pioneer Hi-Bred (US)	3,300	15	3,976	14	4,641	14
3 Syngenta (Switzerland)	2,018	9	2,442	9	2,564	8
4 Groupe Limagrain (France)	1,226	6	1,254	4	1,384	4
5 Land O'Lakes (US)	917	4	1,185	4	1,105[d]	3
6 KWS AG (Germany)	702	3	839	3	969	3
7 BayerCropScience (Germany)	524	2	664	2	700	2
8 Dow (US)	n/a	–	499	< 2	635	2
9 Sakata (Japan)[c]	396	< 2	n/a	–	n/a	–
10 DLF-Trifolium (Denmark)	391	< 2	$383	< 2	398	< 2
11 Takii (Japan)[c]	347	< 2	n/a	–	n/a	–
Total	14,785	67	$17,611	62	19,693	61

a Data from Action Group on Erosion, Technology, and Concentration.
b Data from company annual reports (foreign currencies converted to US$ using average annual exchange rate).
c Neither Sakata nor Takii make public their seed revenues and, therefore, are not included beyond 2007.
d As of 2009, Land O'Lakes no longer disaggregates its seed sales. The figure presented is based on the 2008 figure of seeds, comprising 33.7 percent of the "crop input" segment. This figure likely slightly underestimates seed sales for 2009.

and soybeans are handled somewhat differently, and more opaquely, in that the price of a bag of seed is not disaggregated according to seed price and technology fee.

Farmers who chose to grow any of Monsanto's Roundup Ready products were, until recently, also contractually compelled to use only Monsanto's Roundup herbicide. This stipulation was particularly important for the company since its patent on Roundup expired in 2000. By contractually bundling its herbicide-tolerant seeds and Roundup formulations, Monsanto was able to resist price pressure from generic producers once glyphosate went off patent, and, thus, maintain its dominant position in the agrochemical market. As one observer notes, the success of such tactics helps explain why large pesticide companies were prompted to move into agricultural biotechnology (Kuyek 2002, 77-78).

In addition to the stipulation that only Roundup or another authorized herbicide could be used, the technology use agreements (now rebranded as "technology stewardship agreements") that accompany all of Monsanto's seed varieties, and which farmers purchasing the seed must sign, contain the following terms: growers must implement an insect resistance management program (applicable only to Bt crops); a grower may use the seed solely for planting a single commercial crop; none of the seed may be given to another person or entity for planting; none of the seed from the resulting crop may be saved for planting in subsequent years; seed may not be used or allowed to be used for crop breeding, research, generation of herbicide registration data, or seed production; Monsanto expressly limits its liability for any loss incurred from using the product to the price of the seed and dictates how farmers must proceed if they launch a claim; farmers must provide Monsanto complete access rights to their fields and records to conduct what the company refers to as technology protection audits; and Monsanto reserves the right to enjoin growers found in breach of contract from ever again using or selling its seed (Monsanto Canada n.d.).

Although, in recent years, Monsanto appears to have relaxed the Roundup requirement by permitting growers to use authorized non-Roundup glyphosate herbicides, this change is more show than substance. Part of the technology fees included in the price of Monsanto's genetically engineered seeds covers a crop protection program (branded as "Roundup Rewards") whereby Monsanto will waive the technology fee for replacement seed if a crop fails within sixty days after planting. However, the fee waiver is contingent upon the use of Roundup formulations. In fact, in its technology use guides, the company states explicitly that "Monsanto does not make any representations, warranties or recommendations concerning the use of glyphosate products supplied by other companies which are labelled for use over Roundup Ready [seed]. Monsanto specifically denies all responsibility and disclaims any liability for any damage from the use of these products over-the-top of

Roundup Ready [seed]" (*Monsanto Canada* 2010, 25). Since technology fees comprise a significant portion of the overall cost of genetically engineered seeds, there is clear pressure on growers to purchase and apply only Roundup formulations on their glyphosate-resistant crops.

Given these various business practices, it is perhaps not surprising that the company's chief executive officer, Hugh Grant, reaffirmed to investors as late as September 2009 that the company would meet its goal of doubling gross profits in 2012 from 2007 levels of US$4.3 billion. However, by late 2010, it looked as if such future revenue predictions were wholly unwarranted, as many North American farmers, seemingly aggravated by the continual escalation of Monsanto's genetically engineered seed prices, adopted the company's latest Roundup Ready 2 soybeans and SmartStax corn at levels substantially below forecasted demand. This lower than expected demand, coupled with disappointing yields of SmartStax, forced the company to drastically reduce its prices on both new varieties and endure an almost 42 percent decline in its stock price in the first three-quarters of 2010 (Pollack 2010). In fact, year-over-year net sales declined in 2010 by 10 percent (US$1.2 billion). Gross profit was down an astounding 25 percent (US$1.6 billion), although the bulk of that reduction was attributable to lower sales volumes and sales prices of Roundup and other glyphosate-based herbicides ("Monsanto's Annual Form 10-K" 2011). By the following year, however, this poor fiscal showing seemed to be more of an anomaly than a trend. In 2011, Monsanto increased its net sales by 13 percent to US$11.8 billion, and gross profit was up by 20 percent to just over US$6 billion. The company's improved sales and revenue were driven by both its seeds and genomics segment and by its agricultural productivity segment. Net sales of corn, soybean, and cotton seeds were up in 2011 by 13, 4, and 39 percent, respectively. Sales of Roundup and other glyphosate-based herbicides increased by 4 percent because of increased demand in Europe and Argentina. Moreover, average net sales prices for these herbicides increased due to lower sales deductions for marketing programs (ibid.). The stabilization of its herbicide business is, no doubt, due in large part to the "Roundup Rewards" system, which, as outlined earlier, provides a very clear incentive to farmers to only use Monsanto's glyphosate formulations in conjunction with the company's genetically engineered seeds.

Despite the *mea culpa* and improved revenue picture, storm clouds of a different sort are appearing on the horizon for Monsanto as both the US Department of Justice and at least seven state attorneys general have announced that the company is being investigated for possible violations of American antitrust laws (Fitzgerald 2010; Leonard 2009). In a related move, the attorney general of West Virginia in October 2010 filed a civil enforcement action in circuit court to stop Monsanto from selling any of its products in the state until the company complies with a subpoena that compels it to co-operate with an investigation launched in June into the failure of Roundup Ready 2 Yield soybeans to produce the yield advantages advertised by the company (claimed to range from between 7 and 11 percent over first generation Roundup Ready soybeans) (Volkmann 2010). In his initial letter to Monsanto, Attorney General Darrell McGraw cited US Department of Agriculture statistics and a number of academic and independent research studies that dispute the company's enhanced yield claims. According to Assistant Attorney General Douglas Davis, Monsanto did finally respond to the subpoena, and, although he did not rule it out, there are currently no plans to bring further civil action against the company.[14]

In addition to exerting substantial pressure on the demand side, Monsanto also engages in significant market manipulation on the supply side. This concerted combination of both push and pull factors that constrain the supply of, and demand for, its competitors' seeds and herbicides has been a vital element in Monsanto's strategies to achieve the market dominance it enjoys today. Many of the supply-side means by which Monsanto has achieved its dominant market position have increasingly come to light over the past few years. According to a class action lawsuit brought against Monsanto in 2007 by Texas Grain, Monsanto is alleged to have pursued ownership interests in a wide variety of potential competitors in order not only to expand and consolidate its existing seed-trait monopolies but also to block development and market entry of alternative herbicide-tolerant seed varieties that would have facilitated increased competition in the non-selective herbicides market.[15] For example, its acquisition of the majority interest in DeKalb Genetics Corporation in 1998 apparently led to the cancellation of an existing project on which DeKalb was working to develop glufosinate-tolerant corn traits.

Similarly, after its purchase of Asgrow in 1997, Monsanto is purported to have caused the company to breach two soybean research and development agreements it had established with Dupont. In its litigation against these breaches of contract, Dupont claimed that such actions constituted a broader plan by Monsanto to impede the development of competing herbicide-tolerant seed traits and undermine competition from other herbicides. In its statement of claim, Dupont alleged that "Asgrow and Monsanto have acted with the joint purpose of eliminating STS [sulfonylurea-resistant] soybean seeds as a competitor with Roundup Ready soybean seeds in order to facilitate Monsanto's scheme to monopolize the soybean seed and soybean herbicide markets."[16] Through its control of Asgrow, Monsanto was also able to terminate a collaborative research agreement between Asgrow and AgrEvo (a Bayer CropScience predecessor) to develop glufosinate-based seed traits.

The situation was similar with Holdens Foundation Seed, another large American seed and technology company, which ceased its support for glufosinate-tolerant corn traits after being acquired by Monsanto in 1997:

"Because foundation seed companies like DeKalb, Holdens and Asgrow played a critical role in the testing, development and commercialization of new types of seeds, Monsanto's acquisition of those companies substantially reduced the number of actual and/or potential sources rivals needed access to in order to compete with Monsanto in the market for genetically modified herbicide-tolerant seed traits."[17] Texas Grain Storage similarly alleges that Monsanto leveraged its market power in seed trait markets to reduce competition in the separate herbicide market, thus permitting it to engage in supra-competitive pricing for its glyphosate herbicide formulations. The magistrate hearing the case refused the plaintiffs' motion for class certification. Counsel for Texas Grain Storage subsequently appealed this ruling, but, as of April 2012, the District Court had not yet ruled on the appeal.

Another prominent prong in Monsanto's strategy to dominate the seed market is a labyrinthine licencing system that it uses to licence its genetically engineered traits to smaller seed companies that typically focus on traditional plant breeding methods in order to develop seeds suited to particular regional environmental and growing conditions. These types of arrangements provide Monsanto with access not only to locally adapted cultivars, into which it inserts its genetically engineered traits, but also to established regional production and distribution infrastructures upon which agricultural producers rely. Obversely, by securing widespread access for itself, Monsanto has simultaneously shut out many of its competitors from important manufacturing and distribution assets and channels. A recent leak of confidential contracts reported by the Associated Press has revealed that, although Monsanto has licenced its proprietary technology to over 200 smaller seed companies, thus allowing those companies to insert Monsanto's genes into their own corn, canola, cotton, and soybean varieties, the restrictions attached to these agreements are very onerous. For example, without the express consent of Monsanto, its licencees are prohibited from breeding new varieties that would combine Monsanto's technology with transgenic material from other companies.

The upshot of this proviso is that Monsanto's competitors are effectively precluded from inserting their patented traits into any of the crops that already contain Monsanto's genes. This prohibition, of course, is doubly profitable for Monsanto since it blocks rivals from competing in the genetically engineered seed market and also impedes competition from other non-glyphosate non-selective herbicides by stacking the market with Monsanto's own herbicide-tolerant seed traits that work in conjunction with glyphosate herbicides. That being said, when it anticipates large returns, Monsanto does engage in licencing and cross-licencing agreements with its major competitors, such as Syngenta, Bayer CropScience, BASF, and Dow AgroSciences.

In addition to this *de facto* ban on stacking traits, another contract provision stipulates that if a licencee should change ownership all of its inventory containing Monsanto technology must be destroyed. The effect of this requirement is a significant reduction in the value and price that any licencee trying to sell its seed business would obtain from anyone other than Monsanto. According to Monsanto spokesperson Lee Quarles, some contracts have permitted licencees to sell their inventory for a specified period and royalty fees have been waived for destroyed seeds. Nonetheless, the Associated Press estimates that this practice has likely helped Monsanto acquire twenty-four independent seed companies in the United States over the past few years (Leonard 2009).

Licencees are also contractually restricted from discussing or otherwise revealing the terms of the contracts they sign with Monsanto. Failure to adhere to this stipulation can result in a cancellation of the deal and the forced destruction of the licencee's seed inventory. Another practice employed by the company earlier in the decade was to discount substantially the price of its seed if independent seed companies agreed to ensure that 70 percent of their corn seed inventory comprised Monsanto products. Now that Monsanto's seed varieties have attained a dominant market position, the company has discontinued this type of discounting (Leonard 2009). As stipulated in a soybean licencing contract reviewed by Bloomberg News, the licencee would receive a 7.5 percent rebate of the royalty payments owed to Monsanto if Roundup Ready varieties accounted for 70 percent of the dealer's annual herbicide-resistant seed sales. At sales levels less than 70 percent, but above 50 percent, the rebate would be reduced by half, and no rebate would be paid if sales of Monsanto's products fell below 50 percent of the licencee's annual volume.

The statement of claim in the class action lawsuit launched against Monsanto in 2007 by Texas Grain Storage alleges that, in many instances, if the specified percentage threshold for even one type of seed variety was breached, Monsanto was permitted contractually to withhold rebates for all of its other seed traits licenced to, and produced and/or sold by, the seed company. So, for example, if the seed company sells too much of a Monsanto rival's soybean seed, it would have to pay higher royalty payments not only for the soybean traits it licences from Monsanto but also for all of the other licenced seed traits (for example, both herbicide-resistant and insect-resistant traits for corn, cotton, canola). The net effect was that Monsanto was able to leverage its monopoly control across all specific seed-trait markets. In its lawsuit, Texas Grain Storage asserts, moreover, that similar rebating practices had applied to dealer sales of Roundup herbicide formulations. For example, through such programs as the "Action Pact Program," dealers were required to achieve Roundup sales volumes at levels equal to, or higher than, Monsanto's existing regional market share for Roundup. Failure to achieve such targets would jeopardize the full range of rebates on Monsanto seed sales and seed-trait technology fees, not just on Roundup sales.[18] According to Harry Stine, president and founder of Stine Seed Company, one of the largest private seed companies in the United States, "the rebates were so large that for all practical purposes you had to do it. In order to get the large rebate they would give you, you had to minimize your sales of other companies' seeds" (as cited in Fitzgerald 2010, para. 20).

The ultimate effect of tight industry control by a handful of biotechnology companies is a decreasing availability of conventional seeds and a correlative escalation in the prices of genetically engineered seeds. These price increases are being fuelled, in part, by the rising prominence of stacked varieties within seed producers' catalogues. Single and even double-stacked varieties are gradually being replaced with triple (or more) trait seeds that fetch higher prices, since more traits mean more royalty payments to the patent holder. Monsanto's strategy to oblige agricultural producers to avail themselves of triple-stack varieties is based on attenuating the number of single and conventional options in its own brands and subsidiaries and also on dramatically escalating the price of single and double-stack varieties (Hubbard 2009).

The latest and thus far highest-stacked variety is Monsanto's new SmartStax corn. Developed through research collaboration and cross-licencing agreements with Dow AgroSciences, this genetically engineered corn contains eight different traits to render it both herbicide (glyphosate and glufosinate) and insect (rootworm and corn borer) resistant. Perhaps the most offensive outcome of this continual ratcheting up of the number of traits genetically engineered into seed varieties and the corresponding reduction in the supply of alternatives is that farmers are compelled to pay for traits that they may not even require (this is particularly the case with SmartStax, as discussed more fully in Chapter 5).

Since recent disaggregated information on seed input prices in Canada is not readily available, the data provided here is from the United States.[19] Although these figures will not map perfectly onto the Canadian market, they do provide a rough approximation and illustration of price-trend comparisons for conventional and genetically engineered corn and soybean seeds over the past decade. It is estimated that Monsanto's genetically engineered traits, either directly or through licence agreements, account for 82 percent of the corn produced in the United States and as much as 93 percent of the soybeans (Fitzgerald 2010). In the case of corn, the premium for genetically engineered seed over conventional varieties has risen from 29 percent in 2001 (with genetically engineered seed costing US$110 per unit[20] and conventional seed costing US$85 per unit) to as high as 69 percent in 2009 (with genetically engineered seed averaging US$235 per unit and conventional seed costing US$139 per unit). The price of genetically engineered corn seed per acre over this period averaged US$53, which represents a 46 percent premium over the average price per acre for conventional corn seed of US$36.

Between 2001 and 2009, the cost of conventional corn seed per acre fluctu-
ated within a range of 6 and 12 percent of gross market income and 14 and
20 percent of operating costs. The comparison ranges for genetically engineered seed averaged between 10 and 16 percent of gross income and 23 and
29 percent of average operating costs per acre (Benbrook 2009b).

As outlined earlier, by 2010, farmer backlash against the skyrocketing
prices of Monsanto's genetically engineered seeds had compelled the company to reduce its prices on several of its products, including Roundup
Ready 2 soybeans and SmartStax corn, which Benbrook (2009b) initially
forecast to be as much as US$70 and US$320 per unit, respectively. In fact,
the average price differentials between genetically engineered corn seed
(US$247 per unit) and conventional corn seed (US$152 per unit) declined
to 63 percent in 2010, meaning that farmers paid about US$55 to plant an
acre of conventional corn and close to US$90 per acre for genetically en-
gineered varieties. As a result of higher conventional corn seed prices
(US$163 per unit) compared to relatively static average prices for genetically
engineered corn seed (US$249 per unit), the price differential narrowed to
53 percent in 2011, when it cost US$59 to plant an acre of conventional corn
and US$91 for an acre of genetically engineered corn (National Agricultural
Statistics Service 2012).

The situation in respect of soybeans is even more telling. With over
90 percent of the soybean seed market dominated by genetically engineered
varieties, companies rapidly escalated retail prices. In 2001, the average price
for a unit of genetically engineered soybean seed was US$24 – just over
33 percent more expensive than a unit of conventional seed at US$18. By
2009, the average premium for genetically engineered soybean seed over
conventional seed had increased to 47 percent (genetically engineered seed
cost US$50 per unit, while a unit of conventional seed cost US$34). On a
per acre basis, conventional soybean seed averaged US$25 between 2001
and 2009, while genetically engineered seed set a farmer back 60 percent
more at about US$40 per acre on average for this period. As a percentage of
gross market income per acre, conventional soybean seed prices moved
within a range of 7 and 11 percent between 2001 and 2009. Conventional
seed costs per acre, as a percentage of average operating costs, ranged from
between 21 and 32 percent during this same period. By comparison, genetically engineered soybean seed costs per acre between 2001 and 2009 ranged
from between 8 and 17 percent of gross income and 32 to 47 percent of
average operating costs (Benbrook 2009b).

In 2010, the average unit price of conventional soybean seed ran about
US$34, as compared to almost US$54 for genetically engineered seed, which
represents a 58 percent price premium. On a per acre basis, farmers planting
genetically engineered soybeans paid almost US$62, while their conventional

counterparts (to the extent that they could source such seed) paid US$39. Although not as dramatically as corn seed prices, these steep price differentials have recently been reduced. The average cost for genetically engineered soybean seed in 2011 declined slightly to US$51 per unit, while conventional seed prices remained more or less the same, which helped reduce the price differential to 52 percent. This translated into a per acre cost of about US$59 and US$39 for genetically engineered and conventional soybean seed, respectively (National Agricultural Statistics Service 2012).

Of course, the biogopolists are quick to justify higher seed prices based on the claimed heavy research and development and regulatory costs associated with discovering a new genetically engineered seed and bringing it to market.[21] Although there is certainly some truth to such claims, cost arguments are perhaps more suitably interpreted as further evidence of the dispensability, not least from an allocative efficiency perspective, of genetically engineered seeds. In fact, the gaping cost differentials between conventional and genetically engineered breeding techniques are truly staggering. While traditionally bred varieties typically come in under US$1 million, Monsanto has stated that a genetically engineered variety requires at least ten years to develop at a cost of between US$100-150 million. Similarly problematic is the fact that the research dollars being expended by the largest agricultural biotechnology companies are for a relatively small number of crops.

By comparison, conventional breeders, assuming they had access to sufficient research and development funds, could introduce between 100 and 150 new varieties in less time (ETC Group 2009). Moreover, there is some evidence that, as the seed industry has consolidated, companies have been sponsoring less research relative to their market share (Schimmelpfennig, Pray, and Brennan 2004). And although biotechnology companies are quick to assert that the putative higher yields of their genetically engineered seeds more than offset any price increases, recent evidence heavily contests such claims. We will elaborate this more fully in Chapter 5, but it is worth noting here that, despite claims advanced by the biotechnology industry, intrinsic yield increases, as well as disease resistance, grain size, maturation period, and responses to biotic and abiotic stresses, are attributable largely to the robustness of the traditionally bred germ plasm rather than to the one or more genetically engineered traits inserted into the seed.

Industry yield claims notwithstanding, the more plausible explanation is that the heavy industry consolidation of the past decade has resulted in an oligopolistic market structure in which the largest actors are able to increase prices while reducing the supply of non-genetically engineered seed varieties. Neil Harl, an agricultural economist at Iowa State University who has studied the seed industry for decades, estimates that Monsanto has "control over as much as 90 percent of (seed genetics). This level of control is almost unbelievable ... The upshot of that is that it's tightening Monsanto's control, and makes it possible for them to increase their prices long term. And we've seen this happening the last five years, and the end is not in sight" (as cited in Leonard 2009, para. 12).

In his analysis of the capitalist control of agriculture, Richard Lewontin (2000a) outlines how the increasing vertical and horizontal integration of multinational agribusinesses, in conjunction with new biotechnologies, has enabled industrial capital to simultaneously impose a suite of high-cost inputs on farmers and exert pressure on them to produce products demanded by a handful of major purchasers with enough market clout to obtain prices for agricultural outputs that serve the interests of capital's bottom line rather than that of farmers. That is, agriculture is increasingly characterized by near monopsony and monopoly at the input and output stages, respectively, of production. In fact, according to data released by Agriculture Canada and Statistics Canada, agricultural input suppliers have captured 99.6 percent of the wealth produced on Canadian farmland over the past twenty-four years. So while farmers have produced an overall average of $388 per acre per year, their net income, exclusive of government payments, has been a mere $1.45 per acre per year. Put another way, farm input and service corporations are receiving 266 times more wealth than the actual agricultural producers in this country (National Farmers Union 2009).

As Karl Kautsky (1988) argued over a century ago, farmers' labour is objectified in their products, for which, as a result of unequal exchange relations, they are paid less than their full value. And despite still owning the land, buildings, and equipment, farmers are not in a position to exploit these means of production in alternate economic ways. The result is that "the farmer becomes a mere operative in a determined chain whose product is alienated from the producer. That is, the farmer becomes proletarianized ... What the farmer has gained is a more stable source of income, at the price of becoming an operative in an assembly line" (Lewontin 2000a, 97, 104; see also Kloppenburg 2004). And although perhaps a stable source of income, it barely exceeds subsistence levels, as the figures discussed earlier illustrate. We are witnessing the emergence of "biolords," the biological version of Roberto Verzola's "cyberlords" – "the propertied class of the information society, [who] control either a body of information or the material infrastructure for creating, distributing or using information" (Verzola 2000, 92). Embodying and combining the biological and the informational, one particular genetically engineered technology, Terminator, holds the promise of further solidifying the power and control exercised by the biolords.

Terminator Technology
Genetic use restriction technology (GURT, popularly known as "Terminator technology") is an umbrella term that encompasses a wide array of interactive or interdependent genes being engineered to work in combination with

some form of chemical or environmental catalyst (for example, ethanol or heat shock) to regulate the expression of the plant's genetically engineered trait, or traits, in the case of stacked seeds. In addition to Monsanto, Syngenta,[22] DuPont/Pioneer (DuPont acquired Pioneer Hi-Bred in 1999), Aventis,[23] and BASF have also developed Terminator technology, which, depending on which company you consult, is known as "gene protection," "gene expression control," or "genetic use restriction."

There are two types of GURTs: varietal (V-GURTs) and trait-specific (T-GURTs). The former controls the reproductive functions of the seed, while the latter are designed to regulate control over the functionality of a particular transgene that has been inserted into the seed, such as herbicide or insect tolerance (ETC Group 2007). These biotechnologies involve the genetic manipulation of seeds to render them infertile after harvest so that farmers have to purchase new seed each year from a handful of international seed suppliers rather than save and reuse them, as millions have been doing for millennia. The ultimate danger from this produced form of organic obsolescence is that multinational seed and agrochemical companies will acquire the capacity to control the food chain and the world's food supply (Heller 2001; Shiva 2001b). According to Vandana Shiva (2001b, 82), "[i]n one broad, brazen stroke of his hand, man will have irretrievably broken the plant-to-seed-to-plant-to-seed cycle – the cycle that supports most life on the planet. The new technologies and system mean no seed and no food unless you buy more seed." It should be noted that, at this point, GURT remains unproven science, and, thus, there exists the possibility that its widespread deployment may be halted not so much by protest but, rather, by its potential for inducing significant environmental risks and its lack of efficacy and practicality. Although Monsanto's previous chief executive officer, Robert Shapiro, penned an open letter in 1999 to the Rockefeller Foundation in which he stated that the company would not commercialize sterile seed technologies, the letter nonetheless left the door open for future development and deployment: "We [Monsanto] are not currently investing resources to develop these technologies, but we do not rule out their future development and use for gene protection or their possible agronomic benefits" (Shapiro 1999).

Industry, in its attempts to mitigate negative public opinion surrounding this technology, asserts that genetically induced seed sterilization offers a safety mechanism to prevent gene flow and cross-fertilization. This apparent new need for genetic innovation confirms the prognosis by many early critics of genetic engineering that such technology will give rise to a "genetic treadmill" in which new technologies have to be developed to respond to the hazards inflicted on the environment and human health by first generation biotechnology. The irony remains lost on the biogopolists. Such industry contentions notwithstanding, there is some concern that the transgenes inserted into Terminator crops, which actually do have a small window of fertility, might escape into other crops or the natural environment. Given the ability of nature to adapt, coupled with the fact that such technology has yet to be tested on any large scale, sterility could spread among seeding plants throughout the environment, which would have catastrophic consequences for the environment and the world's food supply (de la Perrière and Seuret 2000; Shiva 2000). Commenting on the often illusory nature of apparent "human conquest over nature" under capitalist social relations, Friedrich Engels (1940 [1898], 291-92) pointed out long ago that "each such conquest takes its revenge on us. Each of them, it is true, has in the first place the consequences on which we counted, but in the second and third places it has quite different, unforeseen effects which only too often cancel the first."

In truth, the upside for multinational seed companies of this technology is that it will render superfluous the need to pressure farmers to sign contracts that prohibit the replanting of harvested seeds and the consequent expenditure of resources to police such contracts. Moreover, Terminator technology provides multinational seed companies with a means of valorizing genetically engineered seeds in countries where Western standards of intellectual property rights are poorly developed or haphazardly enforced (Kloppenburg 2004). In a study that examined the potential financial impact of Terminator technology on just seven countries (Canada, Brazil, Argentina, Pakistan, Philippines, Ethiopia, and Iran), the ETC Group estimated that the commercialization of Terminator seeds could cost farmers in these nations over US$1.2 billion annually in total. Although Canadian farmers do not engage in seed saving on the same scale as farmers in poorer countries, the bill could still run around US$85 million for farmers in this country (ETC Group 2006c), a not insubstantial sum when one considers that the large multinationals have substantially squeezed farming profit margins at both the input and output stages of production.

Highlighting the connections between biological and knowledge commons, legal scholar Dan Burk (2004) draws an interesting analogy between GURTs and digital rights management systems, both of which employ technological mechanisms to restrict access to, and use of, their underlying technologies, be they genetically engineered seeds or digital media. Access restriction technologies can be particularly insidious if rights holders develop them in such a manner that they curtail the legitimate uses enshrined in, and protected by, public access doctrines, such as fair dealing, or call into question exemptions embodied or read into patent and plant breeders' rights legislation for farmers and researchers. Similar to the content of the debates now raging in regard to technical protocols and their potential to circumscribe access to information, technical standards in the biological realm have the capacity to constrain the behaviour of users and their legitimate use of a technology in ways that overstep the intellectual property rights established through legislation (Lessig 1999; Reidenberg 1998). According to Burk (2004, 1567), "[p]roducers who employ such lock-out technology may in essence become private legislatures, imposing rules of usage without regard to the broader public interest that informs democratic rule making ... The instantiation of a proprietary rule in genetic code, which following Reidenberg, we might call 'lex genetica,' is the first example of regulation through genetic code, but it is unlikely to be the last." Perhaps more practically, even if the enforcement of a GURT patent creates a situation in which the rights holder overreaches in attempting to stop people from "hacking" a technological lock-out system, it probably remains well beyond the ability of laypeople to circumscribe GURTs. No group of biological hackers, akin to those who have succeeded in reverse engineering many digital rights management systems, has yet emerged to provide the wider community with a toolbox of GURT circumvention tools (Burk 2004).[24]

Despite opposition from Canadians (for example, in an effort to emphasize the need for government accountability, a group of about 1,000 people in Ottawa staged a public "trial" of Terminator technology on 20 March 2006), the Canadian government has lobbied on numerous occasions to overturn the *de facto* international moratorium on GURTs that was adopted by the UN *Convention on Biological Diversity (CBD)* in 2000.[25] The Canadian government advocates a "case-by-case risk assessment" for seed sterilization technologies, which, according to the ETC Group, "would open the door to field trialing and commercialisation of sterile seed technology" (ETC Group 2006c, para. 3). The case-by-case approach has been rejected staunchly by a large number of European countries as well as by the CBD. Paragraph 23 of Decision V/5, made by the fifth Conference of the Parties (COP) to the CBD, reads as follows:

> [I]n the current absence of reliable data on genetic use restriction technologies, without which there is an inadequate basis on which to assess their potential risks, and in *accordance with the precautionary approach,* products incorporating such technologies should not be approved by Parties for field testing until appropriate scientific data can justify such testing, and for commercial use until appropriate, authorized and strictly controlled scientific assessments with regard to, inter alia, their ecological and socio-economic impacts and any adverse effects for biological diversity, food security and human health have been carried out in a transparent manner and the conditions for their safe and beneficial use validated. In order to enhance the capacity of all countries to address these issues, Parties should widely disseminate information on scientific assessments, including through the clearing-house mechanism, and share their expertise in this regard.[26]

In spite of lobbying efforts by the Government of Canada to rescind the moratorium on Terminator technology at the eighth COP to the CBD, which was held in Brazil in March 2006, the majority of members voted instead to preserve and strengthen it (Conference of the Parties to the Convention on Biological Diversity 2006). Nonetheless, a number of commentators expect that Terminator technology will continue to be developed by the biotechnology industry and that it will maintain a place on the agenda at upcoming meetings of the COPs to the CBD. According to Pat Mooney, "the only solution is a total ban on the technology once and for all ... The next step is for all national governments to enact national bans on Terminator as Brazil and India have done" (ETC Group 2006b, para. 8). The global peasant farmer movement, La Via Campesina, has demanded that "governments and ... international institutions ... prohibit biopiracy, and patents on life (animal, plants, parts of the human body) including the development of sterile varieties through genetic engineering" (as cited in King and Stabinsky 1999, 86).[27] Similarly, the Consultative Group on International Agricultural Research (CGIAR), an international organization concerned with achieving sustainable food security and reducing poverty in developing countries through research in agriculture, forestry, fisheries, policy, and the environment, declared in 1998 that it would not facilitate the introduction of Terminator technologies into farmers' fields: "The CGIAR will not include in its plant-breeding material any genetic system aimed at preventing the germination of seeds because of the potential risks of a sterilising gene flow through pollen, the non-viability of exchange or sale of seeds, the potential negative impact on genetic diversity and the importance of the breeding programme and reproduction on the farm for agricultural development" (as cited in de la Perrière and Seuret 2000, 31).[28] Instead, the CGIAR advocates policies that promote local seed saving (a practice that it believes contributes to the sustainability of local food systems), protect biodiversity, and serve to maintain or re-establish food security. For these same reasons, the CGIAR contends that international aid programs could be improved if they were to incorporate efforts to ensure local seed independence (Cummings 2008).[29]

The CBD's agreed moratorium on Terminator technology notwithstanding, a significant amount of research is still being conducted on seed sterilization technologies. An emerging hub for such research activity is Europe, where the European Union invested millions of Euros in its Transcontainer Project, which was tasked with developing biological mechanisms to contain transgene flow from genetically engineered plants and trees.[30] This three-year project, which ran from May 2006 to 2009 and had a budget of €5.38 million, involved thirteen public and private research partners and was driven by the underlying assumption of "co-existence," a controversial notion that genetically engineered crops and traditional crops can co-exist (or that it is possible to determine some level of "acceptable" genetically engineered contamination). As one organic farmer points out, "'coexistence' is a nice term, but it turns out that coexistence means we put up with their contamination" (Jenkins 2007, para. 57). Members of the ETC Group dismissed this research project as a publicly funded effort to aid the biotechnology industry in breaching the European public's resolute rejection of genetically engineered foods and crops. Although the website for the project indicates that eventually all of the research will be made publicly available, as of early 2012 only two peer-reviewed article citations were listed.[31]

Similarly troubling is the fact that Terminator technology has been continually refined since its initial development in the early 1990s, with researchers now experimenting with environmental or chemical "switches" that can turn on or off a plant's reproductive abilities (dubbed "zombie seeds" by the ETC Group). Significant research is also being conducted on developing means to excise transgenes from a genetically engineered plant at a particular time in its growth cycle (christened "exorcist seeds" by the ETC Group),[32] as well as ways to kill a plant that contains "conditionally lethal" transgenes (referred to as "pull-the-plug plants" by the ETC Group).[33] One of the most significant benefits to industry from such genetic technologies (along lines that are similar to genetically engineered seed tolerant of particular herbicides) is that they will compel farmers to accept responsibility for trait control. Farmers will be forced to purchase these new proprietary inducers (most likely chemicals) that are required for the plants to either express or repress the desired genetic trait, including fertility. Should excision techniques become more advanced, it is quite probable that companies will lobby vigorously to have their products categorized as being genetic engineering free in order to avoid labelling requirements in those jurisdictions that have them, such as the European Union (ETC Group 2007). As Jim Thomas of the ETC Group argues, gene giants such as Monsanto and Dow AgroSciences are trying to do an end run around the moratorium on Terminator technology: "They're going to go to the next [biodiversity] convention, and argue that sterile seeds are not a problem any more because the sterility is reversible." Thomas goes on to contend that if they succeed it will "open the floodgates" to other sectors of the biotechnology industry that are currently too controversial, such as pharmacrops and genetically engineered plants that produce industrial products (as cited in Patterson 2007, A5).

In general, one might argue that capital's co-optation of generations of socially produced knowledge as it relates to agricultural biotechnology and genetic engineering is robbing living organisms of their natural commonness (free reproducibility), highlighting the paradox of the commercial logic inherent in genetically engineered seeds. On the one hand, it is claimed that

genetically engineered seeds will deliver increased yields and great abundance, while, on the other hand, those who refuse or are unable to pay for
access to them are simultaneously excluded. Perhaps more importantly, if
biotechnology companies are successful in infiltrating the market with
Terminator technology, we can, based on past experience, expect that more
and more seed varieties will be engineered to express these traits.

Since the multinational seed companies exercise such extensive control
over the world seed market, they are in a position to eliminate the supply
of many traditional varieties, thus compelling farmers to purchase Terminator
technology. Not only would this possibility reduce the number of different
types of seeds available to farmers, thus threatening biodiversity and the
biological commons, but it would similarly exercise a ripple effect on the
knowledge commons, as agricultural producers are further removed from
their traditional relationship to seeds and plant breeding, including all of
the knowledge that this relationship entails. Thankfully, there exists a strong
coalition of civil society groups from across the globe, including Canada,
that is dedicated to stopping the introduction of Terminator technology
into farmers' fields.

Terminating Terminator

Building on momentum generated at the March 2006 meeting of the COP
to the CBD, at which the Canadian Biotechnology Action Network (CBAN)
successfully helped to prevent efforts by the Canadian government to overturn the international moratorium on Terminator seeds, CBAN members
decided at their first annual general meeting in October 2006 that they would
take on the Ban Terminator campaign in Canada as their pilot project. The
Ban Terminator project seeks to ensure the continued international moratorium on GURTs and to achieve a complete ban in Canada. As Lucy Sharratt
points out, Ban Terminator was particularly important as a first campaign
because it not only allowed CBAN to insert itself into the debate on biotechnology in this country while developing its organizational capacity, but it
also helped establish the discourse and work on genetic engineering within
a strong social justice framework rather than through a food security concept
or individual health concern perspective. Groups working with CBAN on
this campaign include the National Farmers Union, the Council of Canadians, the ETC Group, the National Council of Women of Canada, Inter
Pares, the Saskatchewan Organic Directorate, Beyond Factory Farming,
Rightoncanada.ca (an Internet and public advocacy campaign of the Rideau
Institute that seeks to put human rights back on Canada's political agenda),
GenEthics of Australia, Oxfam, and the National Family Farm Coalition of
the United States, among others. The steering committee of the Ban
Terminator campaign includes the following groups: the ETC Group, GRAIN,

the Indigenous Peoples Council on Biocolonialism, the Intermediate Technology Development Group, the Pesticide Action Network in Asia and the Pacific, the Third World Network, and La Via Campesina. There is also an international campaign to ban seed sterilization technology in which CBAN and over 500 other movements from around the world participate.

Sharratt strikes an optimistic tone when speaking about the chances of success, pointing in part to the victory achieved in March 2007 to continue the international moratorium on Terminator technology. Reflecting on the national scene, she outlines the Private Member's Bill (C-448) to ban Terminator technology, which was introduced on 31 May 2007 by Alex Atamanenko, agriculture critic and New Democratic Party Member of Parliament for British Columbia's Southern Interior. This bill also contained a consequential amendment to the *Patent Act* that would have prohibited patents for Terminator technologies.[34] With the dissolution of the second session of the thirty-ninth Parliament on 7 September 2008, this bill died on the Order Paper. On 30 March 2009, Atamanenko reintroduced his Private Member's Bill (C-353) to ban the release, sale, importation, and use of Terminator technology. This bill also died on the Order Paper when the Harper government, in a very controversial move, prorogued the second session of the fortieth Parliament on 30 December 2009.

Nonetheless, Sharratt believes that, if CBAN can secure enough resources, it can facilitate a general trend toward success. To illustrate her point using an example, Sharratt outlines how the Canadian government is talking seriously about genetically engineered trees and the combination of genetically engineered trees and Terminator technology, all the while knowing that the Canadian public does not want such trees or, at least, that it would be controversial. If CBAN can gather the resources to flag these issues publicly, it can possibly reverse plans to commercialize those products that have received regulatory approval, but have not yet been marketed, of which genetically engineered alfalfa is another good, and controversial, example (discussed more fully in the following chapter). Sharratt contends that various oppositional movements in Canada have laid such good groundwork through solid research and their engagement and interaction with Canadian government regulatory agencies that, when they appear at meetings, it now matters. She further believes that government regulators are gradually being convinced that they need to be concerned about the critique and the presence of mobilized opposition in this country. CBAN believes that if it can sufficiently expand the general public discourse around Terminator technology to admit clear and strongly articulated economic and social justice concerns, then the regulators' assertions that this technology is safe will be unmasked as being a wholly inadequate response to Canadians' apprehension over genetic engineering and agricultural biotechnology.

Theoretical Outlook 1: Primitive Accumulation and Enclosures

As outlined in the introduction, this book seeks to interrogate critically, and situate theoretically, the developmental trajectory of agricultural bio-technology in Canada. It will do so in a way that not only problematizes the various capital-employed, federal policy-supported practices facilitating the commodification of this technoscience but that also reflects the perspectives of those people and groups mobilizing in response to such corporate capture. Very few of the informants who participated in this research articulated their opposition to biotechnology utilizing the terminology of "enclosure" and "commons." However, these concepts nonetheless resonate strongly with the concerns they expressed about their desire to rescue and reinvigorate agricultural products and processes from expanding corporate encroachment.

Conceptualizing Primitive Accumulation

Karl Marx (1992) provides his most expanded discussion of primitive accumulation in volume 1 of *Capital,* where he develops a critique of the "so-called primitive accumulation" articulated by classical political economists. At its most basic, primitive accumulation can be understood as providing the origin of the separation between producers and the means of production (including nature) – a separation that is responsible for the alienated character of labour and thus for defining the opposition inherent in capitalist social relations. The alienation of labour under capitalist social relations manifests itself in manifold ways, two of which are most germane to the current work. The first consequence of the estrangement of practical human activity – of labour – is a resulting alienated relationship between the worker and the product of labour, which, because of private property and the capital-labour relation, appears as something alien – as a power independent of the actual producer. Since the product of the worker's labour is an alien object that belongs to the capitalist paying her wage, the more she toils under capitalist social relations the more powerful becomes the alien, objective world she brings into being against herself.

Moreover, this estrangement – this loss of the object of production – has implications for the relation between humanity and nature, given that "the worker can create nothing without *nature*, without the *sensuous external world*" (Marx 1975a [1844], 325 [emphasis in original]). The implication of this relationship is that alienation reflects the estrangement of the worker not only from the object of her labour but also from her active role in transforming nature. Considered from this perspective, the alienation of humanity's relation to nature becomes a consequence of the practical organization of human life under capitalist social relations. According to Paul Burkett (1999), it is precisely by conceptualizing the separation of producers from the necessary means of production achieved through practices of primitive accumulation that Marx is able to develop a critique of the anti-ecological tendencies inherent in capitalist relations of production. These tendencies, as we will see later in this chapter, are being exacerbated rather than abated through agricultural biotechnology. On this reading of Marx, we note that he, despite a number of interpretations to the contrary, rejected reductionist, teleological accounts of the natural world. He instead articulated the need to employ materialist and dialectical analyses that admit the possibility of contingency that is free of determinism in an open, evolutionary process of natural history. Put more explicitly, Marx envisioned and stressed the possibility of marrying a materialist conception of history to a materialist conception of nature, implying a connection between the alienation of human labour and the alienation of human beings from the natural world based on material inter-relationships that co-evolve dialectically with one another (Burkett 1999; Foster 2000).

A second and related aspect of the alienation of labour encompasses the relationship of the worker to the act of production within the labour process. Under the control of capitalist production processes, not only is the product of labour objectified in an alien object that holds power over the actual producer, but the corresponding form of productive activity also renders the worker's own labour as something alien and opposed to him, reflecting an estrangement from himself and from his own activity. Rather than offering satisfaction in and of itself, alienated labour is external to the worker, something sold to and thus belonging to someone else. Through its saleability, the relationship of the worker to his activity becomes one of what Marx refers to as "self-estrangement": "[E]strangement manifests itself not only in the result, but also in the *act of production*, within the *activity of production* itself ... So if the product of labour is alienation, production itself must be active alienation, the alienation of activity, the activity of alienation. The estrangement of the object of labour merely summarizes the estrangement, the alienation in the activity of labour itself" (Marx 1975a [1844], 326 [emphasis in original]). The rising prominence of science and abstract knowledge in capitalist production processes would similarly portend the further alienation of direct producers. The alienated "general social powers of labour, including natural forces and scientific knowledge ... appear most emphatically as forces not only alien to the worker, belonging to *capital*, but also directed in the interests of the capitalist in a hostile and overwhelming fashion against the individual worker" (Marx 1994, 29-30 [emphasis in original]).

Indeed, Marx long ago contemplated the connection between the development of science and capitalist social relations, arguing that the latter encourage "the development ... of the natural sciences to their highest point" (Marx

1993, 409). More specific to this chapter, capital exhibits a tendency "to give production a scientific character," whereby "the entire production process appears as not subsumed under the direct skilfulness of the worker, but rather as the technological application of science" (699). Similar to the separation of producers from their physical means of production (including nature), as the social knowledge requisite for social reproduction becomes increasingly vital to capitalist accumulation, *"knowledge thus becomes independent of labour and enters the service of capital."* As Marx goes on to explain in this same passage, "this process belongs in general to the category of the *attainment of an independent position by the conditions of production* vis-à-vis labour. This separation and autonomisation, which is at first of advantage to capital alone, is at the same time a *condition for the development of the Powers Of Science and Knowledge"* (Marx 1994, 57 [emphasis in original]).

Yet capital is adept at exploiting and appropriating science and knowledge in service of its production processes. Under the dominance of capitalist social relations, we witness a further instance of the social separation of the conditions of production from the control of the direct producers in service of capitalist valorization: "[A] *separation of science,* as *science applied* to production, from *direct labour,* whereas at earlier stages of production the restricted measure of knowledge and experience is directly linked with labour itself, does not develop as an autonomous power separated from labour" (Marx 1994, 33-34 [emphasis in original]). Marx's discussion here of the appropriation of science from the direct producers in service of capitalist accumulation mirrors his elaboration of the various forms of abstract knowledge encapsulated in the collective social intelligence of the general intellect that is increasingly objectified in fixed capital. By separating science from direct labour, capital induces a "concentration ... [and] ... development into a science of the knowledge, observations and craft secrets obtained by experience and handed down traditionally, for the purpose of analysing the production process to allow the application of the natural sciences to the material production process" (34). Capital's "application of science ... rests entirely on the separation of the intellectual potentialities of the process from the knowledge, understanding and skill of the individual worker, just as the concentration and development of the conditions of production ... rests on the divestiture – the separation – of the worker from those conditions" (34).[35] Thus, already a century and a half ago, Marx recognized the expanding capitalist appropriation of science and other abstract knowledge constitutive of the general intellect: "It becomes the task of science to be a means for the production of wealth; a means of enrichment ... science appears as a *potentiality alien* to labour, *hostile* to it and *dominant* over it" (32, 34 [emphasis in original]). As the empirical evidence examined in the following chapters will demonstrate, this dual form of alienation inheres in the science of agricultural biotechnology and its attendant technological applications. In fact, biotechnology could be construed as an application of scientific knowledge that accelerates the development of capitalist-controlled forces of production by socially separating this technoscience from the control of direct producers in service of capitalist accumulation imperatives.

While it is true that farmers were historically beyond the purview of the capital wage labour relationship as it was emerging within the industrializing factories of the early to mid-nineteenth century that most occupied Marx when he developed these ideas, they nonetheless resonate with the contemporary situation in respect of agricultural biotechnology. So while there might not be a direct wage relationship between agricultural producers and capital, large transnational firms, particularly Monsanto, have developed alternative mechanisms of control in service of their accumulation imperatives, all of which tend to animate these same forms of alienation in our contemporary conjuncture. For example, the technology use agreements to which agricultural producers must adhere provide Monsanto with a substantial level of control over not only its seeds but also the farming practices of growers. Similarly, as we will examine in more substantial detail in Chapter 4, patents constrain the natural reproducibility of seeds – a critical agricultural input long considered to be a common resource developed through co-operative social labour over centuries – in a way that enforces artificial scarcity upon a natural resource. Finally, and relatedly, genetic contamination and the emerging evolution of herbicide-resistant weeds (as elaborated more fully in Chapter 5) portend very real and detrimental material consequences for the natural resources that provide the basic prerequisites for sustainable agricultural production. By reworking Marx to our current historical conditions, we not only demonstrate the enduring relevance of his ideas but also, in fact, continue in the tradition of Marx himself, who stressed the need to develop and apply materialist analyses suitable to their historical epochs.

Primitive accumulation thus represents a historically specific and class-differentiated relationship of control over the necessary means of social production. Most contemporary scholars engaging in a reinvigoration of primitive accumulation as a theory for comprehending contemporary capitalist development tend to agree on three additional basic points about this concept.[36]

First, primitive accumulation should be understood as a continuous process that remains vital for capitalist accumulation. As Marx (1992, 874) informs us, "the capital-relation presupposes a complete separation between the workers and the ownership of the conditions for the realization of their labour. As soon as capitalist production stands on its own feet, it not only maintains this separation, but reproduces it on a constantly expanding scale." That is, the separation between producers and the means of production, a central category of Marx's critique of political economy, is the constitutive presupposition of accumulation and, thus, common to both primitive accumulation and accumulation in general – capital presupposes this separation. In Marx's own words, "the manner in which the capitalist mode of production *expands* (takes possession of a greater segment of the social area) and subjects to itself spheres of production as yet not subject to it ... entirely reproduces the *manner* in which it arises altogether" (Marx 1994, 327 [emphasis in original]). The *Grundrisse* (1993, 462) similarly weighs in on the issue: "Once this separation is given, the production process can only produce it anew, reproduce it, and reproduce it on an expanded scale." As Marx (1967, 246) again points out in volume 3 of *Capital*, accumulation is really nothing more than primitive accumulation – which he conceptualizes in volume 1 in terms of separation – "raised to the second power."[37] In the third part of his *Theories of Surplus Value*, Marx (1972, 315) is even more explicit about the continuous nature of primitive accumulation, contending that accumulation "reproduces the separation and the independent existence of material wealth as against labour on an ever increasing scale."[38] For this reason, accumulation "merely presents as a *continuous process* what in *primitive accumulation* appears as a distinct historical process" (272 [emphasis in original]). We thus note that Marx's discussion of primitive accumulation contains a basic ontological connection between primitive accumulation and expanded reproduction, such that accumulation in general represents a form of intensified primitive accumulation (Bonefeld 2001, 2002; De Angelis 2001, 2007; Mandel 1975).

Similar in principle, the two forms of accumulation differ in their historical basis and their intensity. Whereas primitive accumulation entails an *ex novo* separation, accumulation proper follows from the expanded reproduction of the separation between producers and the means of production. The *ex novo* separation is produced through extra-economic means that set the context for the opposition between producers and the means of production and that give rise to the particular alienated character of social labour under the capitalist mode of production.[39] A typical historical and enduring mechanism for realizing the *ex novo* separation between producer and the means of production is enclosure. The oft-cited historical example is the enclosure of common lands in England.[40]

More contemporary examples occurred in the last decade of the twentieth century throughout Africa, Latin America, and Southeast Asia as a direct result of various structural adjustment programs imposed on heavily indebted nations that, among other things, transformed traditional land tenure systems to facilitate the privatization of vast expanses of once common lands (De Marcellus 2003; Federici 1992; Midnight Notes Collective 1992; Routledge 2004).[41] Other instances range from water privatization to the enclosure of knowledge and natural resources through overly restrictive intellectual property regimes.[42] Enclosures might also emerge as a by-product of a particular accumulation process or what economists refer to as a "negative externality." In both cases, enclosing the commons augments the disciplinary capacity of capital because such practices render greater numbers of people dependent upon the market in order to reproduce their livelihoods – that is, they enforce the basic element of primitive accumulation.

While there is a temporal element that distinguishes primitive accumulation from accumulation proper – indeed, the *ex novo* separation between producers and the means of production represents an *a priori* historical event – the critical distinction between the two is grounded less in temporality and more in the conditions and exigencies that comprise the separation. As Marx (1993, 459 [emphasis removed]) tells us in the *Grundrisse*, "once developed historically, capital itself creates the conditions of its existence (not as conditions for its arising, but as results of its being)." That is, once produced, capital must reproduce the separation between producers and the means of production (and, indeed, expand this reproduction). In order to normalize capitalist social relations, increasingly larger swaths of the population must be brought into the fold of capitalist commodity production through "the silent compulsion of economic relations [that] sets the seal on the domination of the capitalist over the worker. Direct extra-economic force is still of course used, but only in exceptional cases. In the ordinary run of things, the worker can be left to the 'natural laws of production,' i.e., it is possible to rely on his dependence on capital, which springs from the conditions of production themselves, and is guaranteed in perpetuity by them" (Marx 1992, 899). The outcome of these processes is twofold, transforming labour power and the means of subsistence into commodities in a process that establishes both a labour force from which capital can extract surplus value and a market through which that surplus value can be realized.

Over time, through the commodity form, capital is able to displace exchange relations based on use value by raising exchange value to prominence, which ultimately compels independent producers to accept the yoke of capitalist relations of production. Obviously, this process entails a linear temporal dimension in which primitive accumulation must precede the emergence of the capitalist mode of production as the preponderant system of social reproduction. However, once we recognize that primitive accumulation satisfies a precondition for the expansion of capital accumulation, the temporal element assumes a secular form that encompasses not only the period in which the capitalist mode of production emerges but also the reproduction and expansion of the capitalist mode of production. The implication of this process is that capitalist production entails the production of surplus value as well as the reproduction of social relations of production in an inverted form – social production alienated through private property and the commodity form (Bonefeld 1992; De Angelis 2007).

The second point about primitive accumulation is that it assumes a variety of forms, including the privatization of once public goods, which has the ultimate effect of reorganizing class relations in favour of capital. Here, we are talking about the "social commons," which encompasses those areas of existence that emerged as commons through active social movements in the past and were subsequently formalized through institutional norms and practices. For example, the rights and provisions typically associated with the welfare state, such as health, education, pension, and unemployment benefits, provide access to social wealth without a corresponding labour requirement (De Angelis 2007). In a similar vein, John McMurtry (1998, 24) speaks of a "civil commons," which signals "what people ensure together as a society to protect and further life, as distinct from money aggregates." As both David Harvey (2003) and James McCarthy (2004) point out in respect of this aspect of primitive accumulation, international trade regimes, which impinge on domestic governance, facilitate capital's appropriation of the conditions of production.

International trade agreements, particularly the World Trade Organization and its *Agreement on Trade Related Aspects of Intellectual Property Rights,* and their associated administrative bodies increasingly circumscribe the ability of sovereign states to enact laws and regulations within their territories.[43] These international trade treaties, which are enforced by private adjudication bodies that focus solely on neo-liberal accumulation imperatives, offer multinational corporations a back door to circumvent national regulators who are "forced" to respond, even if only nominally, to broader societal interests. In this constructed environment, the rights of trade and investment enjoy precedence over all other rights (McCarthy 2004). For example, the contemporary intellectual property system functions as an important mechanism for primitive accumulation by stripping indigenous populations of their rights to natural resources and knowledge that have been developed in common over centuries. In what can only be regarded as blatant acts of biopiracy, a few multinational corporations are instead appropriating rights of control over, and access to, such resources and the information and knowledge embodied in these physical artifacts (Shiva 1997, 2001b). In fact, Harvey (2003, 148) speaks of "the wholesale commodification of nature in all its forms," arguing that "the rolling back of regulatory frameworks designed to protect labour and the environment from degradation has entailed the loss of rights. The reversion of common property rights won through years of hard class struggle (the right to a state pension, to welfare, to national health care) to the private domain has been one of the most egregious of all policies of dispossession pursued in the name of neo-liberal orthodoxy."

The third feature of primitive accumulation speaks to its spatial ambition. Despite a general ethnocentrism present in Marx's work (an ethnocentrism that Marx readily admits), he discusses not only the historical, but also the global, elements of the processes of primitive accumulation, through which a privileged minority relentlessly pillaged the means of production from the people of pre-capitalist civilizations around the world: "The discovery of gold and silver in America, the extirpation, enslavement and entombment in mines of the indigenous population of that continent, the beginnings of the conquest and plunder of India, and the conversion of Africa into a preserve for the commercial hunting of blackskins, are all things which characterize the dawn of the era of capitalist production. These idyllic proceedings are the chief moments of primitive accumulation" (Marx 1992, 915). Ensuring an expanded reproduction of capital depends upon enveloping new spheres and peoples within the web of capitalist social relations of (re)production. Although long a feature of capitalist expansion in the global South, primitive accumulation today is assuming an integral role in capitalist accumulation processes in the global North, particularly given the vital importance that information and knowledge play in generating value for multinational corporations.

Having historically extended the territorial reach of capitalist social relations through colonialist expansion and the imposition of private property rights across the globe, primitive accumulation in the twenty-first century has become both more extensive and intensive, affecting an enormously broad range of spatio-social activity. In practice, primitive accumulation motivates efforts by capital to enclose more and more areas of our social and natural being, as evidenced particularly within the agricultural biotechnology sector. Having run up against a number of natural limits to growth in the form of finite territorial resources, biotechnology offers capital a new suite of tools to plumb the depths of biological existence at the genetic level in search of new sources of capital accumulation. Through biotechnology, capital is able to shift its practices of primitive accumulation from an expansive and extensive plundering of the world's geography that itself is being increasingly exhausted toward an intensive and interior exploitation of the natural world (De Angelis 2007; Harvey 2003, 2006; Katz 1998). Put another way, the technoscience of biotechnology represents a contemporary modality of primitive accumulation that, as the material presented later in this volume attests, facilitates intensified enclosures of social co-operation and production.

Industry consolidation, bundling practices, coerced technology use agreements, and the biologically induced sterility being developed by the Monsantos, Syngentas, Bayer CropSciences, and BASFs of the world, which wring maximum surplus value out of agricultural production chains, seem to bear out Marx's contention that accumulation, in general, is a form of intensified primitive accumulation. I contend that Terminator technology qualifies as a new modality of capitalist primitive accumulation that strives to circumscribe natural cycles of reproducibility in a manner that forces agricultural producers to purchase a vital input, freely given by nature for millennia, from an oligopolistic set of supposed life science companies. Writing in the *Grundrisse,* Marx (1993, 527 [emphasis in original]) long ago anticipated the development of precisely such a scenario under capitalist agriculture: "[I]f agriculture itself rests on scientific activities – if it requires machinery, chemical fertilizer acquired through exchange, seeds from distant countries etc., and if rural, patriarchal manufacture has already vanished – which is already implied in the presupposition – then the machine-making factory, external trade, crafts etc. appear as *needs* for agriculture." This suite of technological mechanisms represents the extra-economic prerequisite to biocapitalist production that not only endures in contemporary society but is also being extended across the globe.

I further agree that Terminator technology represents the biological in-stantiation of those digital exclusionary technologies seen in the music and software industries. Similar to the way information today is easily copied, transformed, and circulated beyond the control of intellectual property rights holders, the natural progenitive capacity of seed renders strict control over it and the genetic information it contains quite difficult. However, Terminator technology – or, to carry the analogy further, these "biological rights management systems" – could potentially provide their owners with the ultimate form of control in mediating the relationship between people and not only information and knowledge but also the "stuff of life" in a manner that encloses both the biological and the knowledge commons. Through this powerful technology that corrupts the natural rhythms of na-ture in service of accumulation imperatives, we can discern the emergence of a type of capitalist biopower capable of regulating and controlling life at the genetic level.

Similar to what Marx (1993, 799) outlined in respect of industrial produc-tion, refracted through agricultural biotechnology traditional knowledge appears increasingly as alien and external to agricultural producers "in the same proportion as science is incorporated in it as an independent power." Capitalist controlled agricultural biotechnology represents an application of scientific knowledge designed to accelerate the development of the forces of production, and thus accumulation, in a way that simultaneously separ-ates such knowledge from the control of the direct producers (Burkett 1999).

However, at an even more basic level, Marx (1992, 348), who was heavily influenced by the work of the German agricultural chemist Justus von Liebig, recognized that the contradictions inherent in capitalist property relations promoted a "blind desire for profit that ... exhausted the soil." The result was that farmers began losing their ability to autonomously reconstitute their own means of production: "[I]f agriculture itself rests on scientific activities – if it requires machinery, chemical fertilizer acquired through exchange, seeds from distant countries etc., and if rural, patriarchal manufacture has already vanished – which is already implied in the presupposition – then the machine-making factory, external trade, crafts etc. appear as *needs* for agriculture ... Agriculture no longer finds the natural conditions of its own production within itself, naturally, arisen, spontaneous, and ready to hand, but these exist as an independent industry separate from it" (Marx 1993, 527 [emphasis in original]). As he relates in another work, "[t]he moral of the tale ... is that the capitalist system runs counter to a rational agriculture, or that a rational agriculture is incompatible with the capitalist system (even if the latter promotes technical development in agriculture) and needs either small farmers working for themselves or the control of the associated producers" (Marx 1967, 216; see also Kautsky 1988 [1899]).

Thus, while there might not be a formal separation of agricultural producers from the most basic means of production (land), agricultural production nonetheless is subsumed increasingly within the capitalist mode of production. Through this process, agriculture is transformed from a handicraft, the knowledge of which is passed down through generations, to a science or, perhaps more accurately, a complex of sciences, integrated into capitalist relations of production (Kautsky 1988 [1899]). Such prospects are exacerbated by capitalist control over seed technology, which strips farmers of traditional agrarian knowledge and nudges them even further into debt as prices for these critical inputs soar.

Conceptualizing Enclosures

The Midnight Notes Collective (1992, 318) articulates quite forcefully the conceptual and practical link between enclosures and primitive accumulation: "The Enclosures, however, are not a one time process exhausted at the dawn of capitalism. They are a regular return on the path of [primitive] accumulation and a structural component of class struggle. Any leap in proletarian power demands a dynamic capitalist response: both the expanded appropriation of new resources and new labor power and the extension of capitalist relations, or else capitalism is threatened with extinction." This same group goes on to articulate five different modes of enclosure in what it refers to as the "pentagon of enclosures." First, and similar to historic examples, enclosures rob local communities of their control over the means of subsistence; second, and again exhibiting a historic continuity, the catastrophe of enclosure is executed through seizures of land for debt service and retirement; third, contemporary enclosures, by dispossessing peasants and indigenous people, create a new mobile and migrant population; fourth, modern enclosures presupposed the demise of state socialism in the Soviet Union, China, and other countries in order to increase international competition among workers; and fifth, new enclosures have induced an array of ecological and biological depredations that threaten the vitality of the terrestrial commons.

Consonant with the first element of primitive accumulation – the separation between the actual producers and the means of production – De Angelis (2007) collapses these five points into two main modes of capitalist enclosure: enclosure that is achieved through a concerted strategy, such as privatization, public spending austerity, or structural adjustment, and enclosure that emerges as a by-product of a particular accumulation process. Since enclosures make possible M-C-M', as well as its continued reproduction, they all share the basic universal character of separating people from access to any social wealth that falls outside the purview of competitive markets and money as capital.[44] That is, in line with the first element of primitive accumulation that was outlined earlier, we note that enclosures provide a mechanism for realizing the *ex novo* separation between producer and the means of production.

What I hope to have demonstrated with the preceding discussion is that the capitalist enclosure of science, and, more specifically, agricultural biotechnology, represents a contemporary instance of primitive accumulation that touches on all three elements of this process. As biotechnology developed from the 1970s onward to a stage sufficient for capitalist valorization, capital began exerting a stranglehold over this technoscience in what can be interpreted as yet another area of social existence now brought under capitalist control, thus reinforcing the idea that primitive accumulation remains a continuous social process. Large swaths of social knowledge in respect of agricultural production are being increasingly enclosed by capital, which represents intensified efforts to privatize once public goods. Finally, efforts by capital to bring agricultural biotechnology profitably within its control involve the same spatial ambitions outlined previously in regard to primitive accumulation.

However, the imposition of an *ex novo* separation represents a social process that, in practice, is susceptible to contestation by oppositional social forces seeking to recover those social spaces appropriated by capital and to reinvigorate them as spaces of commons. Capital is thus compelled to wage a two-front war in its battles for enclosure – invading and enclosing new realms of social existence that can be subverted in service of capital's accumulation priorities in the face of resistance and defending those enclosed areas governed by accumulation and commodification imperatives against *ex novo* guerrilla movements struggling to liberate enclosures from capitalist control. The point to take from this discussion is that not only does separation occur *ex novo* but also that *ex novo* opposition can form in response to capitalist enclosure. Enclosures, and the responses they engender, thus represent strategic problems for capital. They pose limits that must be overcome if capital is to be successful in colonizing new areas of social existence or in sustaining those areas already enclosed from attacks by alternative social forces seeking to de-commodify such spheres and transform them back into commons.

One sees, therefore, that limits to capital are both endogenous and exogenous. In the former, capital itself identifies and defines a limit that it must overcome and, in the latter, this limit is defined for capital by the oppositional social forces that strive to liberate an already enclosed space. But regardless of how limits are identified, it is critical to recognize that counter-enclosures (read commons) represent alternatives that seek to circumscribe capitalist accumulation imperatives either by resisting enclosure strategies or by liberating enclosed areas of social life. Commons therefore tend to emerge out of struggles against their negation: "Therefore, around the issue of enclosures and their opposite – commons – we have a foundational entry point of a radical discourse on alternatives" (De Angelis 2007, 139).

As discussed briefly at the outset of the book, leading theorists, such as Charlotte Hess and Elinor Ostrom (2007) and Yochai Benkler (2006), tend to conceptualize commons as specific institutional forms that structure the way resources, be they material or immaterial, are accessed, used, and managed by a group of people beyond the logic of the capitalist market. Yet as we will see in the chapters that follow, the opposition being waged against capitalist practices of enclosure has thus far placed less emphasis on the articulation and development of specific institutional forms. Thus, for present purposes, and following Massimo De Angelis (2007), we need to go further and conceive of "BioCommons" not merely in relation to the resources they embody but also as social practices that produce meanings and values through active engagement between subjects who struggle to maintain or regain social control over social wealth through opposition to capitalist and state practices of enclosure.

As the authors of *The Ecologist* (1993, 6) have written, the commons "is the social and political space where things get done and where people derive a sense of belonging and have an element of control over their lives." This more active conceptualization permits us first to pose a limit to capital that simultaneously throws open to debate the possibility of alternatives and their problematization, which, in turn, reduces their susceptibility to capitalist co-optation. This notion of the commons thus admits in a similar way the dialectical relationship between enclosures and commons. And, as the following chapter aptly demonstrates, a number of successful struggles have been waged against capitalist attempts to enclose agricultural biotechnology in service of accumulation imperatives.

3
Battles to Reclaim and Maintain the BioCommons

Canadian government policy that seeks to exploit biotechnology for its ability to foster international competitiveness and promote economic growth has found a natural complement in the major industry players' efforts to subsume agricultural biotechnology firmly within capitalist relations of production. Yet a variety of anti-enclosure movements and projects have emerged in response to capital's attempts to corral the BioCommons.

Recombinant Bovine Growth Hormone (rBGH)

In her chapter, "No to Bovine Growth Hormone: Ten Years of Resistance in Canada," Lucy Sharratt (2001) describes one of the most protracted battles to date in Canada that erupted over one particular aspect of biotechnology. In 1986, a number of Health Canada scientists claimed that rBGH developed by Monsanto did not pose demonstrable risks to human health, although the assertion was not supported by scientific evidence. By 1988, rBGH was being tested on dairy cows in four provinces, and the resulting milk was introduced into the commercial supply. It was not until dairy processors informed the media of the situation in November 1988 that the public became aware that parts of the Canadian milk supply had been infiltrated by milk "contaminated" with rBGH. The result was immediate and negative. Moreover, it signalled the beginning of substantial public opposition that would continue for ten years until rBGH was denied regulatory approval by Health Canada (Sharratt 2001).

In British Columbia, unease at the prospect of consuming milk containing the growth hormone elicited such an outpouring of concern that telephone trunk lines were jammed with concerned calls to milk processing plants (Sharratt 2001). In Ontario, processors wrote letters to the Ontario Milk Marketing Board until it acquiesced and instructed researchers not to allow milk from their experiments into the commercial supply. The National Farmers Union (NFU) was also vocal in its opposition to rBGH out of fear for both animal safety and the farm industry. Position papers and letters from a diverse constituency of individuals and organizations opposing rBGH streamed into the federal Departments of Agriculture and Health. Brewster Kneen and Lorraine Lapointe, who were instrumental in the struggle in Ontario, organized the "Pure Milk Campaign," the first systematic effort to inform the public and encourage opposition to rBGH. The issue also received wide coverage in *The Ram's Horn*,[1] which helped link it to wider trends in genetic engineering and the growing control that agri-businesses were exerting over agriculture. According to Kneen, the ultimate success of the opposition movement to rBGH in Canada can be attributed to three main factors. First, the emerging countermovement ensured that the term "bovine growth hormone" dominated public consciousness, instead of the sanitized term preferred by Monsanto and its champions, recombinant bovine somatotropin or rBST. Second, attention was focused on ensuring that farmers' concerns remained central to the issue so that apprehension over rBGH did not degenerate into a concern "merely" over food safety. Finally, opponents of rBGH issued a contingency demand that rBGH milk be labelled as such, should Health Canada issue regulatory approval. This latter factor was a particularly strategic one because, in this country, milk is pooled in a single system. Labelling requirements would have required segregation and the development of a second marketing system, which would have been costly for both dairy producers and consumers (Macdonald 2000; Sharratt 2001).

By 1994, shortly before Health Canada was poised to approve rBGH, opposition had grown even further, with hundreds of thousands of signatures having been collected by the Council of Canadians and sent to Parliament. The minister of agriculture issued a moratorium in August 1994 that included a one-year delay on the sale and use of rBGH. Although the moratorium was not renewed, Health Canada did request additional data from Monsanto, which had the same effect as delaying approval. Senate hearings in 1998 would later discover that executives at Monsanto and bureaucrats within Health Canada exerted pressure on scientists in the department to rush approval for rBGH and disregard negative findings. In response to some of the internal conflicts between scientists and managers at Health Canada, the department requested that two independent bodies – the Royal Society of Physicians and Surgeons and the Canadian Veterinary Medical Association – review the available data. Internal Health Canada documents show that Monsanto tried to insert itself into this process by suggesting both agenda items for meetings in which the work of the independent panels was being considered and specific scientific studies that it wanted the panels to consider in their review (Sharratt 2001).

At around the same time that the battle over Monsanto's application for regulatory approval was being waged in this country, studies were being published in the United States that demonstrated a range of concerns related to the use of rBGH for both animal and human health. For example, this synthetic growth hormone appreciably increases the risk of mastitis among dairy cows. The danger is that milk from cows suffering from this ailment can be contaminated with the pus from mastitis and with the antibiotics used to treat the inflammation. According to one study, the United States Food and Drug Administration (FDA) admitted that the molecular structure of rBGH differs significantly from the growth hormone naturally present in cows – although this agency would later change its position, arguing against any difference. According to scientific studies, milk from cows injected with rBGH differs chemically, nutritionally, pharmacologically, and immunologically from natural milk (Epstein 1996). As documented in the film *The World According to Monsanto,* secret documents liberated from the FDA outline serious health effects associated with Monsanto's rBGH.[2]

In particular, cows treated with Monsanto's growth hormone secrete increased amounts of the powerful hormone IGF-1 in their milk. Samuel Epstein, chairperson of the Cancer Prevention Coalition, points out that around sixty studies have established links between IGF-1 and breast, colon, and prostate cancers (Dona and Arvanitoyannis 2009; Epstein 1990a, 1990b, 1996). The data provided by Lilly Industries in support of its application for regulatory approval of rBGH in the European Union showed a ten-fold increase in IGF-1 concentrations in milk produced using the synthetic hormone. However, even more problematic is the fact that the research methodologies employed in most of the studies conducted by industry suffered from serious design flaws that served to underestimate IGF-1 levels in treated milk (Epstein 1996).

While the report written by the Royal Society of Physicians and Surgeons merely made the point that the Bureau of Veterinary Drugs (the directorate within Health Canada responsible for the rBGH review – now reorganized in the department as the Veterinary Drugs Directorate) had sufficient data to reach a decision, the Canadian Veterinary Medical Association confirmed the conclusion of critical Health Canada scientists that rBGH poses risks for bovine health. This panel's findings provided Health Canada with the scientific justification it needed to refuse Monsanto regulatory approval for its drug. Finally, on 15 January 1999, Health Canada announced that Monsanto's application for regulatory approval of rBGH would not be approved "at this time" (Sharratt 2001, 394; Smith 2003).[3] One of the major lessons learned from ten years of opposition to rBGH was that a diverse movement of citizens with support from across the country made it almost impossible for Monsanto to single out and discredit specific individuals or groups.

Genetically Engineered Pigs and Salmon

Although less the result of concerted oppositional effort, a recent successful challenge to a genetically engineered organism was achieved in early April

2012, when the University of Guelph announced that Ontario Pork will no longer finance the university's research program involving the "Enviropig™."[4] According to spokesperson Keith Robbins, Ontario Pork rescinded support for the Enviropig because its annual research budget shrank from about $1.5 million to $500,000. In addition to not "getting the kind of return that we were looking at," Robbins said the "project itself was just taking a very, very long time" (as cited in Schmidt 2012a, paras. 15-16).

Dating back to 1995, the goal of this research project was to reduce the amount of phosphorous contained in the pig's feces. By 1999, the researchers involved in the project produced the first Yorkshire pig that was genetically engineered to digest plant phosphorus more efficiently than conventional Yorkshire pigs. Employing a transgene sequence that includes an E-coli bacteria phytase gene and a mouse promoter gene sequence, scientists genetically engineered these pigs to produce the enzyme phytase in their salivary glands and secrete it in their saliva. As the pig masticates, the phytase mixes with the feed. The phytase enzyme is activated in the acidic environment of the pig's stomach and helps to degrade indigestible phytate in the feed that accounts for 50 to 75 percent of the grain phosphorus. According to the University of Guelph, this enzymatic activation reduces phosphorus content in manure anywhere from between 30 and 70.7 percent, depending upon the pig's age and diet (*Enviropig™: Environmental Benefits* n.d.).

Although phosphorous from animal manure is a plant nutrient and thus a critical element in soil fertility, when the amount produced by animals outstrips the absorption capacity of plants, the excess becomes a pollutant that may leach into tile drains, ponds, streams, rivers, and lakes. High phosphorus content in water is a major contributing factor to extensive algae growth (and, in some cases, algal toxin production) in fresh water systems, which eventually leads to a reduction in oxygen content that is lethal to fish and other aquatic organisms. It also has serious implications for the potability of the water. Moreover, blue green algae (Cyanobacteria), which thrive in phosphorous-rich waters, often produce cyanotoxins that can be lethal to livestock and pets if they ingest the polluted water.

Given the structural changes to the Canadian hog industry over the last two decades, through which the majority of smaller-scale, family-farm operations have been replaced by intensive, factory farm-type operations in which thousands of pigs are raised in giant barns, excess phosphorous production has become a significant production problem that requires factory farmers to transport liquid manure longer distances, dry compost manure, or expand the area of land on which they can spread manure. By reducing the amount of phosphorous produced by hogs, genetic engineering promised a potentially easier and less costly solution that leaves intact the logic of factory pig farming, which is characterized by strict control by a handful of vertically integrated corporations that now dominate the industry at both the production and processing levels. Rather than develop solutions that might respond to the underlying structural problems of factory farming operations, the University of Guelph sought to capitalize on such problems by producing and patenting (the university holds patents on the pig in the United States and China) a genetically engineered pig that it hoped would generate revenue streams through licencing agreements with pig breeding and production companies (*Who "Created" and Owns "Enviropig™"?* n.d.).

Yet another less radical and cost-neutral solution to excess phosphorous production in the form of a phytase feed supplement has been available to hog producers for at least a decade. According to researchers from the Ontario Ministry of Agriculture, Food and Rural Affairs, and the University of Guelph, adding phytase to feed regimens designed to match the animal's requirements and replacing protein with synthetic amino acids can reduce nitrogen and phosphorous excretion in pig manure by up to half (Simpson and de Lange 2004).

The University of Guelph submitted in April 2009 an application to Health Canada seeking regulatory approval for human consumption of its genetically engineered pig. According to Cathy Holtslander and the Canadian Biotechnology Action Network (CBAN), because Health Canada has failed to develop specific guidelines for evaluating the safety of genetically engineered animals for human consumption, the department planned to rely on the United Nations' Codex Alimentarius guidelines and to refer to the US Food and Drug Administration (*Is "Enviropig™" Safe?* n.d.). In fact, because of this regulatory lacuna, it was Environment Canada that, by issuing a Significant New Activity Notice (No. 15676), granted approval to the University of Guelph to reproduce and export the Enviropig. As discussed previously, Environment Canada was almost completely shut out of regulatory oversight of genetically engineered organisms in this country. However, under the *Canadian Environmental Protection Act,* the department does function as a sort of jurisdiction of last resort to regulate all animate products of biotechnology for uses not covered under other federal legislation.[5]

Since pigs can be rendered as a protein source for other animals, the university also applied to the Canadian Food Inspection Agency (CFIA) for approval to use the Enviropig in other livestock feed.[6] CBAN is now calling on the University of Guelph to withdraw its application for regulatory approval pending with Health Canada and to stop attempts at commercializing this technology. Yet according to University of Guelph spokesperson Lori Bona Hunt, the regulatory applications at Health Canada and the US Food and Drug Administration will remain in place: "Those are still under review and ... they are going to remain active until regulatory decisions are made. The science is done, the genetics are done. The genetics will be stored. The only difference would be you wouldn't see a pig" (as cited in Schmidt 2012a, paras. 13-14). Based on this assertion, it is probable that the university would restart the program should an alternative funding source be found.

The other genetically engineered animal currently on the regulatory radar screen is AquaBounty Technologies' AquAdvantage® salmon, which has been designed to grow to maturity in sixteen to eighteen months rather than the three years that is typical for conventional Atlantic salmon. Invented by scientists at Memorial University, who subsequently patented and licenced their invention to AquaBounty, a company based in Waltham, Massachusetts, the process to achieve accelerated growth relies on inserting into conventional Atlantic salmon a DNA sequence from Chinook salmon that codes for a growth hormone as well as regulatory sequences from Chinook salmon and ocean pout. These genetic constructs induce the genetically engineered salmon to produce growth hormones throughout the year, thus causing them to reach market weight twice as fast as their wild brethren. Although at this point it can only be confirmed that the company has applied for regulatory approval in the United States, the case is germane to the Canadian context for several reasons. Canada has a vibrant Atlantic salmon industry and AquaBounty has located part of its operations in this country, which will require Canadian regulatory approval prior to any commencement of full-scale commercial production.

Although the company apparently plans only to sell table-ready, genetically engineered Atlantic salmon into the American market rather than actually produce these fish in the United States, the meat must be approved by the FDA for human consumption. As such, in 2001, AquaBounty filed a New Animal Drug Application with the FDA for approval to market its genetically engineered salmon to American consumers (in the United States, the FDA, rather inexplicably, has elected to categorize genetically engineered animals as "animal drugs" for regulatory purposes). According to its application, the company will produce all of the genetically engineered salmon eggs at its broodstock production facility in Bay Fortune, Prince Edward Island, and then ship the fertilized triploid eggs to Panama for growing out and processing.

After almost ten years, FDA officials published a preliminary conclusion on 25 August 2010, which maintains that the genetically engineered salmon is safe for human consumption and poses no risk to the environment. At the same time that it published its opinion, the FDA announced that it was tasking its Veterinary Medicine Advisory Committee with examining and discussing the science submitted by AquaBounty. As part of this discussion, two days of public hearings were scheduled for September 2010. However, even a casual observer is left wondering whether the true intent behind these sessions was to lend the overall review process a veneer of public participation that would reinforce the FDA's position in favour of approval. The FDA released only a redacted set of documents pertaining to the application and that release occurred a mere two weeks prior to the hearings. Moreover, the Veterinary Medicine Advisory Committee, only one of whose members had experience with fish, was not mandated to provide any recommendations.

In any event, all did not go as presumably envisioned. A number of committee members, and even more public presenters, articulated substantial concerns about the scientific quality of the data AquaBounty had submitted in support of its regulatory application. During the discussion period, some committee members went so far as to demand that the FDA obtain more comprehensive food and environmental safety reviews, including a full environmental impact statement rather than the less rigorous environmental assessment provided by the company. Reflecting on the substance of many of the exchanges at these public meetings, Jaydee Hanson, senior policy analyst at the Center for Food Safety, pointed out that "the committee raised many of its own concerns regarding small sample sizes, incomplete data, culling practices, and poor scientific assessments over the past days, and pointed to the complexity and numerous uncertainties that remain" (Center for Food Safety 2010, para. 5).

In particular, critics of AquAdvantage salmon homed in on issues around allergenicity, elevated levels of growth hormones (including IGF-1), and the dangers of escape into the wild. For example, although AquaBounty detected a 19 percent increase in allergenicity of triploid fish (the ones to be grown out for human consumption) over the control group, it found no statistically significant difference, which forms the basis of its conclusion that the meat from its genetically engineered salmon poses no heightened risks of allergens. However, the experiment relied on a very small sample size of six fish in each of the experimental and control groups. Moreover, the company admits that its process of transforming the fish into triploid sterile females is only about 95 percent accurate. This means that up to 5 percent of the fish might remain diploid, which the company determined to be 40 percent more allergenic than conventional fish – a statistically significant result.

The Canadian government has been quite generous with funds to try and perfect the induction of triploidy, awarding the company $2.9 million in January 2009 from the Atlantic Innovation Fund "to improve the culture of reproductively sterile Atlantic salmon. The future objective is the safe commercial launch of triploid salmon with Atlantic Canada identified as the source for associated commercial benefits, and worldwide distribution of the product" (*Government of Canada Invests*, 2009, para. 13). The US Department of Agriculture also announced in September 2011 that it would award the company US$494,000 to improve its process of inducing sterilization but, later, rescinded the grant following substantial public opposition (Levuax 2012).

In addition to the greatly expanded allergenicity posed by diploids, they are not rendered sterile, which poses a substantial environmental danger should they escape into waterways and breed with wild salmon. Although AquaBounty maintains that its inland grow-out facility in Panama is sufficiently remote to avoid environmental catastrophe in case of fish escape, experts remain skeptical of the science presented by AquaBounty. According to Anne Kapuscinski, professor of environmental studies at Dartmouth College and who specializes in environmental risk assessment in aquaculture systems:

> There were a lot of things in the way they [AquaBounty] presented their approach to risk assessment and a lot of what we saw in the Environmental Assessment that was released to the public [that] suggested that they are not on the cutting edge of the state of the art of risk assessment science. They need to be on that cutting edge when we're dealing with a precedent-setting case, and one that is challenging because we're dealing with an animal that's barely domesticated. Any salmon that escapes from human control, if it's in habitat that it can survive in, is going to continue to live. And we know that farmed salmon is a global commodity. Let's imagine that tomorrow the FDA approved this application. The company is not going to make a profit by simply growing small numbers of salmon in Panama. That's just a proof of concept facility. This company is going to make money by selling millions upon millions of eggs to big salmon farms all over the place. (As cited in Levuax 2012, para. 8)

In its application to the FDA, AquaBounty included two studies that assessed potential risks from increased levels of growth hormone in AquAdvantage salmon, both of which have been critiqued for lack of methodological rigor. The first study, a peer-reviewed experiment conducted in 1992, was based on a very small sample size: five genetically engineered salmon, seven non-genetically engineered siblings, and five control fish. Moreover, rather than examine full-grown fish, this experiment included fish that were quite small. The genetically engineered salmon averaged only forty-seven grams, while the non-genetically engineered siblings and control fish averaged about ten grams. Since people do not eat salmon that small, this study provides no definitive evidence about any risks associated with consuming these genetically engineered fish.

Although the second study corrected for these shortcomings by including a larger sample size of market-sized fish (thirty genetically engineered salmon, thirty-three sponsor controls, and ten farmed controls), it employed methods that were not sufficiently sensitive and thus failed to detect growth hormone in any fish in either the experimental or control groups. Similarly, insensitive tests were used to look for elevated levels of IGF-1. As senior scientist at Consumers Union and attestant at the public hearings, Michael Hansen (2010, 11-12) points out, "to base a conclusion of no additional risk on exactly six engineered fish, when those data themselves suggest a possible problem, is not responsible science or responsible risk assessment. FDA owes it to the thousands of Americans who are allergic to finfish to demand more data on the allergenicity of these engineered salmon from AquaBounty."

Shortly after the public hearings, the consumer advocacy group Food and Water Watch made public a number of e-mails and internal documents released to it by the US Department of Interior's Fish and Wildlife Service pursuant to an access to information request. These records make clear that although required under section 7 of the US *Endangered Species Act*, the FDA, as late as October 2010, had failed to consult with the Fish and Wildlife Service, or the National Marine Fisheries Service (as also stipulated by law), to assess whether approval of AquAdvantage salmon might impact wild Atlantic salmon, an endangered species.[7] Moreover, as the material obtained by Food and Water Watch clearly reveals, despite the failure of the FDA to discharge its legislated consultative duties, Food and Wildlife Service scientists proactively wrote to the FDA, expressing their belief that "[t]here are several unknowns and uncertainties regarding possible genetic, ecological, and environmental effects of AquAdvantage salmon that must be elucidated before an environmental risk assessment can be thoroughly evaluated and approved. This, along with a situation where regulatory oversight is adequate at best, suggests that approval of AquaBounty Technologies' request for commercial rearing of AquAdvantage salmon is premature" (as cited in Hauter 2010, 2-3).

These, and a litany of other concerns about flawed conclusions reached in respect of genetically engineered salmon, fell on deaf ears at the FDA, as demonstrated by the assessment articulated in its environmental assessment published on 25 August 2010, which claimed that "the production, grow-out, and disposal of AquAdvantage Salmon under the conditions described in this Environmental Assessment are highly unlikely to cause any significant effects on the environment, inclusive of the global commons, foreign nations not a party to this action, and stocks of wild Atlantic salmon" (US Food and Drug Administration 2010, 11). However, the wave of critiques that washed over the FDA during the public consultations in September 2010, followed a month later by the public revelation from a coalition of consumer, environmental, animal welfare groups and commercial and recreational fisheries associations and food retailers that over 360,000 public comments had been sent to the FDA demanding that it reject AquaBounty's application or, failing that, require mandatory labelling for genetically engineered fish, may have given the FDA some pause. As of mid-2012, it had yet to publish a final decision on AquaBounty's application.

According to Ronald Stotish, president and chief executive officer of AquaBounty, at the same time that the FDA appeared poised to approve AquAdvantage salmon in the United States, the company was entering the initial stages of applying for regulatory approval from Health Canada to sell its genetically engineered salmon as a novel food in this country (as cited in Schmidt 2010). Since Health Canada refuses to disclose even whether it has received an application for approval to market engineered salmon for human consumption in this country, the current status of any possible application remains unclear. In any event, the task of assessing an application for approval to commercially produce genetically engineered fish eggs falls to Environment Canada, which, as will be elaborated more fully in Chapter 5, is the regulator of last resort for genetically engineered products that have no other regulatory home. Similar to Health Canada, this department is not required to disclose publicly whether or not an application has been received. Aside from having negligible experience with the regulation of genetically engineered products, Environment Canada would be required to render its decision on any application from AquaBounty within a mere 120 days of submission. And this rapid process provides for no public participation or hearings. The only way that AquaBounty has thus far been able to operate on Prince Edward Island is by taking advantage of a research and development exemption contained in the Canadian *Environmental Protection Act* that permits operations without triggering an environmental assessment.

However, according to a reporter from the *Vancouver Sun* and based on government records obtained under an access to information request, scientists at Environment Canada have questioned the ability of the department to adequately protect wild fish stocks if it authorizes the commercial-scale production of genetically engineered salmon eggs (Schmidt 2011). As outlined in the released internal documents, officials at Environment Canada, in anticipation of a formal application by AquaBounty, identified a regulatory dilemma in respect of the department's environmental protection mandate. Pursuing a narrow interpretation of its responsibility to safeguard the environment would require the department to assess only the production and transportation of genetically engineered salmon eggs from AquaBounty's facility on Prince Edward Island to Panama.

Conversely, a broader interpretation of the department's, and thus Canada's, legal obligations under the Canadian *Environmental Protection Act* would impose an additional duty to assess the wider potential effects that genetically engineered salmon could have on both this country and the international environment should the fish ever escape the grow-out facilities in Panama and migrate into international and Canadian waters. Indeed, as Schmidt (2011) outlines, some scientists at the Department of Fisheries and Oceans have suggested that there exists the potential risk of fish migrating back to Canada and subsequently impacting this country's fish stocks. Despite

Canada's duty under the act to evaluate risks to the global environment, analysts at Environment Canada concluded that this broader approach, which would require monitoring of the grow-out operations in Panama, would be logistically challenging and almost certainly beyond the current capacity of Canadian authorities (Schmidt 2011).

This more cautious perspective on how to proceed with any regulatory assessment was markedly different from the analysis conducted at Agriculture and Agri-Food Canada in consultation with staff at the Department of Foreign Affairs and International Trade, which was shared with Health Canada. Based on internal documents obtained through another access to information request, analysts at these two departments have suggested that because Canada would experience difficulty keeping genetically engineered salmon out of its food supply if approved for sale in the United States, the easiest solution, and one that would purportedly avoid bilateral trade complications, would be to simply follow the American lead and permit the fish into the Canadian market. According to one of the talking points included as part of the material released by the government to *Postmedia News*, "[w]e [regulators] want to work closely with the U.S. to ensure our approval processes for GE animals compliment [sic] one another and that we avoid any potential bilateral complications. Canada-U.S. regulators work closely together on an ongoing basis, but perhaps there is merit in seeking specific opportunities for them to meet and talk about GE animals" (as cited in Schmidt 2012b, C7).

Despite being unclear whether AquaBounty has submitted any application for regulatory approval to Health Canada or Environment Canada, opponents of genetically engineered salmon have mobilized against what could become the first genetically engineered animal approved for human consumption in the world. At the forefront of action in Canada is CBAN, which, in addition to preparing and disseminating background documents and press releases about the dangers of genetically engineered salmon, has coordinated a joint statement of opposition to genetically engineered fish that has been signed by sixty-four fisheries and oceans, environmental, and social justice groups from across the country that demand protection for the health and future of our food system and our aquatic ecosystems. Titled "No GE Fish Research, Production, Consumption in, and Export from, Canada," the signatories to this statement:

- [s]tate [their] categorical objection to the introduction of genetically engineered (GE) fish into our food system;
- [o]bject to the raising of GE fish, whether in open netpens or land-based facilities, object to the production of GE fish eggs in Canada and their export from Canada to other countries, and object to the use of public resources for the research and development of GE fish;

- [c]all upon Environment Canada to reject any request for permission to commercially produce genetically engineered fish or fish eggs;
- [c]all upon Health Canada to reject any application for approval of genetically engineered fish for human consumption;
- [c]all upon the federal government to stop any current risk assessments of GE fish.

Genetically Engineered Wheat

On 11 September 1997, Agriculture and Agri-Food Canada signed a Collaborative Research Agreement (Roundup Ready® Wheat) with Monsanto Canada, according to which, "AAFC [Agriculture and Agri-Food Canada] and Monsanto desire a collaboration so that AAFC can develop commercial Roundup resistant wheat cultivars for production in Canada and Monsanto can generate revenue on the sale of seed of these cultivars and the sale of access to the Roundup resistance technology" (section H). The agreement established an estimated $2.45 million commitment from the Government of Canada over its eight-year lifespan ($1.6 million in facilities and expertise and time of government scientists and an $850,000 Matching Investment Initiative component) and required that research would be conducted at the Agriculture and Agri-Food Canada Cereal Research Centre in Winnipeg as well as in field space either owned or leased by the federal government.[8] The agreement further stipulated that any patentable inventions developed during the project that represented improvements on pre-existing products for which Monsanto holds the patent would become the intellectual property of Monsanto. Unfortunately, details about the intellectual property treatment of germplasm were almost completely excised from the copy of the agreement released under an access to information request.[9] The agreement also set out that Monsanto retained the right to block publication of any results derived from the project for up to two years. Although the exact terms were not made public, the deal included a royalty-sharing arrangement that observers estimated would net the federal government between 1 and 10 percent of Monsanto's total genetically engineered wheat sales if this new seed were to be commercialized. When questioned about this point of the agreement, John Culley, director of intellectual property for Agriculture and Agri-Food Canada, refused to divulge an exact figure of the profit-sharing agreement reached with Monsanto (Bueckert 2004b).

While the terms of Monsanto's obligations under the contract are deemed confidential by the department and thus hidden from public scrutiny, Jim Bole, director of cereal research at Agriculture and Agri-Food Canada at the time, informed a reporter that Monsanto had invested $1.3 million in the project. As part of the deal, the department conducted field trials and provided Monsanto with access to wheat germplasm that had been developed over the years by public scientists and plant breeders – germplasm considered by many to be some of the best in the world (Warick 2003). Opening the doors to part of Canada's common genetic heritage yielded relatively quick benefits for Monsanto, which, on 23 December 2002, submitted an application to the CFIA seeking regulatory approval of its Roundup Ready wheat for unconfined environmental release and for livestock feed. A number of people involved in the fight against genetically engineered wheat surmise that Monsanto purposely chose the Christmas season to submit its application hoping that a distracted public would minimize protest. On 31 July of the same year, Monsanto applied to Health Canada for approval to market its genetically engineered wheat for human consumption.

From 2000 onward, a variety of groups including the NFU, Greenpeace, the Council of Canadians, Manitoba's Keystone Agricultural Producers, the Agricultural Producers of Saskatchewan, the Saskatchewan Organic Directorate, the Saskatchewan Association of Rural Municipalities, and the Canadian Wheat Board, among others, mobilized a concerted and decisive effort against genetically engineered wheat. As part of its class action suit against Monsanto and Bayer CropScience (elaborated more fully in Chapter 4), the Organic Agriculture Protection Fund asked the Court of Queen's Bench in Saskatchewan for an injunction against the marketing of genetically engineered wheat seed. Members of both the Saskatchewan Organic Directorate and the Organic Agriculture Protection Fund offered testimony before the House of Commons Standing Committee on Agriculture and Agri-Food and the Senate Standing Committee on Agriculture and Forestry, voicing their opposition to genetically engineered wheat. Glen Neufeld, president of Saskatoon's Sunrise Foods International, which exports $10 million annually of Saskatchewan-grown organic grain, succinctly summed the fears of many people on the prairies: "It [introduction of genetically engineered wheat] would be devastating. Once it comes, it's all over. I don't hear farmers or anybody asking for this. We are going to be screwed. Somebody needs to shut Monsanto down. Monsanto doesn't really care about the industry. They care about making the next big buck" (as cited in Warick 2003, E1). On 31 July 2001, these groups presented a letter signed by over 300 organizations to then Prime Minister Jean Chrétien demanding that the federal government respond to the wishes of Canadians and put a halt to the development and introduction of genetically engineered wheat in this country: "We represent diverse constituencies and interests, but we are unified in asking that you act immediately to prevent the introduction of genetically modified (GM) wheat into Canadian food and fields unless the concerns of Canadian farmers, industry, and consumers are addressed adequately" (*NGO Letter to Chrétien on GE Wheat* 2001).

The Canadian Wheat Board, which until 1 August 2012 was the sole body responsible for Canadian wheat exports, was quick to articulate its opposition to the unconfined release of genetically engineered wheat in western Canada, except where certain stringent conditions could be met.[10] As outlined on its website, these conditions include: widespread market acceptance; the establishment of achievable contamination tolerance levels; the development of an effective segregation system;[11] the availability of rapid, accurate, and inexpensive detection technology; and a positive cost-benefit throughout the wheat value chain with particular emphasis on farmer income. The Canadian Wheat Board also supports the inclusion of cost-benefit analysis as part of the regulatory process prior to the unconfined release of genetically engineered wheat or barley varieties to ensure that the interests of farmers and customers are fully considered.

To bolster its position, the Canadian Wheat Board commissioned in 2003 leading plant scientists from the University of Manitoba to assess the environmental impacts of unconfined release of Monsanto's genetically engineered wheat. The multi-faceted report considered the introduction of genetically engineered wheat in the following contexts: the trend in western Canada toward reduced tillage, the historical experience with Monsanto's genetically engineered canola, the potential for genetic drift through pollen flow and selection pressure, the economic impact of genetically engineered wheat on low-disturbance direct seeding,[12] the risk of evolution of glyphosate-resistant weeds, and the effect on seed-saving practices. It also examined how genetically engineered wheat would affect the ability of Canadian farmers to segregate genetically engineered crops from conventional varieties, including how segregation might impact our international obligations under the *Cartagena Protocol on Biosafety* to the UN *Convention on Biological Diversity*.[13]

The overall assessment made by the authors of the report is that unconfined release of Monsanto's genetically engineered wheat would negatively affect the environment and impinge on the ability of farmers to conserve natural resources. Moreover, the report adds that the environmental impact of Roundup Ready wheat will be even higher than that of Roundup Ready canola because of the particular nature of the genetic trait involved and the way it relates to western Canadian farm practices: "Adding cost, complexity and herbicide load to pre-seeding weed control in low-disturbance direct seeding cropping systems will threaten the sustainability of these systems for *all* farmers in western Canada ... Controlling volunteer Roundup Ready wheat volunteers will cost even more than controlling Roundup Ready canola volunteers, and *this cost will be borne by both adopters and non-adopters of Roundup Ready wheat*" (Van Acker, Brûlé-Babel, and Friesen 2003, 2, 25 [emphasis added]). On the basis of these research findings, the Canadian Wheat Board wrote to Monsanto Canada in 2003 asking it to withdraw its applications for regulatory approval of its genetically engineered wheat in the interests of the Canadian wheat industry. Monsanto refused.

The Sierra Club of Canada also involved itself in the debate by producing and disseminating fact sheets about how genetically engineered foods are regulated in this country and about the potential implications of genetically engineered wheat for the environment, economy, farmers, and consumers. The efforts of this organization helped educate the broader public about the multiple facets encompassed by agricultural biotechnology. Contrary to the statements made by Monsanto and others, the Sierra Club of Canada referenced the growing evidence about the need for *increased* herbicide applications on genetically engineered crops. This group also illuminated the stranglehold Monsanto places on farmers who choose to plant Roundup Ready products, forcing them to use only its herbicide and contractually disallowing them from saving and replanting seeds. The Sierra Club of Canada also helped inform the public about the evolution of superweeds that are resistant to Roundup herbicide and the resultant chemical treadmill this threatens to create, where increasingly stronger herbicides must be produced and applied to the earth when growing food crops. In terms of concrete action, the Sierra Club of Canada invited Canadians to engage in a variety of strategies to stop the introduction of genetically engineered wheat, including: writing letters to members of parliament (MPs), the ministers of health and agriculture, and the prime minister; signing a petition that called on Parliament to place an immediate ban on the environmental and commercial release of genetically engineered wheat; avoiding genetically engineered food by purchasing organic; and not investing in biotechnology products and companies (Sierra Club of Canada n.d.).

The NFU developed and disseminated a fact sheet outlining ten of the top reasons why Canadian farmers and consumers do not need genetically engineered wheat. At the top of the list is the market loss that would occur if genetically engineered wheat were permitted into Canadian fields. Given the lack of effective segregation or testing systems, the overwhelming majority (82 percent) of international wheat customers have indicated that they would stop purchasing *all* wheat from Canada. Second, similar to the experience with genetically engineered canola, the NFU predicts the loss of wheat as an organic crop if genetically engineered varieties were given regulatory approval for unconfined release. Third, the rapid decrease in demand for Canadian wheat that would follow from the first point would serve to depress wheat prices for farmers (unfortunately, predictions of negative financial impact were recently borne out in markets for flax, another important Canadian export commodity worth $320 million annually).[14] On the other hand, if Canada refused approval for genetically engineered wheat, but other large producers such as the United States allowed it, then Canadian farmers might achieve competitive price advantages in the world market.

The fourth and fifth reasons why the NFU believes we needed to reject genetically engineered wheat relate to its unnecessary health and environmental impacts. The sixth point, echoing the research results found by Van Acker and his colleagues (2003), points to the inevitable increase in Roundup

Ready wheat volunteers and the consequent increase in weed control costs, which were estimated at $400 million annually. In the seventh point, the NFU adopts a skeptical position *vis-à-vis* segregation, contending that it is expensive and quite likely to be unsuccessful in the long term, a position shared by Ken Ritter, the chairperson of the Canadian Wheat Board at the time. In the eighth point, the NFU articulates the demand for a mandatory labelling regime, arguing that without one it would be illegitimate to introduce genetically engineered foods. The ninth point argues that genetically engineered crop varieties represent a blatant attempt by multinational corporations such as Monsanto, Cargill, and ConAgra to bring more of the food supply chain under their control in order to increase profitability. These sentiments were echoed by the president of the UN General Assembly, H.E.M. Miguel d'Escoto Brockmann (2008, para. 14), in his opening remarks to the High-Level Event on the Millenium Development Goals on 25 September 2008, in which he condemned large corporations, particularly Monsanto and Cargill, for profiteering on the back of the world's poor:

> The essential purpose of food, which is to nourish people, has been subordinated to the economic aims of a handful of multinational companies that monopolize all aspects of food production, from seeds to major distribution chains, and they have been the prime beneficiaries of the world [food] crisis. A look at the figures for 2007, when the world food crisis began, shows that corporations such as Monsanto and Cargill, which control the cereals market, saw their profits increase by 45 and 60 percent respectively.[15]

Concluding its top-ten list, the NFU points out that farmers have been told, whether correct or not, that there is a glut of grain on the world markets. If that is the case, then producers certainly do not require genetically engineered wheat to grow more grain and depress world prices even further (National Farmers Union 2003). This argument does not necessarily accept the higher yield claims articulated in respect of genetically engineered wheat (and other crop varieties), but it is rather an immanent critique designed to undermine genetically engineered wheat proponents on the basis of their own claims. As Fred Tait of the NFU pointed out at the time, "the introduction of (genetically modified) GM wheat will result in similar [as with the introduction of genetically engineered canola in the late 1990s] problems of cross pollination. Following the introduction of GM wheat, farmers will be faced with the choice of growing GM wheat, GM contaminated wheat, or not growing wheat. Consumers will have the option of eating products that contain Canadian GM wheat, or locating wheat products manufactured from imported GM-free wheat" (*Broad Alliance Formed in Canada* 2001, para. 7).

Another prong in the NFU strategy against Monsanto's genetically engineered wheat was the organization in 2003 of a nationwide anti-genetic engineering tour designed to garner wider public attention to the issues surrounding genetic engineering and agricultural biotechnology. Part of the message being disseminated during the tour was that farmers could demonstrate to Monsanto their unhappiness with genetically engineered wheat by boycotting Roundup and purchasing another brand of glyphosate. Such actions were made easier with the expiration of the patent on Roundup and the subsequent introduction of generic glyphosate herbicides (although not necessarily for farmers planting Monsanto's herbicide-resistant seeds, as noted in the previous chapter).

Greenpeace Canada, working in collaboration with Greenpeace International, launched an online petition drive by which people from around the globe could ask the Government of Canada not to approve genetically engineered wheat. In addition to the petition, by early May 2004, the Prime Minister's Office had received more than 5,000 faxes from global citizens, the overwhelming majority of whom made clear that they did not want genetically engineered wheat in their food (Greenpeace Canada 2004). In November 2002, Greenpeace Canada also released a study that it had commissioned to investigate the consequences of introducing genetically engineered wheat into Canadian fields. The overall critical assessment in the report is based on six key findings regarding the commercialization of genetically engineered wheat, some of which mirror the conclusions reached by the Sierra Club of Canada and the NFU. First, citing the purity assurances required by most Canadian wheat importers, the authors of the report spell out the market loss Canadian wheat farmers would suffer if genetically engineered wheat were allowed into our wheat supply. In part, such economic loss would occur because of the second point made, namely, the impossibility of developing a segregation system sufficient to guarantee the purity of non-genetically engineered wheat.[16] Third, the report contends that genetically engineered wheat would require the government to revise, or perhaps even completely abandon, our varietal registration system, which would have the ultimate effect of undermining Canada's reputation for quality. The fourth claim speaks to the increased management burdens that genetically engineered wheat would place on farmers as a result of genetically engineered volunteers, the higher dependence on chemical inputs as weeds develop tolerance to glyphosate, and the contractual limitations on farmers' ability to save and re-use seed. According to the fifth claim in the report, the professed benefits of this technology rest on scant empirical evidence. Indeed, as we examine more fully in later chapters, many of these corporate myths are now being systematically debunked as more long-term research emerges about agricultural biotechnology. The final conclusion drawn is that, because of our flawed regulatory system, the public would have no surety about the safety of genetically engineered wheat for either the environment or for humans (MacRae, Penfound, and Margulis 2002). Greenpeace also published a more general report at the international level, outlining the environmental issues related to genetically engineered wheat, ranging from gene flow to volunteer weeds, herbicide toxicity, and other indirect effects (Mayer 2002).

On 9 January 2004, Agriculture and Agri-Food Canada announced that it was withdrawing from its partnership with Monsanto. When queried as to whether the decision reflected the concern about losing Canadian wheat export markets, Jim Bole of Agriculture and Agri-Food Canada replied in the affirmative. He added that "Agriculture Canada would probably no longer anticipate a return on their investment" (as cited in Bueckert 2004a, para. 9). Monsanto offered a different spin on events, contending that the partnership had achieved its aims and thus had run its course. Company spokesperson Trish Jordan also made it clear that Monsanto would continue "trying to find ways to make this [commercialization of genetically engineered wheat] doable and come up with solutions rather than just stopping all work altogether" (as cited in Bueckert 2004a, para. 6). In commenting on the departmental decision to end the partnership with Monsanto to develop genetically engineered wheat, Bole candidly admitted that the "biotechnology revolution" was oversold: "We expected to be growing crops at this time with many traits that would be of great value to consumers and producers. But the regulatory area has been much more stringent than anyone anticipated and market acceptance hasn't been as positive as we would have anticipated" (as cited in Bueckert 2004a, para. 10). While many people and groups mobilizing resistance against agricultural biotechnology in this country would take issue with the characterization of our regulatory regime as overly stringent, Bole's second point is well taken.

On the heels of this decision, the Organic Agriculture Protection Fund issued a press release that both welcomed the end of federal government involvement in the commercialization of genetically engineered wheat and called upon Monsanto to abandon the project by withdrawing its application for regulatory approval, halting all field trials, and destroying all existing genetically engineered wheat seed (Organic Agriculture Protection Fund 2004). On 5 February 2004, the Saskatchewan Association of Rural Municipalities followed suit and sent a formal letter to Monsanto that outlined similar demands. Monsanto refused both requests. A year earlier at its annual convention, the Saskatchewan Association of Rural Municipalities passed a resolution that demanded the disclosure of current genetically engineered wheat test plots and an immediate moratorium on any further testing. As early as 2001, this organization called on the federal government to ban genetically engineered wheat in this country, including importation. In fact, a large number of Saskatchewan municipalities declared themselves to be genetically engineered wheat-free zones (Organic Agriculture Protection Fund n.d.).

In the run-up to the federal elections in late June 2004, Greenpeace, the NFU, the Saskatchewan Organic Directorate, and Canadian Organic Growers launched an advertising campaign on 19 March in a bid to stop the commercial introduction of Monsanto's genetically engineered wheat into Canada. Under the headline "The Greatest Threat to Wheat Farming Isn't Hail or Drought. It's Roundup Ready™ Wheat," the advertisement called on farmers to contact then Minister of Finance Ralph Goodale (a prominent western Canadian cabinet minister) or the minister responsible for the Canadian Wheat Board, Reg Alcock, to articulate their opposition to genetically engineered wheat. According to Darrin Qualman, former director of research for the NFU, "[w]e are hoping that these ads will help mobilize farmers to challenge the government in the run-up to the federal election. Ottawa needs to know that farmers do not support GM wheat" (as cited in "Opposition to GE Wheat Growing" 2004, para. 4). Arnold Taylor, chairperson of the Organic Agriculture Protection Fund and president of the prairie regional chapter of Canadian Organic Growers, is even more forceful in his assessment of the motivation behind the advertising action: "GM wheat should be stopped in its tracks and that's what this advertising campaign is all about. If the Liberals want support in the West, they need to say no to Roundup Ready wheat" (ibid., para. 6). The advertisement ran in the *Ottawa Hill Times* on 22 March and in the *Farmers Independent Weekly,* the *Saskatoon Star Phoenix,* and the *Regina Leader Post* on 25 March. The largest farm newspaper in western Canada, the *Western Producer,* refused to run the advertisement because of some of the groups involved in the campaign.

This stance of the newspaper certainly adds some weight to the critique levelled by Taylor and others (articulated more fully in Chapter 6) about the constricted scope of media coverage of opposition to agricultural biotechnology in this country. Taylor believes that the media, for the most part, fail to understand the mindset around organic farmers. Moreover, he recognizes that companies such as Monsanto and Bayer CropScience are the largest advertisers for newspapers such as the *Western Producer,* which has implications for the amount and depth of coverage that organic farming issues will receive. As he says, "[y]ou know, I'm not naïve enough to think that Monsanto doesn't say to the owner of the *Western Producer* that, you know you're advertising is at stake here. Let's just kind of downplay this. That's all they have to do and you know it's thousands of dollars" (Interview).

At the same time that these print advertisements were appearing, the Canadian Wheat Board released information demonstrating the devastating impact genetically engineered wheat would have on Canadian wheat farmers and the Canadian economy more broadly – at the time, the Canadian wheat export market was worth $3 billion annually. On average, the Canadian Wheat Board held 20 percent of the world market for wheat and 65 percent for durum wheat. In 2004, 87 percent of all countries that purchased

Canadian wheat required certification that the wheat they were receiving was not genetically engineered, up from 82 percent in 2002. The Council of Canadians developed and posted on its website a table outlining the state of market acceptance and non-acceptance of genetically engineered wheat (*State of Market Acceptance* 2004). On 22 March 2004, the Council of Canadians and representatives from Japanese consumer and industry groups held a press conference in Ottawa to announce the intention of the Japanese representatives to present a petition to then Minister of Agriculture and Agri-Food Bob Speller from 400,000 Japanese consumers, indicating that they would not buy Canadian wheat products unless the Government of Canada were to ban genetically engineered wheat. In addition to concerns about the almost complete loss of our export markets that would result from the introduction of genetically engineered wheat on the prairies, polling conducted by Decima on behalf of Greenpeace in March 2003 revealed that 60 percent of Canadians would avoid food products that contained genetically engineered wheat ("Farmers, Greenpeace Launch Anti-GM Wheat Ads" 2004).

Survey data collected by interview participant Ian Mauro also helped in the fight against the introduction of Monsanto's genetically engineered wheat into Canadian markets. Monsanto Canada said it would withdraw its regulatory application if farmers indicated they did not want genetically engineered wheat, all the while believing that such data were not available, which is normally a relatively safe assumption. What Monsanto did not know was that researchers at the University of Manitoba were finishing up a major survey of 2,000 Canadian prairie farmers that asked about precisely this issue. The findings demonstrated an overwhelming majority (83 percent) of respondents were strongly opposed to genetically engineered wheat (Mauro, McLachlan, and Van Acker 2009).[17] In addition to catching Monsanto off-guard, this episode indicated the vital importance of disseminating this type of information in the public domain.

In a more colourful action, Friends of the Earth in Quebec and the Council of Canadians kicked off a campaign in early May 2004 asking Canadians to mail then prime minister Paul Martin a slice of bread in protest over the government's support of genetically engineered wheat (Leahy 2004). At around the same time, Greenpeace Canada launched a quarantine action against the Agriculture and Agri-Food Canada research farm in Morden, Manitoba, where field trials of genetically engineered wheat were being conducted under the auspices of the partnership between the government and Monsanto. Activists scaled the research building and unfurled a 120-square-metre banner from the roof that read "Stop Genetically Engineered Wheat." Others locked the front doors of the building and set up quarantine signs. Around the farm itself, signs were posted that read "Biohazard: Stop Genetically Engineered Wheat" (Greenpeace Canada 2003a).

Figure 3.1   Greenpeace banner draped across Agriculture and Agri-Food Canada building at research farm. Photograph by Michael Aporius, courtesy of Greenpeace.

According to Josh Brandon, an interview respondent and genetic engineering campaigner for Greenpeace's Vancouver office, the CFIA had no mechanism that it could use to prohibit genetically engineered wheat because the safety issue of that crop was not really that different from other genetically engineered products that had already been approved, such as canola and corn. The overwhelming logic for not allowing approval of Monsanto's genetically engineered wheat was the fear of destroying the Canadian wheat export market because major importers of Canadian wheat would refuse to accept genetically engineered crops. On 10 May 2004, Monsanto formally withdrew its application for regulatory approval of its Roundup Ready wheat. Brandon believes that the government compelled Monsanto to withdraw its application due to pressure from Greenpeace and other groups opposed to genetically engineered wheat. The company also dropped its submissions for regulatory review in Australia, New Zealand, Russia, South Africa, Colombia, and the United States (Department of Agriculture and Environmental Protection Agency; it did, however, leave its application before the US Food and Drug Administration) (Hall 2004).

Although Monsanto eventually backed down from seeking permission to market this product, some still expected that the company would seek regulatory approval in the future. In fact, in announcing its decision in May

2004 to halt the introduction of genetically engineered wheat in Canada, Monsanto spokespeople stated that the decision was a matter of timing rather than a complete cessation of its attempt to market the product. According to a statement made by Carl Casale, executive vice-president for global strategy and operations at Monsanto's headquarters in the United States, and echoing the message voiced by Trish Jordan at Monsanto Canada, "this decision allows us to *defer* commercial development of Roundup Ready wheat in order to align with the potential commercialization of other biotechnology traits in wheat, estimated to be four to eight years in the future" (as cited in Leahy 2004, para. 3 [emphasis added]). The untapped commercial potential of the Canadian market alone, in which farmers typically save and reuse as much as 76 percent of wheat seed, no doubt furnishes a substantial incentive for Monsanto to bring this crop variety within its patent portfolio (Mauro, McLachlan, and Van Acker 2009).

Indeed, by mid-2009, the genetically engineered wheat issue was once again being pushed onto the public stage. In a clear shot across the bow of public opinion and activist groups, a number of Canadian, American, and Australian wheat organizations released a joint statement on 14 May calling for the commercialization of genetically engineered wheat. Although neither Monsanto nor any other biotechnology company signed the statement, the organizations involved had clearly learned some discursive strategies from the biogopolists, as evidenced by their attempts to wrap the call for approval of genetically engineered wheat in the discourse of feeding the world. The statement uncritically puts forth the usual claims attributed to agricultural biotechnology, including reduced input use, enhanced soil nutrient and water use, and improved tolerance of abiotic stresses. Yet, as will be elaborated in Chapters 5 and 6, a plethora of scientific studies of the main genetically engineered crops currently on the market are increasingly challenging such claims. It remains unclear why genetically engineered wheat should prove to be any different in practice. Indeed, the statement is littered with claims of dubious causality, if not outright misrepresentation. For example, the fifth point of the statement maintains that "biotechnology is a proven technique to deploy traits of interest with a high degree of precision in agricultural crops. Crops derived through biotechnology are subjected to strict regulatory scrutiny before commercialization. Over 10 years of global experience with biotechnology has demonstrated a convincing record of safety and environmental benefits as well as quality and productivity gains" (*Wheat Biotechnology Commercialization* 2009). As the evidence presented in Chapter 5 attests, none of these claims can withstand empirical scrutiny.

Commenting on this latest effort by these wheat organizations, Terry Boehm asserts: "GE wheat is a potential disaster of huge proportions. We refuse to allow Monsanto and industry groups to restart any campaign to commercialize GE wheat" (Canadian Biotechnology Action Network 2009a, para. 3). Lucy Sharratt adds that "Monsanto needs to accept defeat. The industry groups in our three countries are promising to force this product on all of us but today we reiterate our pledge to stop them" (ibid., para. 7). In response to the industry statement, a coalition of fifteen Canadian, American, and Australian consumer and civil society organizations released their own statement on the *Definitive Global Rejection of Genetically Engineered Wheat* (2009) on 1 June 2009 (by February 2010, 233 consumer and farmer groups in twenty-six countries had signed the statement), which focuses on the following joint pledge: "In light of our existing experience with genetic engineering, and recognizing the global consumer rejection of genetically engineered wheat, we restate our definitive opposition to GE wheat and our commitment to stopping the commercialization of GE traits in our wheat crops. We are committed to working with farmers, civil society groups and Indigenous peoples across the globe as we travel the road towards global food sovereignty."

The statement responds to the false claims of those groups seeking the commercialization of genetically engineered wheat, particularly by pointing out how farmers and scientists have successfully adapted and hybridized wheat varieties to local growing conditions for centuries. As the opponents of genetically engineered wheat point out, the only trait being genetically engineered into wheat varieties is for herbicide tolerance and not increased yield. The ultimate goal of the genetically engineered wheat developers is to bring this crop, which has profound cultural and religious significance for a number of cultures around the globe, under the firm control of the dominant seed companies as has occurred with genetically engineered canola, soybean, maize, and cotton. Aside from appealing to additional organizations to sign the statement against genetically engineered wheat, CBAN in Canada encouraged citizens to write to the minister of agriculture to make known their opposition to the introduction of genetically engineered wheat in this country.

Although apparently not directly involved in this latest action, Monsanto was once again ramping up its efforts on the genetically engineered wheat front. In March 2008, the company established the Beachell-Borlaug International Scholars Program, which, endowed with US$10 million, supports young academics working on wheat and rice research. In a more direct move, on 14 July 2009, Monsanto announced that it had acquired WestBred, a Montana-based company that specializes in wheat germplasm. According to the press release outlining the acquisition, "the investment will bolster the future growth of Monsanto's seeds and traits platform and allow farmers to benefit from the company's experience in drought-, disease- and pest-tolerance innovations" ("Monsanto Company Invests in Developing New

Technologies for Wheat with Acquisition of WestBred Business" 2009, para. 1). Carl Casale, in direct reference to the statement issued by the wheat growers' organizations, added that "the US wheat industry has come together to call for new technology investment, and we believe we have game-changing technologies – like our drought-tolerance and improved-yield traits – that can meaningfully address major challenges wheat growers face every season" (ibid., para. 2). Finally, on 2 November 2009, Monsanto issued a press release confirming that its purchase of WestBred was a precursor to the company's move back into the wheat seed market. Even without the powers of clairvoyance, one can safely predict that another skirmish on the genetically engineered wheat battlefield is not that far off.

Genetically Engineered Alfalfa

Alfalfa, an important forage crop employed in a variety of functions in agricultural production systems, is the third largest crop in Canada in terms of area planted, with 4.5 million hectares in production. The three prairie provinces account for approximately 75 percent of production, with an additional 20 percent grown in Ontario and Quebec. Canada also maintains a significant alfalfa processing industry that ranks within the world's top five in terms of alfalfa pellet and cube exports. Although genetically engineered alfalfa has been approved for unconfined environmental release by the CFIA, it cannot yet be commercialized because alfalfa requires variety registration, for which Monsanto has thus far not applied. In a 23 March 2006 position paper, the Saskatchewan Organic Directorate recommended that, in order to ensure food and feed safety and support organic farming, the Government of Canada should rescind approval for the environmental release of genetically engineered alfalfa, ban imports of genetically engineered alfalfa or alfalfa contaminated by genetically engineered alfalfa into Canada, and prohibit the testing of genetically engineered alfalfa in Canada.[18]

Given the loss of canola as an organic crop in this country due to genetic contamination, the Saskatchewan Organic Directorate is particularly concerned about the detrimental impact that the commercial release of genetically engineered alfalfa would have on organic farmers. According to their position brief, "[a]ll organic farmers use legumes as a soil-building component in their crop rotations. Alfalfa is a perfect legume for nitrogen fixation in the crop rotation for the majority of organic farms. To lose alfalfa in organic farm crop rotation would severely hamper our ability to maintain soil fertility and prevent soil erosion, which would harm the future of our soil's health and sustainability" (Saskatchewan Organic Directorate 2006, para. 3).[19] Moreover, alfalfa has a very competitive nature that allows it to thrive without typically requiring herbicides, another reason why it is so important as a rotational crop in both organic and conventional agriculture. Similarly, because of its ability to out compete weeds, it is a vital crop in a farmer's arsenal for cleaning up arable land from various weed incursions. Widespread contamination from genetically engineered alfalfa would also exercise knock-on effects on the organic dairy, meat, and honey industries given the importance of organic alfalfa as a major forage and feed product for various organic livestock. The Saskatchewan Organic Directorate was therefore quite pleased with the US injunction on the sale of Roundup Ready alfalfa seed, particularly given the high probability that this case could set a precedent that would provide Monsanto and Canadian authorities with the regulatory rationale for approving variety registration in this country – the last hurdle before genetically engineered alfalfa may be grown here.

In the US District Court, Justice Charles R. Breyer of the Northern District of California ruled on 13 February 2007 that the US Department of Agriculture's Animal and Plant Health Inspection Service (APHIS) violated the law when it did not properly assess the environmental impact of genetically engineered alfalfa before deregulating Monsanto's Roundup Ready alfalfa (in the United States, genetically engineered crops must be deregulated before they can be marketed and planted).[20] The lawsuit against the federal agency was brought by a coalition of groups led by the Center for Food Safety, which is based in Washington, DC. In his decision, Justice Breyer cited concerns about gene transfer through pollen to organic or non-genetically engineered alfalfa, which could impinge on export markets given that a number of foreign countries refuse to accept genetically engineered crops. Breyer also censured APHIS for failing to consider the possibility that the use of Roundup Ready alfalfa could result in increased application of glyphosate-based herbicides, which, in turn, could lead to evolving weed resistance to glyphosate (Pollack 2007).

In response to Justice Breyer's ruling, a Monsanto spokesperson stated that "we're going to do everything we think is appropriate to defend growers' right to choose this technology. Our goal is to restore that choice for farmers" (as cited in Jenkins 2007, para. 38). The disingenuity of this statement is not lost on a number of commentators, including interview informant Arnold Taylor, who points out that gene flow from genetically engineered plants is contaminating traditional crops to such an extent that farmers are losing the ability to choose to grow non-genetically engineered crops. These natural pressures serve to exacerbate the detrimental impact of increasing corporate consolidation and subsequent bundling practices, restrictive licencing agreements, and alleged price fixing on farmers and seed markets, as examined in the previous chapter.

In a later decision on 12 March 2007, Justice Breyer vacated APHIS's 2005 deregulation of genetically engineered alfalfa and ordered an immediate injunction on sales of such seed. Justice Breyer made this injunction permanent in a 3 May 2007 decision that enjoined any further planting of genetically engineered alfalfa until the agency completed an environmental impact statement (Pollack 2007). On 2 September 2008, the US Court of Appeals for the Ninth Circuit upheld Justice Breyer's ban on planting genetically engineered alfalfa pending the outcome of an environmental impact statement. The court's opinion, written by Circuit Judge Mary M. Schroeder, articulates the belief that genetically engineered alfalfa poses potentially irreversible harm to organic and conventional crop varieties, to the environment, and to the financial status of farmers. In weighing the two sides of the issue, the Court of Appeals stated that "Monsanto and Forage Genetics contend that the District Court disregarded their financial losses, but the District Court considered those economic losses and simply concluded that the harm to growers and consumers who wanted non-genetically engineered alfalfa outweighed the financial hardships to Monsanto and Forage Genetics and their growers."[21]

Monsanto subsequently appealed the decision at the Supreme Court of the United States (US Supreme Court), asking the court to lift the injunction on genetically engineered alfalfa, permit the sale and planting of genetically engineered alfalfa, and not allow contamination from genetically engineered crops to be considered "irreparable harm." In a seven-to-one decision delivered in June 2010, the majority of the US Supreme Court found that, although the lower court did not err in vacating APHIS's decision to completely deregulate Roundup Ready alfalfa, it had abused its discretion in enjoining the agency from partially deregulating the crop and in invoking a nationwide injunction that prohibited planting the crop pending a complete environmental review.[22] In practice, however, the US Supreme Court's decision maintains the vacatur of the complete deregulation decision because APHIS stated that it had no plans to consider a partial deregulation prior to the completion of the environmental impact statement. Thus, a *de facto* ban on the selling and planting of genetically engineered alfalfa remained in force during the pendency of the environmental review process.

The US Supreme Court did affirm the lower court's finding that contamination from genetically engineered crops is a sufficient cause of environmental and economic harm to challenge future deregulation decisions for genetically engineered crops.[23] On the heels of this decision, fifty members of the US House of Representatives and six US senators sent a letter to Agriculture Secretary Thomas Vilsack asking that the department not deregulate genetically engineered alfalfa.

In its draft environmental impact statement released in December 2009, APHIS claims that, although application levels of glyphosate may increase with the introduction of genetically engineered alfalfa, these will be offset by reductions in the use of other more toxic herbicides. Despite the projected increases in glyphosate, the statement contends that this will likely not lead to the development of glyphosate-resistant weeds nor significantly impact human health and safety, land use, or the physical environment. Yet, as we will see in Chapters 5 and 6, emerging scientific research provides evidence of precisely the opposite, namely, that the evolving glyphosate resistance among greater numbers of weed species will require heavier doses of glyphosate and/or the application of other more toxic herbicides. The draft statement similarly dismisses the claim that the widespread release of genetically engineered alfalfa will, in aggregate, pose negative economic effects for organic and conventional alfalfa growers. In fact, the statement's authors assert an absence of sensitivity to genetically engineered alfalfa in domestic organic markets, a claim resolutely rejected by the Center for Food Safety's staff attorney George Kimbrell: "When the initial National Organic Program rule was published, USDA [the US Department of Agriculture] received 275,000 public comments from people demanding that genetic engineering be excluded from organic food. This is evidence that people do care about GE contamination" (as cited in Roseboro 2010, para. 4). Despite such overwhelming opposition, in early 2011, US Agriculture Secretary Tom Vilsack announced that he would authorize the unrestricted commercial cultivation of genetically engineered alfalfa.

Various organizations on this side of the border also waded into the debate. The NFU, Beyond Factory Farming, and the Saskatchewan Organic Directorate, in co-operation with CBAN, all sent formal submissions to APHIS outlining their opposition to the conclusions outlined in the agency's draft environmental impact statement. As these organizations point out, any American decision to permit the environmental release of glyphosate-tolerant alfalfa will impact Canadian producers as a result of the inevitable contamination of non-genetically engineered alfalfa crops and seed. Although Canada is a large alfalfa producer, we nonetheless import American hay. More problematic, despite substantial indigenous production of alfalfa seed, Canada still relies on American imports to meet domestic seed requirements. Disputing claims that stewardship plans can be implemented to mitigate the probability of gene flow, these groups point to actual experience with both genetically engineered canola and, more recently, genetically engineered flax to remind APHIS that gene flow is not only inevitable but also unpredictable and ultimately detrimental to farmers with regard to market impact and long-term costs. Contra the APHIS position, Canadian producer groups contend that both domestic and export markets will reject alfalfa products contaminated with genetically engineered substances, that the current institutional mechanisms will fail in their attempts to prevent contamination, and that no measures can be implemented that would avert or mitigate unintended presence.

The submission of the Saskatchewan Organic Directorate accuses APHIS of failing to properly grasp real world organic farming practices, which, in turn, undermines the central claim made by the agency that gene flow can be mitigated to such an extent that it will not significantly impact the human environment. In particular, the brief outlines six specific factors that have a high probability of contaminating non-genetically engineered alfalfa. First, there exists a myriad of ways in which seed mixing may occur (for example, by failing to properly clean out hoppers and bins between crops, spilling seed while hauling, and enabling volunteer growth due to seed shattering, high winds blowing swaths during harvest, floating swaths in flood conditions during harvest, or seed spread from storage bins by birds and rodents). Second, there is significant potential for pollinator-mediated gene flow, against which the suggested counter measures outlined by Monsanto in the stewardship agreement that accompanies its seeds are completely inadequate. Third, a number of environmental influences increase the probability of unplanned large-scale blooming of alfalfa crops before cutting, which could offer a window of opportunity for bees and other pollinating insects to transfer pollen between genetically engineered and non-genetically engineered alfalfa seed crops. Moreover, areas with longer growing seasons can yield two or three cuts of hay, thus increasing these windows of opportunity. Fourth, genetically engineered volunteer alfalfa will be a problem because alfalfa seed can persist and germinate in fields for several years. Fifth, feral alfalfa, which is widespread in alfalfa production regions, will likely increase the probability of gene flow between fields and, thus, ultimately facilitate contamination of conventional and organic seed.[24] Feral crop populations can be particularly prone to gene flow because of their high levels of outcrossing, prolific seed production, dispersal, and dormancy, desultory germination, speedy vegetative growth, and tolerance to competition, as well as biotic and abiotic stresses. According to one group of agricultural researchers, "feral alfalfa populations can and will act as bridges for long-distance gene flow and facilitate the adventitious presence [contamination] of novel traits in the environment. As such, feral populations will become a potential barrier for achieving coexistence of transgenic and nontransgenic alfalfa fields" (Bagavathiannan and Van Acker 2009, 70). Finally, because the composition of alfalfa seeds allows them to survive intact and pass through the ruminant gut, livestock manure and hoof action might also facilitate the spread of genetically engineered seed and subsequent germination.

Taylor points to the success that farmers and others have achieved in the battle with Monsanto over its genetically engineered wheat. However, the current and growing fear is that Monsanto and other companies will renew their efforts to expand their control over agricultural markets through genetically engineered alfalfa. As part of its efforts to oppose the commercialization of genetically engineered crops that threaten the sustainability of organic farming, the Organic Agriculture Protection Fund has written to Monsanto and Syngenta in order to put them on notice that, if they do try to introduce genetically engineered wheat or genetically engineered alfalfa in Canada, organic farmers will respond with all legal mechanisms at their disposal, including attempts to obtain an injunction until the liability issue has been settled. As Taylor puts it, these companies need to understand that "we are here and we are not going away" – important words that industry would do well to heed given that Saskatchewan is home to 1,200 organic farmers who till over 485,000 hectares of organic land (Interview).

This warning was put into action in early 2009, when the Saskatchewan Organic Directorate launched its "No to GMO Alfalfa" campaign. As part of its initial strategy, the directorate asked concerned groups to sign a petition that commits them to the following three statements: "We oppose the sale, trade and production of GMO Alfalfa in Canada. We ask the Canadian Food Inspection Agency (CFIA) to reassess its approval for environmental release of GMO Alfalfa. We want the public to understand the hazards, costs and market losses that would result if GMO Alfalfa were released into our environment." The next steps include striking a campaign team to mobilize farmers and other citizens against the introduction of genetically engineered alfalfa into the Canadian market and environment. As of March 2012, 127 groups, including farmer associations, farm businesses, sprouting and seed companies, food retailers, and public interest groups, had signed the "No to GMO Alfalfa" letter.

In addition to these types of grassroots organizing, opposition to genetically engineered alfalfa and the scope of its regulatory assessment has also been mobilized in Parliament. Alex Atamanenko, agriculture critic and New Democratic Party MP for British Columbia's Southern Interior, introduced a Private Member's Bill (C-474) on 2 November 2009 that sought to amend the *Seeds Regulations* to require that an analysis of potential harm to export markets be conducted before the sale of any new genetically engineered seed is permitted.[25] Despite strong opposition from the Conservatives, tepid support from the Liberals helped the bill pass second reading in April 2010, at which time it was referred to the House of Commons Agriculture Committee. In June, Jim Lintott, chairman of the Manitoba Forage Council, told the parliamentary Agriculture Committee that "[f]orage producers support Bill C-474 because we believe it would have the ability to protect the alfalfa industry from the truly dangerous effects of introducing GMO varieties that are not approved by our customers" (as cited in Sharratt 2010, para. 10). According to Kelvin Einarson, director and secretary treasurer of the Manitoba Forage Seed Association, "Bill C-474 is the first step in offering some protection in the future for Canadian family farms. Market acceptance must be made part of the evaluation process and incorporated into the Seeds Regulation Act" (ibid.).

In the face of a staunchly pro-genetic engineering government and substantial industry lobbying against the bill, the Agriculture Committee hearings were shut down in late October when Conservative MPs defeated a motion to extend debate on the bill within the committee for an additional thirty days (many Liberals were suspiciously absent from the House for this vote, including then party leader Michael Ignatieff). With the support of the Liberals, the Harper government voted down the bill on 9 February 2011 in the House of Commons.

Genetically Engineered Sugar Beets

In early 2009, CBAN initiated a public relations campaign against genetically engineered sugar beets in Canada aimed at convincing Lantic – Canada's largest and only sugar company to process sugar beets – not to accept genetically engineered sugar beets. Coinciding roughly with Valentine's Day, CBAN offered concerned citizens the opportunity on its website to send e-mails and valentine messages to the chief executive officer of Lantic asking him to keep sugar non-genetically engineered. Despite over 4,000 e-mails, letters, and cards, Rogers Sugar/Lantic agreed to accept genetically engineered sugar beets. In response to this decision, Alberta farmers began planting genetically engineered sugar beets for the first time in April 2009. Ontario farmers in Lambton County and Chatham-Kent planted genetically engineered sugar beets in 2008, but their complete production was sent to the United States for processing by Michigan Sugar. Given the position of Rogers Sugar/Lantic, CBAN is now focusing attention on Cadbury, a major sugar purchaser, hoping to convince this company to refuse to purchase sugar processed from genetically engineered crops. CBAN is also asking consumers to boycott Rogers Sugar and Lantic sugar products.

Similar actions are being mounted in the United States, where, according to a registry developed for this campaign, over 100 grocery chains and food producers have committed to neither sell nor use genetically engineered beet sugar. By signing the registry, signatories commit to the following principles: "Our company does not support the introduction of Genetically Modified (GM) Sugar from GM sugar beets; We will seek wherever possible to avoid using GM beet sugar in our products; and, We ask the Sugar Beet industry to not introduce GM beet sugar into our food supply."[26] The July/August 2009 issue of the *Organic and Non-GMO Report* indicates that genetically engineered sugar beet plants have been discovered in soil mix sold to gardeners in Oregon. Oregon's Willamette Valley is a production site for Monsanto's Roundup Ready sugar beet seeds. This discovery, confirmed by Carol Mallory-Smith, a professor of weed science at Oregon State University, highlights the extreme difficulties of containing genetically engineered seeds and their progeny and helps dispute industry and government claims about "co-existence" ("GM Sugar Beets Found in Soil Mix" 2009).

Mirroring the recent success with respect to genetically engineered alfalfa, the federal District Court for the Northern District of California ruled in September 2009 that APHIS had violated the *National Environmental Protection Act* by failing to properly assess the environmental impacts before it made the decision to deregulate Monsanto's Roundup Ready sugar beets.[27] The suit was brought against APHIS in January 2008 by the Center for Food Safety and Earth Justice on behalf of a coalition of farmer and consumer groups, including the Organic Seed Alliance, the Sierra Club, and High Mowing Organic Seeds. In ordering APHIS to conduct an environmental impact statement, Justice Jeffrey S. White's ruling is critical of the agency for failing to consider the effects that gene transmission would have on conventional farmers and consumers, as well as on organic beet seed markets. The ruling goes on to express concern about the effects that gene escape could have on related crops such as red table beets and Swiss chard.

Perhaps most interestingly, and something upon which Canadian legislators and courts remain muted, Justice White admonished American regulators to consider the economic effects, contending that they "are relevant and must be addressed in the environmental review when they are 'interrelated' with 'natural or physical environmental effects.'"[28] Finally, reflecting previous concerns about the increasing consolidation of seed markets, Justice White wrote that "there is no support in the record for APHIS's conclusion that non-transgenic sugar beet will likely still be sold and will be available to those who wish to plant it and that farmers purchasing seed will know whether it is transgenic because it will be marked and labeled as glyphosate tolerant. Therefore, the Court finds that APHIS's finding of no significant impact was not supported by a convincing statement of reasons and thus was unreasonable."[29] The ruling further outlines that, at a later date, the court would consider remedies, including injunctive relief similar to that granted in the genetically engineered alfalfa case. In March 2010, Justice White denied a request by the coalition of plaintiffs for a preliminary injunction pending the outcome of the environmental impact statement, but he did make clear that this decision should not be interpreted as a dismissal of a possible future injunction.

On 13 August 2010, Justice White vacated APHIS's deregulation decision and denied the latter's request for a stay of the vacature. Yet in spite of the vacature order, APHIS granted in early September several applications to permit production of genetically engineered sugar beet stecklings. The Center for Food Safety and its allies who brought the original suit against the deregulation decision promptly filed another lawsuit challenging the issuance of the permits, including a motion for a temporary restraining order. On 30 November, Justice White granted the plaintiffs' motion for a preliminary injunction and ordered that all of the stecklings planted pursuant to the permits issued by APHIS be removed from the ground. This order was stayed pending appeal, which took place in February 2011 at the US Court of Appeals in San Francisco. In its decision on 25 February 2011, the appellate court vacated the preliminary injunction and ordered that the APHIS permits be given full effect. At around the same time, APHIS announced that it would partially deregulate genetically engineered sugar beets so that American farmers could plant them before the agency completed the court-ordered environmental impact statement.

Labelling of Genetically Engineered Foods

> It all cries out for a truth-in-advertising reality check. And the
> reality is that the federal government and the biotech industry –
> it's almost impossible now to distinguish between the two –
> do not want you to know what you are consuming. It's a so-far
> successful policy of obfuscation and delay on the issue of labelling
> GMOs at the expense of Canadian consumers and exporters.
>
> – Lyle Stewart, "A Question of Credibility,"
> *Montreal Gazette*, 24 May 2002

The labelling of genetically engineered foods, which, as some observers have noted, might be considered a type of information policy, continues to be an issue pressed by Greenpeace with the support of the Council of Canadians (Magat and Viscusi 1992). In April 2001, the New Democratic Party introduced a bill in the British Columbia Legislative Assembly that would have required mandatory labelling of genetically engineered food. However, the bill died on the order table when elections were called, and the new government did not reintroduce it. In October 2001, a federal Liberal Private Member's Bill (C-287) that would have legally mandated labels for genetically engineered foods was defeated in Parliament, despite the support of more than eighty civil society groups and opinion polls that demonstrated overwhelming public support (between 88 and 90 percent) for such labelling requirements.[30] According to Aaron Freeman, a reporter for the *Hill Times*, when MPs arrived in the House of Commons to vote for the bill on 17 October they found pamphlets published by different food industry associations on their desks that urged them to "vote against Bill C-287 and support Canada's Agri-food business" (Freeman 2001, B4). Part of the lobbying effort against the federal bill was mounted by Loblaws, which also operates as Zehrs, No Frills, Provigo (Quebec), Real Canadian Superstore (Western Canada), and Atlantic Superstore (Atlantic Canada).[31] Aside from defeating the bill, the Liberal government halted within six months a promised study into the issue by the House of Commons Standing Committee on Health.

Although Bill C-287, which would have implemented a mandatory labelling system for genetically engineered food in this country, was defeated in October 2001, Eric Darier, interview informant and agriculture co-ordinator for Greenpeace Canada, argues that it was an interesting battle because the bill came very close to being adopted. In fact, half of the Cabinet at the time made sure to be present in the House of Commons to vote against the proposed bill, which, as Darier suggests, demonstrated both the high level of concern among the government of the day and the degree to which the Liberal Party was split on the issue. Subsequent to this parliamentary defeat, the debate on mandatory labelling took two different directions. The federal government moved slowly on a voluntary labelling scheme, which was precisely what industry advocated since ultimately it would be unenforceable – a Canadian standard for voluntary labelling of genetically engineered foods was released in 2003. However, as Sharon Labchuk of the Prince Edward Island Coalition for a GMO-Free Province adds, "of course no companies have ever voluntarily labelled their foods as containing GE ingredients. It is only mandatory labelling that will give consumers choice in the grocery store" (*Bill to Label Genetically Engineered Foods* 2008, para. 4).

The Canadian Biotechnology Advisory Committee (CBAC) voiced similar opposition to any mandatory labelling system in its 2002 final report on the regulation of genetically engineered foods, despite the fact that its survey research indicated that 85 percent of Canadians favoured a mandatory, over a voluntary, labelling regime for genetically engineered products (Pollara Earnscliffe 2003). Similarly, the majority of respondents who commented on the CBAC's interim report on regulation wanted to see Canada adopt a mandatory labelling regime.[32] In order to justify its position on labelling, which was clearly at odds with the demands of the broader population, CBAC's report articulated the following concerns: possible market damage if consumers associate genetic-engineering labels with a perception that a particular product may not be safe; the potentially onerous cost structure; the logistical difficulties associated with segregating conventional and genetically engineered crops; and the possibility that mandatory labelling would cause Canada to contravene some of its international trade agreements, thus rendering Canadian exports subject to trade sanctions (Canadian Biotechnology Advisory Committee 2002a). In a particularly condescending fashion, the CBAC report suggested that "many people do not have a clear understanding of GM foods and could be confused or misled by a label indicating GM content" (38).[33] Aside from the paternalistic character of such an assertion, one wonders whether the irony was lost on the drafters of this document, given the CBAC's mandate, which, in part, tasked this committee with providing Canadians with "easy-to-understand information on biotechnology issues" (Canadian Biotechnology Advisory Committee 2005a).

In articulating its opposition to labelling in 2002, Agriculture and Agri-Food Canada employed the "substantial equivalence" rationale (that genetically engineered products are not substantially different from conventional ones) as support for its position that mandatory labelling requirements for genetically engineered crops and their food and feed products are marketing decisions and not health or nutrition issues (Wilson 2002). As one commentator points out, a statement issued by the department in that same year contained a direct admission that mandatory labelling was defeated out of concern about annoying the United States, which is the largest importer of Canadian food exports and which does not support mandatory labels for genetically engineered foods (McGiffen 2005). According to the departmental response, "a disjointed approach with the US on voluntary versus mandatory labelling could place both trade and investment at risk" (as cited in Wilson 2002, para. 6). Wilson (2002, para. 9) asserts that "some trade officials have warned that a mandatory labelling regime in Canada likely would be challenged by the US as a new trade barrier that contravenes NAFTA rules." Lyle Stewart (2002a, B3), one of the few journalists in this country to produce critical coverage of biotechnology issues in the early part of this decade, points to the dangers "these trade agreements represent not only for our simple right to know, but for the health of our bodies, environment and democracy." With respect to the trade argument, Nadège Adam (2002, para. 7), former biotechnology campaigner for the Council of Canadians, contends that "the federal government is only trying to find a weak justification for its crawling before the US Government and the Monsantos of the world. Labelling of GM food, in response to the public's wishes and with respect to the precautionary principle, can certainly be showed to be a 'legitimate objective' in the sense of NAFTA. The labelling would affect all GM food, not only from American origin. NAFTA is only a convenient excuse for the government's inaction."

The cost argument – that neo-liberal shibboleth that presumably carries significant weight in the discussions against a mandatory labelling scheme – has been revealed recently to be an industry smokescreen based on extremely inflated estimates. A study commissioned by the Quebec Department of Agriculture, Fisheries and Food determined that a mandatory labelling program would cost the Quebec food industry $28 million and the provincial government $1.7 million annually. To put those figures in perspective, the Quebec food industry is worth $30 billion annually. These cost figures are much lower than those put forth by the food industry – $950 million for the whole country and $200 million in Quebec for both industry and governments. Even before this homegrown study was conducted, John Fagan of Genetic ID, Incorporated, of Fairfield, Iowa, testified before the House of Commons Standing Committee on Health that testing and identifying genetically engineered crops, a prerequisite step for a labelling regime, would impose only negligible costs of about 0.6 percent for large-scale shipments. Even with smaller shipments of around twenty tonnes of food grade soy, the increased cost would amount to 1.6 percent (Stewart 2002a). The argument advanced by the food industry that labelling costs would have to be passed on to consumers is also disputed on the basis of experience in countries with mandatory labelling requirements such as the member states of the European Union, Japan, China, Australia, and New Zealand, where companies have changed the way they operate in order to secure operational savings that offset labelling costs (Lalonde 2007).

In the absence of mandatory labelling, those producers who reject genetically engineered ingredients must, in order to protect their markets, label their products as being free of genetically engineered organisms. As a result, the lack of a mandatory labelling requirement for genetically engineered products actually imposes a cost burden on non-genetic engineering producers. Perhaps most ironic is that an industry that is quick to make appeals to "the market" as the determining factor for introducing biotechnologies continues to ardently oppose labelling requirements for genetically engineered foods, the only real market mechanism available for consumers to make an informed choice (Magdoff, Foster, and Buttel 2000a). As one reporter observes, "it is also in the interests of farmers, and consumers, that GM products be clearly labelled – a minimal requirement that would at least offer buyers a choice. If the cabinet refuses this straightforward demand, it is fair to ask whose side they are on. And what do producers have to hide?"(Riley 1999, A14).

Attempts at maintaining the labelling issue on the public agenda have included the organization of information pickets at major grocery stores by the Council of Canadians and a demonstration by 1,000 protesters in Montreal outside the meeting for the UN *Cartagena Protocol on Biosafety* that took place in 2000.[34] Adopting a more direct course of action, a number of Greenpeace activists, particularly in Vancouver, went into supermarkets and placed their own labels on products that contained genetically engineered organisms. According to interview informant Herb Barbolet, founder of the British Columbia non-profit society Farm Folk/City Folk, such direct action did garner public attention for a short period. Members of the Citizens' Voluntary Labelling Collective carried out similar activities in Montreal in April 2000 and in Charlottetown the following month. Some success has been achieved. McCain and Seagrams have discontinued purchasing genetically engineered potatoes and grains for use in their respective products.

As noted previously, the main group struggling for the institution of a mandatory labelling regime for genetically engineered products in this country is Greenpeace. According to Darier, labelling is an important tactical strategy because Greenpeace found that once labelling or mandatory labelling was instituted in Europe and Australia, consumers avoided genetically engineered foods, which ultimately resulted in such products being removed from the food system. In turn, because there is no incentive to grow them, such crops are then also removed from the environment – although the lasting dangers associated with genetic contamination remain unclear. Greenpeace believes that introducing some kind of transparency in consumer labelling is usually enough to push industry, or at least a segment of industry, to begin removing genetically engineered ingredients from its products. Given such positive experience in other countries, Greenpeace Canada has made the strategic decision to exert pressure on Canadian governments to implement mandatory labelling as the main form of leverage in the struggle to remove genetically engineered organisms from our food supply chain. Indeed, the most recent survey data available suggest that this strategy holds the promise of success in this country. According to Decima Research, 60 percent of Canadians believe genetically engineered foods provide more risks than benefits, 53 percent are uncomfortable purchasing food that contains genetically engineered ingredients, and only 19 percent would buy a food item if they knew it was genetically engineered (Decima Research 2004). In response to the consumer information gaps that follow from the lack of a mandatory labelling regime, Greenpeace publishes a document titled *How to Avoid Genetically Engineered Food: A Greenpeace Shoppers Guide,* which provides information on over 1,000 products typically found on grocery store shelves in this country, including whether or not they contain genetically engineered ingredients (Greenpeace Canada 2003b).

Following the 2001 federal parliamentary defeat of a mandatory labelling system, Greenpeace decided to refocus its efforts and concentrate on those places where it believed it could get some traction on the issue. Based on its own analysis of support levels for mandatory labelling and, thus, its prognosis for success, Greenpeace elected to focus on Quebec and British Columbia as the two likeliest provinces where provincial legislation has a chance of being promulgated. This strategy is designed to capitalize on one of the peculiarities of a federal state: if one level of government refuses to act, it is quite possible to convince another level to do so (a tactic, Darier points out, that industry is quite adept at employing). For example, the Quebec Liberals have promised mandatory labelling, and, in fact, an all-party Agriculture Committee of the Quebec National Assembly unanimously recommended in 2004 that the province implement a mandatory labelling regime based on the European system. That being said, the recommendation has yet to be operationalized, although Greenpeace continues to exert pressure for action on this policy file. The situation is similar in British Columbia, where, although the campaign is younger, the objective is to get that province to move on mandatory labelling for genetically engineered food. The efforts in British Columbia tend to rely on traditional petition action to mobilize public opinion around the issue. Some initial success was achieved when Gregor Robertson, NDP Member of the Legislative Assembly for Vancouver-Fairview, introduced a Private Member's Bill (M 226-2007, *Right to Know Act*) in the British Columbia legislature that sought to make labelling mandatory for genetically engin-eered foods. Unfortunately, the bill never advanced beyond First Reading.

More recent federal legislative attempts have also failed to achieve any lasting success. As a result of winning a random parliamentary draw, Gilles-André Perron, Bloc Québécois MP for Rivière-des-Mille-Îles, was permitted in February 2008 to introduce his Private Member's Bill (C-517) in the House of Commons. This bill would have legislated the mandatory labelling of genetically engineered foods. The bill was debated during a second reading on 3 April 2008 and 5 May 2008, during which time Perron pointed out that the voluntary labelling scheme over the past four years had been a dismal failure. Drawing upon past public opinion polls in which an overwhelming majority of Canadians indicated a desire for labels on genetically engineered foods, he argued that labels are a necessary means to allow consumers to make informed purchasing decisions based on cultural, personal, or religious perspectives. In a contradictory response, Conservative MP Bruce Stanton admitted the interest of Canadians in food labelling but then articulated the rather worn-out talking point about the rigorousness of Canada's regula-tory regime, which, according to his logic, negates the necessity of additional labelling.

In an even more baffling feat of nonsense, Liberal MP Robert Thibault maintained that, if all products containing genetically engineered compon-ents must be labelled, then "we would be labelling so much on the shelves of our stores that the labelling would become meaningless."[35] In cutting through the contradictory rhetoric offered by both the Conservatives and Liberals, NDP MP Nathan Cullen provided an immanent critique of their positions in favour of labelling: "We feel that if genetically modified foods are a safe thing, if the government feels it has the science and the evidence on its side to say that it is safe, 100 percent guaranteed, then the labelling of such products, the identification of those products, should not be a prob-lem. Consumers will then have a choice between a product that has been genetically modified or one that has not."[36] Both CBAN and Greenpeace actively engaged their memberships to write to MPs to express support for the bill once it came up for a final vote in the House of Commons, but in a vote on 7 May 2008 the bill was defeated from progressing further in Parliament.[37]

Although the demands remain unheeded, Darier points out that main-taining pressure on governments about the issues is itself an important oppositional tactic. In fact, he contends that although there has not yet been "success" on the labelling issue, it should be considered relative to the broader struggle. For example, in 2000, the president of the CFIA articulated plans to have 500 new genetically engineered products in Canada within five years. Yet, even in 2012, the numbers of approved products remain more or less the same and are certainly far removed from the prognostications made by government and industry proponents of biotechnology. Darier credits the various campaigns organized against genetic engineering with the success in slowing down the rates of approval for these products. He also points to the successes that opposition struggles have had in convincing industry to withdraw from various biotechnology projects, the most recent success being the decision by Monsanto, under enormous international pressure and market rejection, to scrap its plans to introduce genetically engineered wheat into Canada (although, as outlined earlier, a renewed battle over genetically engineered wheat looms large on the horizon). Overall then, more than fifteen years after the first authorization of a genetically engineered product, the main seed varieties being commercialized in this country continue to be corn, soy, and canola. Although, as of 2008, Monsanto's genetically engineered sugar beets are being marketed here. So, from the perspective of Greenpeace, there has been some success in retarding the progression of biotechnology and the introduction of new genetically engineered crops in Canada.

Theoretical Outlook 2: Conceptualizing the BioCommons

As the preceding evidence attests, struggle is inherent to capitalist social relations, whether at a micro level that remains obscured from widespread view or at a macro level of open confrontation. As Karl Marx articulated with his concept of the circuit of capital, capital must go beyond the exploitation of the immediate workplace to continually infiltrate and subvert new social sites and activities to its own accumulation imperatives. In what can be understood as recurrent processes of primitive accumulation that temporally seek to encompass the full range of the "life time" of waged and unwaged labour beyond the confines of the workday, capital simultaneously engages in spatial colonizing tendencies that strive to penetrate not only new territories across the globe but also areas of social life that reach back to the genetic level of organic existence. However, it is precisely these new sites that open up additional points of resistance. Put more explicitly, the diffusion of capitalist command throughout the social domain brings with it an attendant antagonism that is not only potentially ubiquitous in nature but is also dispersed and generalized in terms of its sites of struggle. The plunder and enclosure of a range of biological artifacts (biological commons) and information (informational/knowledge commons), ranging from the protein sequences contained in deoxyribonucleic acid to particular genes, genetically engineered seeds, and entire plant species, through the capitalist control of biotechnology, is remaking nature-society relations in a way that ignites resistance precisely because it has implications for people's existence and for their relationships to themselves, to others, and to nature.

Articulated in terms of the theoretical framework informing this book, the BioCommons is one of the responses emerging to counter enclosures executed as part of broader capitalist primitive accumulation strategies. And while this application relies more on metaphor than on actual institutional arrangements, it provides an anticapitalist, albeit more implicit than explicit, commitment to the public interest that opens to critical scrutiny the multiple issues and interests encompassed within the battles to control the informational and material resources imbricated in agricultural biotechnology. Recourse to the concept of the BioCommons helps group the seemingly disparate facets of this technoscience to make explicit their interconnections in ways that open discursive and intellectual spaces to oppose their enclosure by capital. Similarly, contemplating such issues through the lens of the BioCommons helps elevate individuals to a role above mere consumer in the marketplace in order to shift the focus of debate around this technoscience onto the rights, needs, and responsibilities of citizens.

Along these lines, much can be learned from Massimo De Angelis's (2007) decidedly sanguine notion of the "beginning of history," which represents a contemporary opportunity to reappropriate and defend the commons, including relations between humans, objects, and the natural world as a means of superseding the individualizing and normalizing tendencies of capital. That is, the beginning of history postulates a struggle between the life-colonizing force of capital that positions individual against individual in pursuit of the capitalist *telos* of accumulation and the life-reclaiming forces of people and movements that strive to construct value practices independent of capital, despite claims that alternatives are no longer possible. We therefore need to make visible those value practices that are situated beyond the value practices of capital. In part, this demands that we first conceptualize the capitalist market regime as an ethical system that involves value judgments and corresponding relations to the "other," although this "other" is often unseen.

As John McMurtry (1998, 13) reminds us: "Economists explicitly deny that any value judgment is at work in their analyses, even though they presuppose a value system in every step of the analysis they make." All market decisions express the values of the market system. Since we, as individuals, are embedded in this system, and to the extent that we accept its codified and normalizing language and conform to its parameters, we are only able to attain cognitive clarity by conceptually exiting the value system given by the market and by refusing to accept its normalization. Despite the discourse of many neo-liberals that would deny an outside to the economic calculus that apparently guides social co-production, social struggle demands that we consider scrutinizing these outside dimensions as alternative value practices among a range of possible actions and processes that can compete with those of capital in meeting the needs of the direct producers rather than of capital. Through the production of the BioCommons, we might begin to "step outside" of the market and develop value practices based on social co-operation that eschew the possessive individualism inherent in neo-liberal capitalist social relations. This more active conceptualization allows us to pose a limit to capital that simultaneously throws open to debate the possibility of alternatives and their problematization, which, in turn, might reduce their susceptibility to capitalist co-optation. This notion of the commons thus similarly admits the dialectical relationship between enclosures and commons.

However, one must remain cognizant of the fact that struggles against enclosure often only open up questions of the commons rather than automatically ensuring their emergence. Capital typically refuses to remain idle when challenged by such resistance, often strategizing to develop means of subsuming struggle into novel forms of accumulation. Indeed, as Marx (1962a, 250-51) told us so long ago, because capital possesses an uncanny ability to reassert itself with renewed vigour after suffering partial defeats at the hand of its adversary, struggles by labour must be informed by an eminent level of reflexive criticality, such that those actors struggling must

criticise themselves constantly, interrupt themselves continually in their own course, come back to the apparently accomplished in order to begin it afresh, deride with unmerciful thoroughness the inadequacies, weaknesses and paltrinesses of their first attempts, seem to throw down their adversary only in order that he may draw new strength from the earth and rise again, more gigantic, before them, recoil ever and anon from the indefinite prodigiousness of their own aims, until a situation has been created which makes all turning back impossible, and the conditions themselves cry out: *Hic Rhodus, hic salta!*

As a more contemporary scholar puts it, "the problem of alternatives therefore becomes a problem of how we disentangle from this dialectic, of how within the social body conflict is not tied back in to capital's *conatus*, but instead becomes a force for the social constitution of value practices that are *autonomous* and independent from those of capital" (De Angelis 2007, 42 [emphasis in original]).

As noted earlier, the majority of current opposition groups in this country have adopted a tack that refuses biotechnology. I qualify this statement because not all of those individuals interviewed reject biotechnology out of hand. However, what does unify all of them is their *refusal to accept capitalist control of biotechnology and its practical applications*. Whether in the realm of the communicational or biological, many of the people involved in such resistance share a common goal of demanding that these emerging technologies, with their potential to impact and transform organic existence, be subject to democratic control from below in a manner that might yield a more equitable distribution of their attendant benefits and disadvantages than is currently the case under capitalist control.

4
Intellectual Property Rights: Facilitating Capital's Command over Biotechnology

> The international effort to convert the genetic blueprints of millions of years of evolution to privately held intellectual property represents both the completion of a half-millennium of commercial history and the closing of the last remaining frontier of the natural world.
>
> – Jeremy Rifkin, *The Biotech Century: Harnessing the Gene and Remaking the World*

In terms of the conceptual framework informing the present work, such efforts also represent the recourse capital makes to laws and other political economic institutional structures that facilitate enclosure of genetically engineered products and the knowledge embodied therein in service to primitive accumulation imperatives. Similar to concerns about the chilling effect of overly stringent copyright protection on the dissemination of information, a number of scholars are questioning the current heavy use of patents within the biotechnology industry. Aside from considerations about the validity of various patents issued on basic biological "building blocks," several commentators worry that the resulting patent thickets will impede research and its dissemination (Eisenberg 1997b; Heller and Eisenberg 1998; Klein 2007; Long and Johnson 1997; Maskus and Reichman 2004).[1] Perhaps more germane to the theoretical position being applied here, it bears emphasizing what sometimes tends to be overlooked (or only implicitly accepted) in much literature on intellectual property, namely, that the value generated through the creation and exploitation of intellectual property relies on an *a priori* exploitation of labour power (Rossiter 2006; Schiller 2007; Trosow 2003).

   With these types of issues in mind, the intent of this chapter is to illustrate the various ways in which capital is responding to Lester Thurow's (1997, 101) recent admonition that "the Industrial Revolution began with an enclosure movement that abolished common land in England. The world now needs a socially managed enclosure movement for intellectual property rights." We will examine how capital applies the contemporary intellectual property protection regime – the umbrella term that includes copyright, patents, trademarks, trade secrets, and industrial design – to facilitate its accumulation processes in ways that not only overstep the conceptual justification for intellectual property but also offend a number of the normative and practical elements of agriculture that have existed for millennia.

Attention in this chapter first turns toward providing a basic outline of intellectual property, including a discussion of its underlying rationale. Having established the general nature of the system, focus will narrow to consider concerns about patents on genetically engineered organisms, including critiques levelled against the genetic reductionist paradigm that provides support for patents on genes. These issues will then be assessed in the context of two of the arguably most famous intellectual property protection battles over genetically engineered organisms waged to date in this country – the Harvard Oncomouse decision and the lawsuit brought against Percy Schmeiser by Monsanto for patent infringement of its genetically engineered canola seed. The penultimate section of the chapter juxtaposes the Schmeiser case with the attempt by the Organic Agriculture Protection Fund to sue Monsanto and Bayer CropScience in order to highlight not only the disconnect between rights and duties that attach to intellectual property protection but also the double enclosing effect of patented genetically engineered crops in Canada.

A Conceptual Overview and Critique of the Intellectual Property System

Bruce Doern and Markus Sharaput (2000) have linked the growing importance of intellectual property as a policy area in Canada to global change in general. As we saw in Chapter 1, Canadian economic policy increasingly seeks to move beyond the traditional exploitation of natural resources toward an expanded culture of invention and innovation. Intellectual property protection today is considered vital not only as an instrument by which to spur domestic innovation but also as a means to gain entrance to world markets of innovation, without which it becomes progressively more arduous for an economy and society to flourish (ibid.). This type of thinking drives Canada's latest iteration of science and technology policy, which contends that facilitating knowledge transfer from institutions of higher education to the private sector requires the development of intellectual property arrangements standardized across countries (Government of Canada 2007). Industry Canada articulates precisely this theme, arguing in the context of biotechnology that "Canada must adapt its delivery of intellectual property services to the competitive conditions of a global, technology-intensive, fast-paced industry" (Industry Canada 2000, 30).

As a direct consequence, intellectual property policy has been increasing in dominance. This expanded prominence has led to a consolidation of power by trade and industry departments such as Industry Canada and the Department of Foreign Affairs and International Trade, where the focus of mandarins is on intellectual property protection; trade imperatives, including trade-related intellectual property norms; and the innovation policy paradigm – the Canadian Biotechnology Strategy (CBS) being a prime example. The burgeoning predominance of "trade" ministries in the drafting of intellectual property policy is quite telling of the extent to which the dissemination function and concepts of the public domain have been rendered increasingly less germane to intellectual property policy debates in Ottawa (Doern and Sharaput 2000). In part, this emphasis responds to demands at the international level where institutions and agreements such as the World Trade Organization (WTO) and the *Agreement on Trade Related Aspects of Intellectual Property Rights (TRIPS)* embody a set of norms based on the belief that information and knowledge must be treated as any other commodity.[2] It is precisely this type of logic that proponents of patent protection for biological products and processes invoke to legitimate their demands for strict intellectual property rights. As we have seen, it is also a logic that infuses Canadian biotechnology policy in its attempts to position biotechnology as one of the latest catalysts for domestic economic growth.

The predominant intellectual property protection device applied throughout the biotechnology industry is the patent. As set out in subsection 27(1) of the *Patent Act,* a patent shall be issued to the inventor of an invention, or the inventor's legal representative, if an application is filed in Canada in accordance with the act and it meets all the other requirements specified by the act.[3] According to section 2 of the act, which sets out definitions, an invention "means any *new* and *useful art, process, machine, manufacture or composition of matter, or any new and useful improvement* in any art, process, machine, manufacture or composition of matter."[4] Based on this definition, we can derive three substantive elements that must be met in order for something to conform to the statutory definition of an "invention" and thus attract patent protection: novelty, utility, and patentable subject matter (that is, the subject must fall within the definition of an invention in order to be patentable).[5] Subsection 27(8) exempts from patentability "any mere scientific principle or abstract theorem." A fourth condition for patentability inferred from the requirement of novelty is that of non-obviousness (also referred to as inventiveness), which is similarly established by section 28.3 of the act. Section 42 of the *Patent Act* stipulates the scope and rights that attach to patent ownership: "Every patent granted under this Act shall contain the title or name of the invention, with a reference to the specification, and shall, subject to this Act, grant to the patentee and the patentee's legal representatives for the term of the patent, from the granting of the patent, the exclusive right, privilege and liberty of making, constructing and using the invention and selling it to others to be used, subject to adjudication in respect thereof before any court of competent jurisdiction." Patents thus convey on their owners the exclusive right, privilege, and liberty to make, construct, and use the invention as well as the right to sell it to others to be used. The duration of patent protection in Canada, and most other signatories to *TRIPS*, is twenty years from the date of filing the patent application.

Canada's intellectual property regime, like that in most other developed nations, is premised on utilitarian considerations of efficiency that seek to balance the incentives presumed necessary to promote the creation of intellectual works with a corresponding public interest in accessing those works. Advocates of strong protection champion the notion that the state is obliged to endow creators with monopoly rights to ensure creative development and consequent societal progress. Drawing in part on John Locke's (1988 [1690]) "just deserts" thesis, this perspective on intellectual property reifies economic rationalism as a natural human characteristic and therefore fits well within the expansionary logic of capital in its drive to stimulate consumer demand and pursue new markets.[6] The economic rationalism inherent in this conceptualization of intellectual property creation is reflected in the substantiating myth of the system, namely, that of the particular, spontaneous genius of an individual creator. However, as Marx made clear, "all scientific labour, all discovery and all invention ... depends partly on the co-operation of the living, and partly on the utilisation of the labours of those who have gone before" (Marx 1967, 104). According to one critical contemporary observer, "the patenting of these ... [genetically engineered agricultural] products ... can no longer be interpreted as an economic recognition of an individual innovation, but should rather be seen as a political action to give companies an exclusive right to introduce new social relations in global food systems" (Ruivenkamp 2005, 13).

Lockean logic provides the basis for the assumption that private property, as an institution, is a necessary mechanism to ensure the efficient distribution of the social benefits and costs that attach to the mobilization of useful resources. Property rights, including those in intellectual assets, are thus construed as being a required instrument to facilitate the allocation of resources among people (North 1990). This emphasis on property rights and efficient allocation suggests that markets offer a suitable mechanism for exchanging and transferring such rights among those able to make best use of them. Viewed from this perspective, the efficient use of intellectual output is considered most appropriately regulated by markets, thus opening the way for the commodification of information and knowledge. Indeed, the very fact that intellectual property rights can be transferred helps give rise to the exchange value of informational and cultural commodities. As one noted Nobel laureate proclaims, "[if] information is not property, the incentives to create it will be lacking. Patents and copyrights are social innovations designed to create artificial scarcities where none exist naturally ... These scarcities are intended to create the needed incentives for acquiring information" (Arrow 1996, 125). As information becomes a key commodity in the market, mechanisms to ensure proprietary control of the information commodity assume added consequence in the context of advanced capitalist social relations. In fact, this logic becomes self-reinforcing in its insistence that the efficient use of knowledge is achieved only through markets and that markets can only develop if information and knowledge are transformed into property (May and Sell 2006).

To the extent that the intellectual property system functions according to these traditional justifications, it is considered to provide a social good by stimulating intellectual creativity and an expanded rate of progress (Bettig 1996; Boyle 1996; Vaver 1991). However, despite the claims by proponents of intellectual property protection for information and knowledge, scant empirical evidence is offered to support the argument that innovation would grind to a standstill absent such proprietary rights. Specific to genetic knowledge and research, this rationale tends to break down. Even before the Supreme Court of the United States (US Supreme Court) decided in *Diamond v Chakrabarty* that genetically engineered organisms are patentable, there was a prolific amount of research and corresponding publication of results in the fields of molecular biology and genetics.[7] Part of the reason for this activity could be that economic gain alone might not always motivate scientists. Perhaps more importantly, the subject matter of genetics and molecular biology occupies such an elemental level of discovery that overprotection could constrain future innovation, thus vitiating one of the oft-cited salutary effects of intellectual property systems (Long and Johnson 1997).

As several prominent intellectual property theorists have pointed out, the practical result of an emphasis on patent protection is a privatization of upstream biomedical research through patent thickets. This could potentially limit innovation further downstream as users are required to navigate multiple regulatory and financial roadblocks in order to secure the inputs necessary for research and product development (Eisenberg 1992, 1997a, 1997b; Heller and Eisenberg 1998; Maskus and Reichman 2004). Similarly, a 2002 report released by the World Health Organization, titled *Genomics and World Health,* concluded that "the current position regarding DNA patenting is retarding rather than stimulating both scientific and economic progress. The monopolies awarded by patents on genes as novel chemicals, therefore, are not in the public interest" (as cited in Shand 2005, 42). In 1995, the

Human Genome Organization (HUGO) also stated that it opposes patent protection for complementary DNA (cDNA) on the grounds that this would hinder the free flow of basic scientific information necessary to facilitate research: "HUGO is worried that the patenting of partial and uncharacterized cDNA sequences will reward those who make routine discoveries but penalize those who determine biological function or application. Such an outcome would impede the development of diagnostics and therapeutics, which is clearly not in the public interest. HUGO is also dedicated to the early release of genome information, thus accelerating widespread investigation of functional aspects of genes" (as cited in Peters 2003, 131-32).

A problem at the other end of the spectrum is the scope of some patents that seek to acquire monopoly protection for entire species. For example, Agracetus received an American patent in 1992 on all genetically engineered cotton and a European patent in 1994 on all transgenic soybean plants, leading one commentator to liken such a practice to an attempt by Henry Ford to seek patent protection for all automobiles (van Wijk 1995). We thus note an immanent paradox of expanding intellectual property rights – creative production relies on a bountiful public domain or commons, which the inherent logic of the system would stymie through stringent intellectual property protections (Boyle 2003; Trosow 2003).

The traditional normative justifications for issuing intellectual property rights notwithstanding, such rights tend to be distributed unequally in practice, resulting in a system that conveys the greatest share of benefits not to society but, rather, to the rights holders, who are not always the creators of the intellectual and creative products. Copyright and patents have become the pre-eminent mechanisms by which to monopolize the production and dissemination of informational, cultural, and industrial artifacts. The outcome has been an assiduous appropriation of almost all tangible forms of creativity within the context of the intellectual property regime and the marketplace.[8] Such monopolization is particularly egregious in industrial sectors that are heavily subsidized through publicly financed research, which characterizes biotechnology. As Peter Drahos and John Braithwaite (2002, 165) argue, "[p]atents, instead of being a reward for inventors who place private information into the public domain, have become a means of recycling public information as private monopolies." Functional arguments for strict intellectual property protection that are steeped in concerns about allocative efficiency fail to admit any of the myriad social and political facets of the commodification debate. Intellectual property rights thus function as particularly forceful forms of social power (Anawalt 2003).

An additional problem with a system of scientific innovation driven by a stringent intellectual property regime is that its commercial focus will tend to concentrate investment activity and attendant research agendas on areas capable of yielding readily monopolizable forms of knowledge and practical applications that promise to deliver high economic rents. Karl Marx (1994, 34 [emphasis in original]) recognized precisely this situation long ago when he opined that "in so far as the sciences are used as a means of enrichment by capital, and thereby become themselves a means of enrichment for those who develop them, the MEN OF SCIENCE compete with each other to discover *practical applications* for their science. Moreover, invention becomes a *métier* by itself." This is particularly the case in contemporary agricultural biotechnology and pharmaceutical research and development (Benkler 2006; Drahos and Braithwaite 2002; King 1997; King and Stabinsky 1999; Moser 1995). Intellectual property might thus be construed as "an architecture of control ... [that] refuses the social relations that make possible the development of intellectual action, and it therefore refuses the potential for social transformation because of the way knowledge is enclosed within a property relation" (Rossiter 2006, 19). Similarly, once such knowledge is captured by intellectual property devices, its price as a research input hinders local researchers' efforts to adapt and improve products to local conditions, thus contributing to developing countries' cycle of dependence on multinational corporations.

In practice, this situation has meant that public sector plant breeding research and a number of the fundamental enabling technologies (transformation methods, selectable markers, constitutive promoters, and so on) that have been employed in genetic engineering have been licenced exclusively to biotechnology companies. These companies have appropriated this publicly produced knowledge, in turn, to develop their product lines that have subsequently received intellectual property protection through patents (Graff et al. 2003).[9] As we saw in Chapter 3, in entering a partnership with Monsanto to develop genetically engineered wheat, the Canadian government made available invaluable germplasm developed by generations of farmer and scientist collaboration in return for the prospect of sharing in the revenue streams that would flow from the patenting and marketing of the genetically engineered wheat. As discussed previously, Monsanto similarly has made adept use of bundling practices that tie its off-patent herbicide formulations to its patent-protected genetically engineered seed varieties.

In our contemporary conjuncture, the traditional bargain designed into intellectual property policy between monopoly rights meant to reward innovators and obligations on rights holders meant to contribute to the public domain is increasingly being eroded in favour of strict protections. As researchers concentrate on exploiting their discoveries through intellectual property devices, there is often a tendency to defer publication of findings until the patent is issued, which circumscribes the dissemination of new knowledge, thus defeating one of the fundamental premises for granting patents in the first place (Blumenthal et al. 1996; Caulfield and Feasby 1998; Cohen 1995).[10] For example, citations are a trusted mechanism within academia to apportion credit and support the legitimacy of one's own work. In the context of patents, however, citations vitiate the non-obvious nature of a particular invention. Scientists seeking patent protection therefore tend only to cite work that patent examiners are likely to know, making sure that such works do not establish any close relationship between the subject matter of the patent application. This process of "constructing nonobviousness" is perhaps most subversive of the collaborative understanding of scientific praxis (Packer and Webster 1996, 439).

Researchers who have conducted a study on the effects of the *Patent and Trademark Law Amendments Act* (also known as the *Bayh-Dole Act*) on American university research and technology transfer conclude that patents and licencing arrangements are not as effective as publication and other less restrictive approaches to information dissemination in ensuring that publicly funded research benefits society (Mowery et al. 1999).[11] Testifying before a 1981 US congressional hearing on the commercialization of academic biomedical research, Jonathan King summed up part of the problem as follows: "Once patenting becomes the mode, then individuals have a vested interest in keeping strains and techniques secret until the patent is granted. This may be up to a year; but even if less, it establishes a destructive element in scientific relations, secrecy and barriers to information flow, which retard overall biomedical progress. I will tell you that the atmosphere around biology department coffee pots has changed in the last few years" (as cited in Loeppky 2005, 138).

In addition to hindering collaborative work among researchers, including science graduate students who are under pressure to publish in a timely fashion, confidentiality agreements that surround patent applications could have negative implications for the reporting of adverse events that occur during clinical and field trials (Caulfield and Feasby 1998). Moreover, genetic information tends to have a short half-life so that it could potentially lose much of its usefulness by the time it makes its way into the public domain (Eisenberg 1997a). Specific to academic agricultural research, a recent survey study of agricultural biologists concludes that these researchers are experiencing time delays in their work when the particular research tools they require are protected by intellectual property devices. Such delay is because patented tools such as vectors, markers, cell lines, antibodies, transgenic seeds and plants, germplasm, genes and proteins, and databases are increasingly less likely to be freely exchanged among scientists without having to first navigate administrative hurdles such as material transfer agreements, which are insisted upon by university administrations and corporations eager to protect their intellectual property portfolios (Lei, Juneja, and Wright 2009). Ultimately, the race to obtain patent protection is creating a vicious cycle in which even those researchers opposed to intellectual property rights for biological products and processes are nonetheless compelled to seek protection for their work in an effort to establish their own revenue streams that can finance their continued research (May 2004; May and Sell 2006; McNally and Wheale 1998).

While it is true that the patent system does require disclosure of the information required to make and use a patented invention, the patent application process can be quite lengthy and obscure. Kathryn Packer and Andrew Webster's (1996, 441) study demonstrates that very few scientists actually make use of patents as a source of information for their research: "When asked whether she used patents as a source of information [a researcher] said, 'Having written patents and knowing the way you write them is to make it impossible to reproduce the work, no.'" It should also be kept in mind that the rationale for obtaining patent protection is to stake a property right in order to prevent competitors from reproducing and marketing the invention. Thus, patent applications tend to be drafted in an opaque manner. According to one scientist who examined a patent, the documentation was "deliberately obscure and I think the reason it was obscure was because the thing doesn't work" (as cited, 441). In fact, a number of scientists have advanced the claim that patents often contain "claims that *resist translation*" and often lay out a causal chain that is "over the top" in terms of plausibility (442 [emphasis in original]). Packer and Webster declare that "in some cases, an education in the patent system can seem to be rather like a crash course in strategic lying" (442). These two authors determined in their empirical study that patents are often employed as weapons to manage the competition involved in external relations rather than as a mechanism for disseminating information.

A 2005 report by the Expert Working Party on Human Genetic Materials, Intellectual Property and the Health Sector (Expert Working Party), which was established by the Canadian Biotechnology Advisory Committee, voices similar concerns about the obstructing effects that strict intellectual property protection can have on research. It traces such deterrent problems to excessively broad patents, the lack of a legislated experimental use exemption against patent infringement (the report considers subsection 55.2(6) of the *Patent Act* to be somewhat ambiguous),[12] refusal of patentees to licence their patents, onerous licencing fees, and excessive transaction costs associated with negotiating licences caught up in patent thickets. In response, the report calls for more rigorous application of the four criteria for patentability and increased scientific expertise among Federal Court justices (one suggestion is to establish an Intellectual Property Division of the Federal Court) (Expert Working Party on Human Genetic Materials Intellectual Property and the Health Sector 2005). Some of the Expert Working Party's recommendations, though disappointing, are not particularly surprising given that part of its mandate included the following statement: "The objective of an effective and balanced intellectual property regime is to act as an important stimulus for innovation, by protecting and nourishing creativity and investment, to the mutual advantage of producers and users of such innovation, and in a manner conducive to economic and social benefits" (vi).

The main message of this report is that Canada's patent regime should be strengthened, sped up, and made more flexible in order to benefit inventors, investors, and producers. The report therefore rejects calls to move deoxyribonucleic acid (DNA) beyond the scope of patentable subject matter, particularly since this would place Canada offside its major trading partners. The report urges the Canadian Intellectual Property Office to adopt interpretive guidelines for the patentability criteria in respect of biotechnology innovations based on American guidelines.[13] The report drafters also rebuff the notion that morality or *public ordre* considerations should figure into patent examinations, instead arguing that other methods of social control should be employed – such methods, however, are not particularly well defined in the report. Fearing a reduction of innovation in the field, the report also argues against excluding diagnostic methods from patentability, although no empirical evidence is mustered in support of this conclusion. Finally, the report, citing a lack of use of sections 19 and 65 of the *Patent Act,* contends that there is no current need to reintroduce compulsory licencing provisions into the legislation. Sections 19 and 65 of the *Patent Act* permit governments and other potential licencees, respectively, to apply to the commissioner of patents for permission to use a patented invention without the approval of the patentee in cases where they have been unsuccessful in securing a licence on reasonable terms (Expert Working Party on Human Genetic Materials Intellectual Property and the Health Sector 2005). Reading the report, however, one is left wondering whether its drafters engaged in any analysis of how "reasonable" is measured and whether this vagueness has exercised any effect on decisions to bring actions based on these two sections of the act. After all, patent holders are often large corporations with deep pockets able to withstand costly litigation, which often serves as a deterrent to challenges.

The few recommendations made in the report that might be considered to support the public interest side of the intellectual property protection equation include the call to establish an experimental use exemption that would shield researchers from infringement claims and the idea of developing a "more open and responsive" process for requesting a reexamination of issued patents (Expert Working Party on Human Genetic Materials Intellectual Property and the Health Sector 2005, x). The latter suggestion would be a welcome change from the current situation in which an overburdened and sometimes underqualified review staff has moved us toward an intellectual property terrain in which examiners tend to issue patents, even if not wholly substantiated, believing that any potential problems can be sorted out by appeal tribunals and the courts (May 2004).

Yet it is precisely once the courts become involved that the power differentials between supporters of the BioCommons and intellectual property rights holders, typically deep-pocketed multinational corporations, become most apparent. Patent litigation is incredibly time and resource intensive, which serves as a very effective barrier against launching patent challenges. With the addition of some qualifications and supplementation, the CBAC accepted the substantive recommendations made by the Expert Working Party when drafting its own report about intellectual property protection for human genetic materials that it presented to Industry Canada and Health Canada in 2006 (Canadian Biotechnology Advisory Committee 2006a). It bears pointing out that the CBAC long advocated for the patentability of higher life forms such as seeds, plants, and animals as long as they met the utility, novelty, and non-obviousness criteria (Willison and MacLeod 2002).

The contemporary privatization of life forms such as human, animal, and plant genomes; cell lines; and agricultural plants and livestock through intellectual property instruments is a qualitatively new dimension of capital's expropriation of social resources (King and Stabinsky 1999). In part, the genetic reductionist paradigm, which posits the gene as a discrete and easily transferable unit of information, provides discursive support for gene patenting based on the putatively exact science of genetic engineering. Indeed, the dominant, though increasingly challenged, genetic reductionist paradigm, which relies on representations of "genes as information" and "genetic codes," has significantly helped to facilitate the mobilization of private property norms and market-based mechanisms to subvert biotechnology into the service of capitalist accumulation. As outlined briefly in the introductory chapter, genetic essentialism emerged in part through an uncritical adoption of various metaphors from cybernetics, information theory, and communication theory into molecular biology. Based on the principles of information exchange inherent in these theories, genetic reductionism assumes that one or more specific genes express a particular trait. At their most basic level, genes are considered to be functional packets of information that not only can be mapped to determine their precise characteristics but that also can be readily enumerated, transplanted from one organism to another through genetic engineering, or otherwise manipulated by adding to, or subtracting from, them in a manner that affects their expression. However, this type of essentialist theorizing about genetics tends to reify conceptions of living organisms as mere conveyors of pre-programmed information necessary for development, thus neglecting the entire range of environmental and chance mediations that might affect the way a gene expresses itself phenotypically. As a result, biological research risks becoming "a kind of artificial life research, [in which] the paradigmatic habitat for life – the program – [bears] no necessary relationship to messy, thick organisms" (Haraway 1997, 245-46; Kay 2000; Smith 2003).

As scholars critical of genetic reductionism point out, our expanding understanding of molecular science seriously questions the validity of the notion that genes are discrete organic objects involved in the seemingly linear chain of causality that is heredity. Indeed, the belief that one gene codes for one protein is nonsense given that the human body produces some 200,000 proteins using only between 20,000 and 30,000 genes. Genes, in fact, are context dependent. Their expression, or lack thereof, follows on the basis of a complex set of interactions between various areas within the genome, between the information contained in DNA and ribonucleic acid as well as information from other cellular molecules; between cells and other physiological systems, including the entire organism itself; and between organisms and their broader environment, which also includes social co-determinants.[14] Molecular scientists are coming to recognize that cells possess capacities for self-regulation and self-repair and that gene expression is influenced by a web of feedback mechanisms between genes, proteins, other cellular elements, and the organism itself.[15]

The emergent properties inherent in organisms indicate a high level of complexity that cannot be deduced in a linear manner from their component parts (Beurton, Falk, and Rheinberger 2000; Bourgaize, Jewell, and Buiser 2000; Fox Keller 2000; Hubbard and Wald 1993; Rose 1997). Moreover, organisms with identical genomes, whether the result of nature or engineering, are susceptible to different developmental and behavioural trajectories when environed by different conditions. Evidence for this position has been offered by plant researchers who have determined that the growth of cloned plants varies when diverse fluctuations (for example, in temperature, elevation, and so on) are introduced into their immediate environment (Lewontin 2000b). Richard Lewontin (1982, 163 [emphasis in original]) writes: "Organisms within their individual lifetimes and in the course of their evolution as a species do not *adapt* to environments; they *construct* them. They are not simply *objects* of the laws of nature, altering themselves to bend to the inevitable, but active *subjects* transforming nature according to its laws."

The reductionist view that an analysis of the parts will yield a complete understanding of the whole must give way to a dialectical method that adopts an opposite approach. That is, the various genetic elements of an organism can only be comprehended through examination of the whole organism as it develops and evolves through interaction with its environment, including the ways in which such organic development and evolution affect the environment (McAfee 2003). Nonetheless, the research into such molecular/environmental interactions tends to remain underdeveloped since the resulting findings do not usually yield forms of knowledge or technology that are easily commodified. Perhaps more importantly, this type of holistic conceptualiztion threatens to undermine the legal fundament upon which the issuance of patents for genes and their component parts has rested to date – namely, the idea that there is a predictable, sequential chain of information from DNA to protein.

Canadian Case Law on Biotechnological Patents

Canadian jurisprudence in respect of biotechnology presents a somewhat mixed picture in that the Supreme Court of Canada rejected Harvard University's patent application on the Oncomouse but affirmed Monsanto's patent on its Roundup Ready canola. This section of the chapter is devoted to elucidating these two decisions, including some of the inconsistencies between them.

President and Fellows of Harvard College v Canada (Commissioner of Patents)[16]

Harvard University applied for a Canadian patent on its invention titled "transgenic animals." According to the application, a cancer-promoting gene (oncogene) was injected into fertilized mouse eggs as close as possible to the one-cell stage. The eggs were then implanted into a female host mouse and permitted to develop to term. After the offspring of the host mouse were delivered, they were tested for the presence of the oncogene, and those that contained it were referred to as "founder" mice. Founder mice were then mated with mice that had not been genetically altered. Fifty percent of the offspring would have all of their cells affected by the oncogene, making them suitable for animal carcinogenic studies. In its patent application, Harvard College sought to protect both the process by which the oncomice were produced and the end product of the process – that is, the founder mice and the offspring whose cells contain the oncogene. The process and product claims extended to *all* non-human mammals genetically engineered to develop cancer.[17]

In his decision, the patent examiner allowed the process claims, agreeing that they were produced completely under the control of the inventor and were readily reproducible, thus qualifying as patentable subject matter as a manufacture or composition of matter within the statutory definition of "invention." However, the commissioner found that the product claims on the actual mice were too greatly influenced by the laws of nature, which removed from the inventor complete control over all of the characteristics of the resulting mice. These claims were ultimately rejected on the grounds that whole organisms reach beyond the scope of the definition of "invention" under section 2 of the *Patent Act*.[18] The appellant commissioner subsequently confirmed the refusal of the product claims.

Harvard proceeded to appeal the decision to the Federal Court's Trial Division, which upheld the ruling of the Patent Office that the Oncomouse did not meet the definition of an "invention" under the *Patent Act*. In reaching this conclusion, Justice Marc Nadon applied the following four indicia to his interpretation of "invention" in this case: (1) the inventors lacked control over the invention beyond the presence of the oncogene; (2) the creation of the Oncomouse, beyond the human intervention involved, relied to a significant degree on the laws of nature; (3) these laws of nature introduced such a degree of variation into the process in terms of which offspring would actually be carriers of the oncogene that reproducibility was significantly vitiated; and, finally, (4) as a matter of policy, higher life forms are not patentable.

The respondent's (Harvard) further appeal to the Federal Court of Appeal was allowed, with the majority of that court ruling that the Oncomouse fits the definition of "composition of matter," since the fertilized mouse egg is a form of biological matter, the egg and the DNA together are a composition of matter, and the resulting mouse is the product of that composition of matter that evolves from a single cell to a multi-cellular level. Moreover, the majority of the Federal Court of Appeal reasoned that the progeny of any oncomice were also to be considered compositions of matter for the purposes of the act because they possess a genetic trait not present in nature and they are linked through heredity to the original composition of matter. Justice Marshall Rothstein, who penned the Federal Court of Appeal's majority opinion, contended that the scope of "composition of matter" should not be interpreted as excluding living things and that decisions about whether patents may issue for life forms must reject the animate-inanimate dichotomy and instead distinguish between discovery and invention. The Patent Office was thus instructed to grant the refused claims.[19]

The commissioner of patents subsequently appealed that decision to the Supreme Court of Canada. In December 2002, the Court overturned the ruling made by the Federal Court of Appeal, upholding the decision made by the commissioner of patents not to allow the product patent claims, arguing that, based on the wording of the *Patent Act*, non-human mammals are beyond the scope of the act. The Court did, however, refrain from commenting on whether life forms should be patentable or not, arguing that is a decision that can only be made by Parliament:

> The sole question in this appeal is whether the words "manufacture" and "composition of matter," within the context of the Patent Act, are sufficiently broad to include higher life forms. It is irrelevant whether this Court believes that higher life forms such as the oncomouse ought to be patentable. The words of the Patent Act "are to be read in their entire context and in their grammatical and ordinary sense harmoniously with the scheme of the Act, the object of the Act, and the intention of Parliament."[20]

In drafting the majority opinion, Justice Michel Bastarache was clear in characterizing the decision before the Court as one of interpretation and not public policy:

> Patenting higher life forms would involve a radical departure from the traditional patent regime. Moreover, the patentability of such life forms is a highly contentious matter that raises a number of extremely complex issues. If higher life forms are to be patentable, it must be under the clear and unequivocal direction of Parliament. For the reasons discussed above, I conclude that the current Act does not clearly indicate that higher life forms are patentable. Far from it. Rather, I believe that the best reading of the words of the Act supports the opposite conclusion – that higher life forms such as the oncomouse are not currently patentable in Canada.[21]

While accepting that the definition of "invention" should be interpreted broadly, Justice Bastarache nonetheless refused an unlimited definition:

> In drafting the *Patent Act,* Parliament chose to adopt an exhaustive defin- ition that limits invention to any "art, process, machine, manufacture or composition of matter." Parliament did not define "invention" as "anything new and useful made by man." By choosing to define invention in this way, Parliament signalled a clear intention to include certain subject matter as patentable and to exclude other subject matter as being outside the confines of the Act. This should be kept in mind when determining whether the words "manufacture" and "composition of matter" include higher life forms.[22]

In response to the admonition that Justice Bastarache articulated in the last sentence of the previous quote, he added that "'composition of matter' does not include a higher life form such as the oncomouse" and that "the words 'machine' and 'manufacture' do not imply a conscious, sentient living crea- ture. This provides *prima facie* support for the conclusion that the phrase 'composition of matter' is best read as not including such life forms."[23]

The rhetoric in the opinion that Justice Ian Binnie drafted for the minority is quite telling of the extent to which neo-liberal perspectives permeate Canada's judiciary. The opinion began by proclaiming that "the biotechnol- ogy revolution ... has been fuelled by extraordinary human ingenuity and financed in significant part by private investment." Justice Binnie went on to stress the importance of monopoly patent rights, maintaining that "in- novation is said to be the lifeblood of a modern economy. We neglect reward- ing it at our own peril."[24] He continued by pointing out the revenue potential of the global biotechnology industry, including Canada's place within that market, ultimately arguing that Canada must be onside with its major trading partners or risk losing out on economic opportunities: "The mobility of capital and technology makes it desirable that comparable jurisdictions with comparable intellectual property legislation arrive (to the extent permitted by the specifics of their own laws) at similar legal results."[25] Aside from trotting out tired statements about the ostensibly necessary link between pecuniary incentives and innovation, the discourse of Justice Binnie sought to firmly situate debate about biotechnological innovations within the dominant context of capitalist social relations and private enterprise. Such discursive positioning is particularly dangerous because, as Teresa Scassa (2003, 117) writes, "the values and perspectives of judges ... are ones that ultimately inform the interpretive outcome. It is in this way that judicial values influence interpretation." It would be precisely this perspective that carried the day in the subsequent Supreme Court of Canada decision in *Monsanto Canada Inc. et al. v Schmeiser et al.*, which undergirds the privatization functions of the intellectual property system in rendering increasing areas of biological existence as exclusive and alienable cogs in the machinery of capitalist value production.

Monsanto Canada Inc. et al. v Schmeiser et al.[26]
Monsanto has long made it publicly clear that it will launch civil suits against farmers who refuse to adhere to the terms it imposes on the sales of its genetically engineered seeds. For example, in June 1998, the company published the following warning to farmers in various newspapers:

> When a farmer stores and sows biotech seeds (genetically modified seeds) patented by Monsanto, he should understand that he is in the wrong. This holds true even if he has not signed any contract at the time of procuring seeds (that is, if he recycles or if he buys seeds illegally from a neighbour). He is pirating ... Moreover, this pirating of seeds could cost the farmer hundreds of dollars per acre by way of damages, interest and legal costs, apart from having to undergo the inspection of his fields and records over many years. (As cited in de la Perrière and Seuret 2000, 12)

In order to prosecute such claims, multinational corporations have devoted significant effort to assaying crops (commonly referred to as "genome control" by seed producers) and substantial laboratory resources in developing detection tools to aid in enforcing their technology use agreements (Lewontin 2000a). According to a report issued by the Center for Food Safety, based in Washington, DC, Monsanto has an annual budget of US$10 million that includes a toll-free number for people to inform on others suspected of "seed piracy" and a staff complement of seventy-five people who are dedicated to investigating and prosecuting farmers. Donald Bartlett and James Steele (2008, para. 9) write:

They fan out into fields and farm towns, where they secretly videotape and photograph farmers, store owners, and co-ops; infiltrate community meetings; and gather information from informants about farming activities. Farmers say that some Monsanto agents pretend to be surveyors. Others confront farmers on their land and try to pressure them to sign papers giving Monsanto access to their private records. Farmers call them the "seed police" and use words such as "Gestapo" and "Mafia" to describe their tactics.

In addition to such invasive practices, Monsanto also maintains and sends to all of its agricultural product suppliers across North America a "blacklist" that contains the names of farmers to whom Monsanto refuses to sell its products. The farmers on this list presumably also include those who resist intimidation by the company's "seed police" force (Lindner 2009).

True to its word, Monsanto has successfully sued a number of Canadian and American farmers. As of 2005, Monsanto had been awarded over US$15 million in the United States against almost 100 farmers, although that number underestimates the true amount collected by the company since it does not include those sums obtained in the large number of lawsuits settled out of court (Center for Food Safety 2004). According to one of its blog postings, as of August 2012 Monsanto had filed 145 patent-infringement lawsuits against farmers in the United States since 1997 (Monsanto 2012a).[27] Monsanto has not revealed how much money the courts have awarded it from farmers, nor has it published comparable information for Canada. Nonetheless, in this country, a number of farmers have been slapped with punitive damages for growing Roundup Ready crops without a licence. In June 2007, a federal court judge ruled that Edward Wouters, of Northspruce Farms, had to pay Monsanto more than $107,000 for knowingly growing, harvesting, and selling 392 acres of Roundup Ready soy (Coad 2007). This amount included an award of $97,554.30 ($248.86 per acre) payable to Monsanto on account of the profits made by Northspruce Farms, pre- and post-judgment interest, and $5,000 in costs. In addition to the monetary award, Edward Wouters, Northspruce Farms, and all persons under their control were enjoined from planting, growing, harvesting, selling, marketing, or distributing Monsanto's genetically engineered soybeans until the end of the patent. Wouters was also ordered to deliver up all seeds and plants in his possession that may contain the Roundup Ready gene.

In November 2007, Monsanto successfully sued another Ontario farmer, Paul Beneteau, for infringing its Roundup Ready soybean patent. Beneteau was ordered by the Federal Court of Canada to pay Monsanto punitive damages in the amount of $8,800 (fifty-five acres at $160 per acre) and turn over to the company any of the genetically engineered soybeans still in his possession (Romahn 2007). In his decision from 26 March 2009, Federal Court Justice Russel Zinn ordered Charles Rivett to pay Monsanto $40,137.94

for planting, cultivating, and harvesting 947 acres of Roundup Ready soybean without a licence. Rivett was also subsequently levied almost $18,000 in court costs and fees. Justice Zinn once again ruled in favour of Monsanto on 10 July 2009, when he found Lawrence Janssens, Ronald Janssens, and Alan Kerkhof guilty of violating Monsanto's Roundup Ready soybean patent when they planted the seed on fifty acres in 2004 without a licence. The three subsequently saved and cleaned 337 bushels of the genetically engineered seed, which they proceeded to plant and harvest in 2005. The three were ordered to pay Monsanto $14,537.50 for the profits they obtained using the company's technology, plus pre- and post-judgment interest, as well as court costs and fees totalling $13,209.55.

Monsanto has also invoked its "violator exclusion" policy against all of these farmers by removing them from the company's authorized user list. In effect, this action bars these producers from access to any of Monsanto's technologies sold by the company or any of its licencees. It should be noted that Monsanto employs this policy only against those farmers who refuse to settle alleged instances of infringement out of court. Despite these more recent patent infringement cases, arguably the most well-known example of Monsanto's penchant for litigation against farmers remains the lawsuit launched against Percy Schmeiser.

Percy Schmeiser, who has been farming for over fifty years in Bruno, Saskatchewan, engages in the practice of saving and crossbreeding seed for use in future plantings. In 1997, he planted a crop of canola using seed he had saved from the previous year – a year in which five neighbouring farmers planted Roundup Ready canola.[28] During the course of growing year 1997, Schmeiser sprayed Roundup herbicide around power poles and ditches near the road that were infested with "volunteer" canola and weeds.[29] Upon noticing several days later that a number of the volunteer canola plants had survived the spraying, Schmeiser proceeded to test for more intensive volunteer canola by spraying an adjacent three-acre patch with Roundup. Approximately 60 percent of the plants on this tract of land survived, which the courts would later accept as evidence Schmeiser should have known that the plants contained Monsanto's patented gene. Schmeiser would allege at trial that these plants were volunteer canola. Schmeiser subsequently harvested the canola from the patch he had sprayed with Roundup, although he did not sell it, and stored it separately on his farm.

In the same year, Monsanto inspectors, acting on an anonymous tip from a neighbouring farmer, took samples from the public road allowances around Schmeiser's farm and confirmed that he had Roundup Ready canola growing in his fields without the company's permission. In March 1998, Monsanto representatives informed Schmeiser of their belief that he had grown Roundup Ready canola without a licence. Schmeiser then proceeded to have the plants he had harvested from the three-acre patch treated at a local seed plant for use as seed, which, together with some other bin-run seed and fertilizer, were subsequently sown on 1,030 acres across nine of his fields in 1998. Monsanto stepped up its sample taking and analysis activities in 1998. Samples were taken from the seed treatment facility Schmeiser had used to have his seed cleaned. According to an independent analysis, between 95 and 98 percent of Schmeiser's canola was Roundup resistant.

Monsanto then initiated and prevailed in a Federal Court lawsuit against Schmeiser for patent violation.[30] Counsel for Schmeiser attempted to argue that Monsanto's patent was invalid on the grounds that the subject matter was not patentable subject matter under the *Patent Act* and that the claimed invention with regard to replication of the gene was not caused by human intervention but, rather, by natural means and that it therefore could not be contained or controlled. The rationale offered was that the enactment of the *Plant Breeders' Rights Act* provides clear indication of Parliament's intent "that intellectual property rights pertaining to new plant varieties are to be governed by legislation other than the *Patent Act* and only to the extent permitted under the former Act."[31]

In his judgment, Justice W. Andrew McKay of the Federal Court's Trial Division wrote that the *Plant Breeders' Rights Act* in no way precludes "registration under the *Patent Act* of inventions that relate to plants, and that may lead to new varieties or characteristics of plants."[32] Justice McKay went on to reason that replication of the gene through natural events similarly did not preclude registration under the *Patent Act* because the invention covered by patent protection is the creation of the gene and the process for inserting it. By following this logic, Justice McKay would similarly contend that the *Harvard College* case, which had provided a common law precedent against patents on higher life forms, supported his contention, since the process and genetically engineered plasmid claims were not in dispute in the Monsanto case: "Here [the case at bar] it is the gene and the process for its insertion which can be reproduced and controlled by the inventor, and the cell derived from that process, that is the subject of the invention. The decision of the Trial Judge and of the Court of Appeal in the *Harvard College* case implicitly support the grant of the patent to Monsanto."[33]

Schmeiser's counsel furthermore sought to argue that any infringement was unintentional and that the source of the original contamination was unclear. Justice McKay made short shrift of the former claim, pointing out that "'infringement occurs when the essence of an invention is taken,' regardless of the intention of the infringer."[34] In respect of the latter, Justice McKay wrote that the evidence suggested Schmeiser "knew or ought to have known" that the seeds he saved and planted the following year contained Monsanto's patented invention, thus establishing infringement.[35] Finally, a principal defence raised on behalf of Schmeiser was that he had not made use of the patented invention because he did not spray the 1998 canola crop with Roundup. Again, Justice McKay rejected this reasoning: "In my opinion, whether or not that crop was sprayed with Roundup during its growing period is not important. Growth of the seed, reproducing the patented gene and cell, and sale of the harvested crop constitutes taking the essence of the plaintiffs' invention, using it, without permission. In so doing the defendants infringed upon the patent interests of the plaintiffs."[36]

In addition to disregarding the fact that Schmeiser had not sprayed his crops with Roundup herbicide, the Federal Court Trial Division reached its decision paying little heed to the facts that Schmeiser did not sell the resulting seed and ultimately received no financial gain from the Roundup Ready plants in his fields. Perhaps most troubling, the court concluded that, even though Monsanto could not control how the gene was dispersed throughout the countryside, it was nonetheless titled to patent protection under the *Patent Act*.[37] Incidentally, establishing corporate liability for the agricultural, economic, and environmental ramifications of genetically engineered seeds is precisely the issue around which the Organic Agriculture Protection Fund suit revolved (as elaborated more fully later in this chapter). Schmeiser was enjoined from planting or selling any seed retained from his 1997 and 1998 canola crops or any other seed saved from plants that were known, or ought to have been known, to be glyphosate-resistant. Moreover, any remaining plants or seeds from the 1997 and 1998 crops in Schmeiser's possession were to be delivered up to Monsanto. Schmeiser was also ordered to pay $19,832 in profits to Monsanto and $153,000 to cover the latter's legal costs.

Although the Federal Court of Appeal unanimously rejected Schmeiser's appeal, on 8 May 2003 he was granted leave to appeal to the Supreme Court of Canada, which heard the case in January 2004.[38] At least one observer contends that Schmeiser's failure at the Federal Court of Appeal would normally have meant the end of the case, but the Supreme Court of Canada's decision in *Harvard College* opened the door to its consideration of the *Schmeiser* case.[39] Precisely this point was made by the dissenting judges in the *Schmeiser* case, who asserted that patents on genes or cells, when reproduced naturally as part of an entire organism could, in fact, confer *de facto* intellectual property rights over whole organisms:

> The heart of the issue is whether the Federal Court of Appeal's decision can stand *in light of this Court's ruling that plants as higher life forms are unpatentable*. A purposive construction that limits the scope of the respondents' claims to their "essential elements" leads to the conclusion that the gene claims and the plant cell claims should not be construed to grant exclusive rights over the plant and all of its offspring. This interpretation is fair and predictable because it ties the respondents to their claims; the respondents specifically disclaim plants. Patents must be interpreted from the point of

view of the person skilled in the art who must also be taken to know the law. A person skilled in the art could not reasonably have expected that patent protection extended to unpatentable plants and their offspring.[40]

Yet, in its May 2004 majority decision, and despite the seeming contradiction with the *Harvard College* decision made in 2002 that refused the patentability of higher life forms, the Court upheld the validity of Monsanto's patent, finding Schmeiser guilty of patent infringement. In their five-to-four decision, written by Chief Justice Beverley McLachlin and Justice Morris Fish, the Court reasoned that Monsanto's patent was valid because it was limited to the genetically engineered genes and the modified cells without including the plant itself:

> The patent is valid. The respondents did not claim protection for the genetically modified plant itself, but rather for the genes and the modified cells that make up the plant. A purposive construction of the patent claims recognizes that the invention will be practiced in plants regenerated from the patented cells, whether the plants are located inside or outside a laboratory. Whether or not patent protection for the gene and the cell extends to activities involving the plant is not relevant to the patent's validity. The appellants have failed to discharge the onus to show that the Commissioner of Patents erred in allowing the patent.[41]

Justices McLachlin and Fish went on to write: "Infringement through use is thus possible even where the patented invention is part of, or composes, a broader unpatented structure or process."[42] Concluding that both the majority and minority opinions in the *Harvard College* case provided support for Monsanto's patent, the majority decision asserted that Schmeiser's reliance on that precedent as a rejection of patents on higher life forms failed to meet his burden of proof in establishing that Monsanto's patent should not have been issued:

> Whether or not patent protection for the gene and the cell extends to activities involving the plant is not relevant to the patent's validity. It relates only to the factual circumstances in which infringement will be found to have taken place ... Monsanto's patent has already been issued, and the onus is thus on Schmeiser to show that the Commissioner erred in allowing the patent ... He has failed to discharge that onus. We therefore conclude that the patent is valid.[43]

As such, by cultivating plants that contained validly patented material vital to the growth of the plant (that is, since the genes exist throughout it), the majority reasoned that Schmeiser necessarily infringed Monsanto's patent. Indeed, throughout the ruling, the majority was at pains to establish a perceived vital connection between the genetically engineered gene and cells to which patent protection attaches and the plants that contain such genes and cells. The argument was thus made that "where a defendant's commercial or business activity involves a thing of which a patented part is a significant or important component, infringement is established. It is no defence to say that the thing actually used was not patented, but only one of its components."[44]

This position was disputed by the minority opinion, which, while affirming the product and process claims of Monsanto's patent, nonetheless pointed out that neither type of claim may extend to higher life forms to which patent protection may not attach. In order to preclude *de facto* protection over the entire plant, the minority reasoned that the plant cell claim must not reach beyond the point where cells begin replicating and differentiating into plant tissues. Otherwise, the claim would extend to every cell in the plant, essentially meaning to the plant itself – something with which the majority seemed comfortable. Similarly, the method claim must end at the point of the regeneration of the genetically engineered founder plant and not extend to methods for propagating that plant. This same claim limitation would also apply to the progeny of the regenerated plant. These are important points given that the scope of "use" of a patented invention, which figures so prominently in findings of infringement, should necessarily be bounded by the scope of the corresponding claims of the patent.[45] The minority thus concluded that Schmeiser could not be held guilty of infringement:

> In the result, the lower courts erred not only in construing the claims to extend to plants and seed, but in construing "use" to include the use of subject matter disclaimed by the patentee, namely, the plant. The appellants as users were entitled to rely on the reasonable expectation that plants, as unpatentable subject matter, fall outside the scope of patent protection. Accordingly, the cultivation of plants containing the patented gene and cell does not constitute an infringement.[46]

The majority of the Court also rejected Schmeiser's claim that he was not guilty of infringement because he did not spray his crops with Roundup Ready herbicide and thus made no use of Monsanto's patented gene. According to their opinion,

> [c]ase law shows that infringement is established where a defendant's commercial or business activity involving a thing of which a patented part is a component necessarily involves use of the patented part. Infringement in this case therefore does not require use of the gene or cell in isolation.

Infringement also does not require that the appellants have used Roundup herbicide as an aid to cultivation ... this argument fails to account for the stand-by or insurance utility of the properties of the patented genes and cells.[47]

The majority went on to assert that "whether or not a farmer sprays with Roundup herbicide, cultivating canola containing the patented genes and cells provides stand-by utility. The farmer benefits from that advantage from the outset: if there is reason to spray in the future, the farmer may choose to do so."[48] One legal commentator disputed this claim, contending that the Court overstated the case of Schmeiser's infringement since the invention at issue was the genetically engineered resistance of canola to glyphosate, which Schmeiser did not make use of since he did not spray his crops with Roundup herbicide (Mgbeoji 2007). Moreover, as the minority opinion asserted, to allege stand-by utility is tantamount to conferring patent protection on the entire plant itself. Citing a CBAC report (2002b, 12), the minority opinion maintained that

[t]he use of biologically replicating organisms as a "vehicle" for genetic patents may overcompensate the patentee both in relation to what was invented, and to other areas of invention ... Because higher life forms can reproduce by themselves, the grant of a patent over a plant, seed or non-human animal covers not only the particular plant, seed or animal sold, but also all its progeny containing the patented invention for all generations until the expiry of the patent term (20 years from the priority date). In addition, much of the value of the higher life form, particularly with respect to animals, derives from the natural characteristics of the original organism and has nothing to do with the invention. In light of these unique characteristics of biological inventions, granting the patent holder exclusive rights that extend not only to the particular organism embodying the invention but also to all subsequent progeny of that organism represents a significant increase in the scope of rights offered to patent holders. *It also represents a greater transfer of economic interests from the agricultural community to the biotechnology industry than exists in other fields of science.*[49]

The Court did, however, reduce the damages awarded to Monsanto in the lower courts. It set aside the requirement that Schmeiser pay Monsanto the profits from his 1998 crop, arguing that he did not earn any additional profit from using Monsanto's genetically engineered seeds above the profit he would have earned by growing non-genetically engineered varieties. More importantly (at least for Schmeiser), given the mixed results of the appeal, each party was ordered to bear their own costs throughout.

In analyzing the decisions made in this case at both the lower courts and Supreme Court levels, a number of observers have reiterated the argument made by the Court dissenters that the majority decision provides Monsanto a backdoor opportunity to acquire patent protection over an entire organism, which is not otherwise permitted under current patent legislation. These same commentators chide the majority ruling for failing to employ alternative legal mechanisms that could have responded to Monsanto's commercial interests without embarking on this particularly slippery path (Gold and Adams 2001; Phillipson 2005). Martin Phillipson (2005), critical of the imbalance between the expanding rights and the corresponding lack of responsibilities associated with patents on genetically engineered organisms, further contends that the Supreme Court of Canada decision in *Schmeiser* vitiates the so-called "farmers privilege" read into the *Plant Breeders' Rights Act,* under which farmers may save and re-use seed in certain prescribed circumstances.[50] As Vandana Shiva (2001a) points out, what has been a millennial duty – saving and exchanging seed – is becoming a crime of intellectual property theft.

The *Schmeiser* case continues to serve as a rallying point around which opposition to genetically engineered seeds is mobilized. In fact, the case, according to one interview informant, has spawned the neologism of "getting Schmeisered," by which is meant being sued by multinational agribusiness for patent infringement. It is worth noting that Schmeiser's very public struggles against Monsanto were recently recognized when he and his wife were awarded a Right Livelihood Award, also known as the "alternative Nobel Prize," which is bestowed annually in a ceremony in the Swedish Parliament to honour and support those "offering practical and exemplary answers to the most urgent challenges facing us today." The award is normally given to four recipients who share the prize money (two million Swedish Krona or about $310,000) to help them with their work. According to the Right Livelihood Foundation, the Schmeisers "have given the world a wake-up call about the dangers to farmers and biodiversity everywhere from the growing dominance and market aggression of companies engaged in the genetic engineering of crops. The Jury honours the Schmeisers for their courage in defending biodiversity and farmers' rights, and challenging the environmental and moral perversity of current interpretations of patent laws."[51]

Attempts at Judicial Redress to Commons Contamination

Based on the various courts' reinforcement of Monsanto's rights of control over its genetically engineered canola, a logical conclusion would be that with such control comes corresponding duties and responsibilities. Precisely this reasoning informs the attempts made by a group of organic agricultural producers in Saskatchewan to pursue class action litigation against Monsanto and Aventis (later amended to Bayer CropScience after it acquired Aventis). In response to the contamination of organic canola by genetically engineered varieties of canola seed, the Saskatchewan Organic Directorate established and mandated a self-sustaining committee known as the Organic Agriculture Protection Fund to pursue the class action civil suit.[52] In addition to raising approximately $250,000 to fund its legal actions, the Organic Agriculture Protection Fund maintains a website that not only outlines progress in the class action suit but also provides a variety of resources related to organic farming and the dangers posed by genetically engineered crops.[53]

The Organic Agriculture Protection Fund, with Larry Hoffman and Dale Beaudoin as the named plaintiffs, attempted to become certified as a class action under the Saskatchewan *Class Actions Act*.[54] As explained by interview informant Cathy Holtslander, who was involved in the efforts against Monsanto and Bayer CropScience, the Organic Agriculture Protection Fund pursued class action because no single member of the Saskatchewan Organic Directorate has the financial ability to single-handedly litigate against a multinational corporation. In addition, if an individual were to lose a court battle, the defendants could be awarded costs, meaning that the plaintiff would be forced to reimburse Monsanto and Bayer CropScience for all the legal costs they incurred defending themselves in the lawsuit. The Saskatchewan *Class Actions Act* was designed to remedy this obstacle by shielding plaintiffs from awards of cost unless there is a fraudulent misuse of the courts. According to Holtslander, the new act provides access to justice by assisting a large group of people, who individually might have only a modest amount of damage that normally would not warrant suing a corporation, but, when aggregated, becomes significant enough to merit class action.

The Organic Agriculture Protection Fund, which submitted its statement of claim on 10 January 2002, was the first group in Saskatchewan that has tried to make use of this relatively new legislation, which was proclaimed on 1 January 2002. The statement of claim was for damages caused by the introduction of genetically engineered canola and for an injunction to stop genetically engineered wheat. The plaintiffs in *Hoffman and Beaudoin v Monsanto Canada* maintained that "as a result of widespread contamination by GM canola few, if any, certified organic grain farmers are now growing canola. The crop, as an important tool in the crop rotations of organic farmers, and as an organic grain commodity, has been lost to certified organic farmers in Saskatchewan."[55] On 2 February 2004, the statement of claim was amended to include compensation for costs incurred by organic farmers to remove genetically engineered canola from their fields and seed supplies:

> Whether growing canola or not, certified organic grain producers have sustained, and will continue to sustain, contamination of their grain field by stray GM canola plants. While certification standards ban the "use" of GMOs, they also mandate that farmers are to take reasonable steps to prevent the contact of prohibited substances with their organic crops. This has resulted in organic producers being required by their certification agencies

to mechanically remove stray canola plants from their grain fields, and not to grow any crops on those fields from which it is difficult to clean canola.[56]

After Monsanto withdrew its application for regulatory approval of its Roundup Ready wheat in June 2004, a further amendment was made to the statement of claim to remove the request for an injunction against genetically engineered wheat.

The statement of claim further alleged that Monsanto and Bayer Crop-Science were liable in negligence, strict liability, nuisance,[57] and trespass.[58] The claim of negligence was based on the defendants' failure to ensure that their genetically engineered canola would not infiltrate or contaminate farmland, to warn farmers about cross-pollination, and to advise growers of farming practices that would limit the spread of genetically engineered canola.[59] The plaintiffs argued that the defendants were liable on the basis of strict liability for engaging in a non-natural use of land and permitting the escape of something likely to do mischief or harm. They also claimed nuisance because of the interference that genetically engineered canola has caused organic farmers in trying to use and enjoy their land. According to the claim, the defendants' introduction and unconfined release in Saskatchewan of genetically engineered canola, which has subsequently trespassed on lands owned by organic farmers, has also given rise to liability based on trespass.[60] The plaintiffs also contended that genetic modifications are "pollutants" within the meaning of the *Environmental Management Protection Act*[61] and that such pollutants have caused harm to organic farmers as a result of their discharge into the environment.[62] The plaintiffs therefore sought to hold the defendants (as "owners or persons in control")[63] liable to organic farmers pursuant to subsection 13(3) of the act for the damages they have incurred as a result of the introduction of genetically engineered canola (the "pollutant") into the Saskatchewan environment.[64] Finally, the plaintiffs asserted that the testing and unconfined release of genetically engineered canola in Saskatchewan was a "development" under the *Environmental Assessment Act*, which requires an environmental impact assessment and ministerial approval prior to environmental release.[65] Having failed to either conduct such an assessment or obtain ministerial approval, the statement of claim maintained that, under section 23 of the act, the defendants are liable for any loss or damage sustained by organic farmers from the "development" without proof of negligence or intention.

The initial hearing on class certification occurred in November 2004 before Justice Gene Ann Smith of the Saskatchewan Court of Queen's Bench. In her ruling from 11 May 2005, Justice Smith denied the plaintiffs' claims, stating that they had failed to meet the criteria for class certification as set out in the *Class Actions Act*, including a failure to demonstrate either an identifiable class or a cause of action in negligence, nuisance, strict liability, or trespass. With respect to negligence, Justice Smith argued that the plaintiffs failed to sufficiently demonstrate proximity and foreseeability: "What is missing from the plaintiffs' claim ... is any specific allegation that the loss and damage to organic farmers in particular which is claimed (*viz.*, loss of the use of canola as a marketable organic commodity and loss of canola for use in crop rotation, plus the clean-up costs and loss of use of fields as a result of GM canola volunteers) was foreseeable."[66] In respect of proximity, Justice Smith's judgment maintained that the plaintiffs "have not alleged any relationship at all, either in the pleadings or in the argument before me, that would give rise to an argument for sufficient relational proximity to support a *prima facie* duty of care."[67]

In her analysis of the nuisance claim, Justice Smith wrote that

> no harm can be said to have been caused by the *mere* sale or marketing of GM canola. The adventitious presence of canola in the crops and on the land of organic farmers required the intervention of neighbouring farmers who cultivated GM canola. While the "release" of the GM varieties of canola by the defendants may have been a necessary condition for the occurrence of the harm alleged, it was far from sufficient, in itself.[68]

The characterization of Monsanto as a *mere* marketer fails to reflect the high level of control exercised by Monsanto over its Roundup Ready seeds through the use of its technology use agreements, which, as we have seen, licence farmers to use the gene technology while explicitly retaining ownership firmly within the control of the company. Moreover, Smith's conclusion was particularly problematic for the plaintiffs since Saskatchewan, like most other jurisdictions in Canada, has so-called "right-to-farm" legislation (the *Agricultural Operations Act*), which, under subsection 3(1), prohibits nuisance suits from being brought against farmers who operate within the parameters of "normally accepted agricultural practices."[69] In interpreting this judicial reasoning, one commentator laments that, through a combination of statute and common law, a claim of nuisance cannot be brought against either the user or the manufacturer of substantially harmful genetically engineered organisms (Phillipson 2005).

A similarly troubling aspect of Justice Smith's ruling, and one that was pervasive throughout her judgment, was her preference for the industry term "adventitious presence" in place of the term "contamination" as employed by the plaintiffs. In what appears somewhat contradictory, she contended that the former term is more neutral but then proceeded to employ the definition as cited by Bayer CropScience in one of its briefs to the court:

A more neutral term, "adventitious presence" (sometimes referred to in the following discussion as "AP") is proposed by the defendants. This term is explained in the brief filed on this application by BCS as follows: "In the context of the production and trade of grain and seed, the term "adventitious presence" (often simply abbreviated as "AP") refers to the unintentional, unavoidable, and incidental commingling of trace amounts of seed, grain, or foreign material (biological or other) or impurities in a quantity of seed or grain. AP can occur through any one of a number of unavoidable means, including mechanical mixing during the harvesting, processing, handling and storage of seed and grain, as a result of inclusion of foreign seed in or around the planted area, as a result of volunteers from the previous year's crop or low seed admixtures, or as a result of cross-pollination with plants near or in the planted area ... (BCS brief of law at tab 1, para. 12).[70]

The insouciant claim by Bayer CropScience that adventitious presence is unintentional, unavoidable, or incidental underscores a level of industry arrogance belied by the realities of contaminated fields, crops, and seed supplies. In fact, the presence of unwanted genetically engineered plants in their fields is anything but adventitious for organic (and no doubt many conventional) farmers. Aside from loss of markets, organic producers incur the cleanup costs associated with the actual removal of the genetically engineered plants growing in their fields and are also subject to additional paperwork and inspections by their organic certification bodies. As Holtslander points out, organic farmers are therefore very clear that genetically engineered organisms and organic agriculture cannot coexist.

The fact that this language was employed by Justice Smith in her decision is particularly telling not only of where the sympathies of the courts lie but also of the degree to which they accept the destruction of non-genetic engineering farming practices that occurs because of the contamination wrought by genetically engineered seeds. Since organic agricultural production expressly prohibits the cultivation of genetically engineered seeds and the co-mingling of organic and genetically engineered plants in the same fields, contamination is, in fact, the appropriate term to characterize the presence of genetically engineered volunteer plants. However, given the pejorative nature of this term, industry and government consistently employ the more benign sounding "adventitious presence," which has the simultaneous discursive effect of trivializing the very real impacts of contamination events on agricultural producers who eschew genetic engineering technologies.

Justice Smith relied on the precedent established in *Rylands v Fletcher* to disallow the plaintiffs' claim alleging strict liability:

Regardless of whether one considers GM canola a "dangerous substance," or the field trials for GM canola an "unnatural" or "non-natural" use of land, it is not reasonably arguable that the commercial release and sale of Roundup Ready canola seed and Liberty Link canola seed constituted an "escape" of a substance, dangerous or otherwise, from property owned or controlled by the defendants in the sense of "escape" required by the rule in *Rylands v Fletcher*. It is my conclusion that the pleadings do not disclose a reasonable cause of action based on the rule in *Rylands v Fletcher*.[71]

Placing the reliance on *Rylands* aside, Justice Smith's logic reveals an unfair incongruity with the rulings established in the *Schmeiser* case. As the reader will recall, in the latter case, all of the courts involved upheld the validity of Monsanto's patent on its genetically engineered canola. Thus, for patent enforcement purposes, Monsanto's transgene was considered sufficiently novel, but, per Smith's ruling, these same transgenes were considered so mundane that they failed to attract issues of liability.

Finally, Justice Smith refused the plaintiffs' claim of trespass, arguing that they were unsuccessful in meeting the directness requirement articulated by Lord Denning in *Southport Corporation v Esso Petroleum Co.*: "It is my conclusion that action in trespass does not lie against the defendants as the inventors and marketers of GM canola for the adventitious presence of GM canola in the crops and on the lands of organic farmers, for even a liberalised requirement ... for direct interference cannot be met in the circumstances of this case."[72] When we juxtapose this reasoning with the decisions in *Schmeiser* that upheld the patent rights of Monsanto over its genetically engineered canola, there emerges a clear disconnect between ownership rights and corresponding duties. In fact, in the case of genetically engineered plants, trespass assumes added gravity since these organisms do not merely enter fields and sit there but, instead, take root, grow, propagate, and spread.

One element of the claims that Justice Smith did decide in favour of the plaintiffs, albeit reluctantly, was for a cause of action under the amended 2002 *Environmental Management and Protection Act*: "Given the literal wording of section 15 I am unable to say that it is plain and obvious that the plaintiffs' claim under this statute cannot succeed. This provision, so interpreted, would not require the plaintiffs to allege and prove that the 'substance' at issue is inherently harmful or unsafe."[73] Nonetheless, the overall judgment, which rejected almost all causes of action claimed by the plaintiffs, seriously calls into question the capacity for common law to provide a judicial remedy to the current asymmetry between rights and obligations in respect of genetically engineered organisms in this country.

In *Schmeiser*, the Federal Court dismissed the defence's claim that releasing Roundup Ready canola into the environment without control over its dispersion negated Monsanto's claim to enforcement of its rights to exclusive use. Indeed, that court went on to place great emphasis on the facts Monsanto introduced to bolster its ownership claims:

> With respect, the conclusion the defendants urge would ignore the evidence of the licensing arrangements developed by Monsanto in a thorough and determined manner to limit the spread of the gene. Those arrangements require agreement of growers not to sell the product derived from seed provided under a TUA [technology use agreement] except to authorized dealers, not to give it away and not to keep it for their own use even for reseeding. It ignores evidence of the plaintiffs' efforts to monitor the authorized growers, and any who might be considered to be growing the product without authorization. It ignores the determined efforts to sample and test the crops of the defendants who were believed to be growing Roundup Ready canola without authorization. It ignores also the evidence of Monsanto's efforts to remove plants from fields of other farmers who complained of undesired spread of Roundup Ready canola to their fields.[74]

Yet when it comes to the question of any corresponding duties that might attach to its ownership rights, Monsanto asserted – and the courts seem to have accepted – that the dissemination of genetically engineered canola is so widespread and uncontrollable that the company should not be held liable. Moreover, both Justice Smith and the Court of Appeal justices deferred to the concept of federal paramountcy, contending that important policy reasons negate any duty of care (even if Smith had found a duty of care to exist) given that the federal government had approved both genetically engineered canola varieties for unconfined release.

Precisely because of this seeming inability of common law to adapt to new realities posed by biotechnology, countries that have authorized the environmental release of genetically engineered seeds must also develop and implement a *sui generis* legislative framework to respond to the challenges posed by these genetic technologies.[75] Not completely unexpectedly given its industry bias, the CBAC opposed such a solution, arguing that "in our view, Canadian law already adequately addresses issues of liability and compensation for damages through the common law of negligence and the civil law of obligations, which are based on principles of accountability and responsibility. Specific provisions for damages caused by products of biotechnology, patented or not, are not required" (Canadian Biotechnology Advisory Committee 2002b, 17). As Terry Zakreski (2007, para. 28) maintains in his memorandum of argument seeking leave to appeal the case at the Supreme Court of Canada, "given the ease at which the courts below swept aside the Applicants' claims, the CBAC's confidence in the ability of Canadian law to adequately address issues of liability and compensation appears to be misplaced."

The Organic Agriculture Protection Fund sought leave to appeal Justice Smith's decision on 25 May 2005, believing that she had engaged too closely with the merits of the case when deciding whether to certify as a class action, something that the *Class Actions Act* specifically disallows at subsection 7(2): "An order certifying an action as a class action is not a determination of the merits of the action." It was similarly asserted that Justice Smith had applied an overly strict interpretation of the requirements of the act that would make it inordinately difficult to ever define any group of people as a class. As one legal observer points out,

> [r]efusing to certify a case as a class action has very serious consequences. The certification decision is not purely procedural, as the motions judge said it was; in fact, it has significant substantive effect. It prevents the attainment of the three goals of class actions: access to justice, judicial economy, and behavior modification in cases of widespread harm. Although technically the case can still go forward as an individual claim, the complexity and cost of arguing the scientific and economic issues mean access to justice would be out of reach of the individual farmer. In terms of behavior modification, even if each of the class members could successfully sue individually, the award would not be an aggregate one, and the message would not have the power of a collected action. If environmental cases are repeatedly refused certification, the potential threat of group action is reduced, leaving those who might inflict widespread harm undeterred. (McLeod-Kilmurray 2007, 197)

Leave to appeal the lower court decision was granted on 30 August 2005, and the actual appeal hearing took place on 11 December 2006 before Justice Cameron, Justice Gerwing, and Justice Sherstobitoff of the Saskatchewan Court of Appeal. In its decision from 2 May 2007, the court dismissed the appeal, thus upholding the lower court's ruling that denied class certification. Holtslander considers this disappointing outcome to have resulted from the Court of Appeal's failure to consider any of the arguments presented by counsel for the Organic Agriculture Protection Fund. According to Holtslander, the court essentially reiterated the lower court's ruling and set an even worse precedent. Justice Cameron, who wrote the Court of Appeal's majority decision, contended that the preclusion of awards of cost in the *Class Actions Act* warrants a high standard for establishing class certification. This higher standard is ostensibly to prevent unscrupulous plaintiffs from seeking class action certification to pressure a company or institution into a settlement "induced not by fear of being found to have engaged in any wrongdoing but by concern over the enormous cost associated with class action litigation. There is an obvious need to guard against such mischief in the interests of furthering, not distorting, the purposes of the *Act*, and of maintaining respect for, and confidence in, the class action regime."[76] Holtslander passionately maintains that this decision is "a real affront to justice, because it is saying that we have to protect these poor corporations from all these uppity people who think that maybe corporations should be responsible." She is quick to add that "those are my words, that's not our official position" (Interview).

On 1 August 2007, the Organic Agriculture Protection Fund filed papers with the Supreme Court of Canada, seeking leave to appeal the decision of the Saskatchewan Court of Appeal. In his memorandum of argument, Zakreski (2007, para. 1), counsel for the Organic Agriculture Protection Fund, stated the following: "This case seeks to ask whether biotechnology companies incur responsibility when their patented genetically modified seed, pollen and plants infiltrate farmland, causing harm. While *Monsanto Canada Inc. v Schmeiser* confirmed that these companies have significant exclusive rights to GMO seed and plants, the question remains whether they have any corresponding duties." Unfortunately, on 13 December 2007, the Supreme Court of Canada dismissed the application without costs, thus declining to involve itself with a case that promised to set an important national, and possibly international, precedent on the potential liability of biotechnology companies for harm caused by their genetically engineered seeds. On 16 April 2008, Larry Hoffman and Dale Beaudoin announced their intention not to proceed with their individual claims against Monsanto and Bayer CropScience, while also noting that they and other organic farmers would continue fighting for the rights to farm free of genetically engineered organisms and to eat non-genetically engineered food. According to Beaudoin, "[w]e are closing a chapter, but not the book. We will challenge Monsanto and Bayer for the liberty, freedom and right to grow GMO free crops. We want to be able to save and use our own seed" (*Individual Action Not the Way to Go* 2008, para. 6).

While there are certainly environmental aspects to the fight being waged by Saskatchewan organic farmers against Monsanto, Bayer CropScience, and any other corporation that seeks to introduce genetically engineered crops into the province, farmers involved have focused the battle on economic issues, asserting that, even with regulatory approval, these large multinationals must accept liability for the damage to organic farmland and markets that follows in the wake of contamination by genetically engineered crops. The National Farmers Union (n.d., art. 9) has similarly waded into the liability debate, arguing that the "federal government must compel companies which own patents on GM seeds or livestock to set up contingency funds to compensate for product liability and legislate efficient and accessible mechanisms to enable liability claims to be effectively pursued." Organic farmers, having already suffered the loss of canola since even the seed stocks are now contaminated, refuse to lose the right to grow any further major crops.

This is also a point of criticism of Canada's regulatory system, which fails to address the potential economic impact of genetically engineered products on current producers. As discussed earlier, Bill C-474 would have amended the *Seed Regulations* to make potential harm to export markets a regulatory criterion when assessing genetically engineered seeds, but heavy industry lobbying, coupled with Conservative and Liberal Party opposition, made passage of this bill impossible. Nonetheless, Arnold Taylor, interview informant and chairperson of the Organic Agriculture Protection Fund, contends that the lawsuit has expanded the clout of organic farmers by impressing upon agribusiness that until the liability issue is adequately resolved it will be highly impractical to introduce new genetically engineered products into the Canadian market, a fact he believes is lost on neither shareholders nor insurers (Interview).

Percy Schmeiser also recently entered the genetic engineering contamination debate. In 2005, while preparing his fields for a mustard crop, Schmeiser found a number of Monsanto's Roundup Ready canola plants. After he informed the company of these volunteer plants in September, Monsanto dispatched a team of investigators to the Schmeiser farm where they confirmed the presence of Roundup Ready canola in his fields. Schmeiser had the plants professionally removed and subsequently demanded that Monsanto reimburse him for the incurred removal costs, contending that stray plants are pollution and that responsibility for remedying the situation remains with the polluter. Monsanto indicated its willingness to cover the costs ($660) of what it referred to as a "specific and local" event on condition that Schmeiser and his wife sign a document that forever releases Monsanto from any lawsuits associated with its products and forbids the Schmeisers from disclosing the terms of the settlement. Schmeiser promptly refused. As he explained,

> [n]o corporation should have the right to introduce GM seeds or plants into the environment and not be responsible for it. It doesn't matter if it was $600, or $600,000. It has now become a very important case, even though it is small, because if we win then it could cost Monsanto millions and millions of dollars across the world. It was almost unbelievable that Monsanto didn't pay, because it came out and admitted it was their GMO [genetically modified organism] on our property. But they said they would refuse to pay unless we signed a non-disclosure statement. No way would we ever give that away to a corporation. (As cited in Adam 2008, 20)

Given Monsanto's intransigence, Schmeiser initiated proceedings in small claims court to recover the removal costs. However, the lawsuit was settled out of court on 19 March 2008. Monsanto agreed to pay all of the clean-up costs of the Roundup Ready canola that contaminated Schmeiser's fields, and it dropped its demands that Schmeiser not disclose the settlement and that Monsanto could not be sued again if further contamination occurred. Schmeiser, according to his website, believes this precedent-setting agreement ensures that farmers will be entitled to reimbursement when their fields become contaminated with volunteer Roundup Ready canola or any other unwanted genetically engineered plants.[77] It will be interesting to determine what, if any, effects this admission by Monsanto will have for future legal struggles along the lines of those waged by the Organic Agriculture Protection Fund to litigate against genetic contamination.[78]

Accounting for Corporate Command through Intellectual Property

> Capital has from the start sought to enclose the commons. From colonization to slavery, from the work day to the home, from activity to the deepest thoughts and feelings, the history of capital is its extension into the human commons.
>
> – Monty Neill, George Caffentzis, and Johnny Machete, *Toward the New Commons: Working-Class Strategies and the Zapatistas*

Corporate control of agricultural biotechnology through such means as patents and technology use agreements represents a new modality of capitalist primitive accumulation, which strives to circumscribe natural cycles of reproducibility and force agricultural producers into purchasing from an oligopolistic set of supposed life science companies a vital input that for millennia was freely given by nature. From the perspective of multinational biotechnology corporations, saving, trading, and reusing seeds present substantial obstacles to one of their main business lines. Through this lens, we can interpret the intellectual property regime, particularly in its expanded contemporary manifestation, as a state-enforced system designed to expedite the private expropriation of some or all of the value produced through the co-operative relationships of biopolitical production (Hardt and Negri 2004). For example, in light of the American decision in *Chakrabarty* and the subsequent US patent issued to Harvard for the Oncomouse case, as well as the Canadian *Schmeiser* case, what were once considered natural elements of common property are now deemed products of individual human labour and ingenuity to which attach private property rights of exclusion and enclosure.

Yet might it not be argued that people who develop ideas and inventions based upon those that have come before them are engaged in personal appropriation of the public domain? Are these creators, similar to Isaac Newton, not standing on the shoulders of those giants who came before them? Is the notion of the "autonomous invention" thus a myth employed to construct discursively and legally an uneven system based on individuating alienable "things" that can circulate as exchange values (Haraway 1997)? Should one not heed the Lockean proviso that removing objects from the commons is permissible only so long as there is "enough and as good left for others"? If the provision of property rights is considered to be the optimal means to spur intellectual innovation, then we must also consider measures and limits on such rights to protect the common pool, or public domain, to prevent a net decrease in the production of informational and cultural artifacts.

Specific to biotechnology, one argument advanced by the proponents of patent protection is that this type of protection safeguards the underlying information rather than the life form. That is, the genes remain a part of the commons and *only* the information that describes them and their use is protected. In part, this contention flows from the traditional notion inherent in intellectual property regimes that the commons remain because of disclosure requirements and limited terms of monopoly protection. However, what these justifications deliberately neglect to consider is that the relatively standardized structure of the biotechnology industry, which relies on a common set of expensive techniques to manipulate biochemical information, effectively renders most current patents a rent that must be paid if further research and innovation is to be conducted (May 2004).

In fact, the transaction costs associated with clearing rights protected by the intellectual property regime is rendering biotechnological research prohibitively more expensive. The contemporary intellectual property system might thus actually give rise to its antithesis and come to retard innovation and scientific progress. That is, intellectual property rights, aside from undervaluing the sources and audiences of intellectual creations, provide protection at the expense of dissemination to such an extent that future creators will find an increasingly limited public and common domain on which to draw for inspiration. This is perhaps the ultimate irony in that the expanded breadth of intellectual property rights threatens to curtail the very process upon which proponents of strict intellectual property protection base their arguments.

These intellectual property enclosures have been made possible by an uncritical acceptance of genetic reductionist representations, which, as discussed earlier, are misaligned with both the theory and practice of the researchers and farmers who engage with human, animal, and plant life forms (Beurton 2000; Falk 2000; McAfee 2003). Yet such depictions of genes as discrete and stable objects that possess knowable and predictable properties offer a theoretical justification for transforming genes and their attendant information into alienable commodities that, through a stringent intellectual property protection system, can be owned and traded in the neo-liberal marketplace. In practice, the attachment of intellectual property protections to particular component parts of an organism has transformed such monopolies into proxies of control over the entire organism, as indicated by the *Schmeiser* decision. Put simply, the genetic reductionist paradigm dovetails quite readily with the economic reductionism inherent in neo-liberalism, helping to sustain capital's encroachment into nature and organic existence. As one commentator lyrically points out, "the naked hubris that posits genes, biological processes, and whole organisms as alienable privatized inventions is quite evidently a multi-faceted theft and ought rightly to be named as such" (Prudham 2007, 414).

Bearing in mind the exposition of capitalist control of biotechnological agriculture presented in Chapter 2, we also need to articulate explicitly the fairly obvious link between Terminator technology and intellectual property. These new biotechnologies can be interpreted as a response to what I, somewhat inelegantly, term the "leakiness" of intellectual property protection. Some form of property protection, be it through patents or plant breeders' rights, has long been available to seed sellers. And while such protections exercise a certain form of control over producers who typically elect to purchase seed anew each planting year out of fear of prosecution, the seed itself refuses such control, spreading its progeny across farmers' fields. In effect, nature spurns the artificial enclosures of the intellectual property system, permitting its seed to leak out around the edges of this artificial enclosing construct. As I have detailed at some length in this and Chapter 2, multinational corporations, particularly Monsanto, are zealously undertaking measures to plug the holes in the contemporary intellectual property regime. Terminator technology threatens to move the fight for control over nature and seeds to a completely new level – to the genetic level in an emerging ensemble of legal and technological measures that would effect a new enclosure movement at the level of the organism never imagined by even the most ruthless of feudal lords.

The intellectual property system facilitates the appropriation of knowledge developed in the commons for centuries, separating labour from its own knowledge, thus serving not only to reduce living knowledge to abstract knowledge but also to actually devalue the former – which increasingly occurs along North/South divisions. Capital's expanding exploitation of social labour and nature brings with it a corresponding substitution of value accumulation imperatives for use value as the driving motivation for production, leading to a situation in which the social conditions that provide the basis for social production come to confront labour as the power of capital:

> The forms of socially developed labour ... appear as *forms of the development of capital*, and therefore the productive powers of labour built up on these forms of social labour – consequently also science and the forces of nature – appear as *productive powers of capital*. In fact, the unity of labour in

co-operation, the combination of labour through the division of labour, the use for productive purposes in machine industry of the forces of nature and science alongside the products of labour – all this confronts the individual labourers themselves as something *extraneous* and *objective*, as a mere form of existence of the means of labour that are independent of them and control them ... And in fact all these applications of science, natural forces and products of labour on a large scale ... appear only as *means for exploitation* of labour, as means of appropriating surplus-labour, and hence confront labour as *powers* belonging to capital. (Marx 1963, 390-92 [emphasis in original])

The prescience and sagacity of Marx's thought to our contemporary situation cannot be emphasized strongly enough when considering the material presented in this book, particularly in respect of the way capital makes adept use of the intellectual property regime to bring agricultural biotechnology firmly within its ambit. In our current conjuncture, intellectual property systems, both nationally and internationally (the latter impinges on the former), function as legal mechanisms or instances of state support of modern enclosures in much the same way that the state facilitated historical terrestrial enclosures. Put more explicitly, primitive accumulation in our contemporary context does not differ in any substantive way from the historical processes that facilitated the rise of capitalist social relations but, instead, merely in the forms, scope, and intensity it assumes. For example, the knowledge enclosures around biotechnology achieved through the intellectual property regime have material implications for biological resources that, until about the last three decades, remained common resources largely beyond the purview of capitalist commodification. This expanding range of biotechnological resources protected by intellectual property rights furnishes capital a vital extra-economic prerequisite to biocapitalist (re)production that not only endures in contemporary society but that also is being extended across the globe. Under the dominance of capitalist social relations, we thus witness a further instance of the social separation of the conditions of production from the control of the direct producers in service of capitalist valorization.

The *Schmeiser* and *Hoffman and Beaudoin* cases provide examples of direct and indirect enclosures that can be understood as contemporary forms of primitive accumulation in the service of capital accumulation. The former case exhibits a deliberate act of *ex novo* enclosure of the canola genome that has been facilitated by the intellectual property regime. The natural reproducibility of seeds, a critical agricultural input long considered to be a common resource developed through co-operative social labour across millennia, is being enclosed through patents to enforce artificial scarcity upon a natural resource. Such enclosure has been made possible because control over a few patented genes and genetic-engineering processes has conferred on Monsanto *de facto* control over the entire plant and, thus, the complete range of previous labour and knowledge contained therein. In what can only be regarded as blatant acts of biopiracy, a few multinational corporations are appropriating rights of control over, and access to, such resources and the information and knowledge embodied in these physical artifacts (Shiva 1997, 2001b). Similarly, the technology use agreements to which agricultural producers must adhere provide Monsanto with a substantial level of control not only over its seeds but also over the farming practices of growers.

The *Hoffman and Beaudoin* case demonstrates how the accumulation imperatives made possible by the first type of enclosure can exercise knock-on enclosing effects on external parties. Genetically engineered canola varieties have effectively infiltrated the market and environment to such an extent that few, if any, pedigreed canola seed growers or grain farmers can warrant their seed as being genetic engineering free. As a result of the contamination caused by genetically engineered canola, conventional and, especially, organic farmers have lost this crop as an important variety within their rotations. These companies were quite aware of, and no doubt banked on, the fact that, without adequate safeguards, conventional and organic farmers would be circumscribed in their ability to use the genetic resources of canola. The net effect of such actions has been the enclosure of canola germplasm by Monsanto and other companies to the detriment of all agricultural producers. Farmers who plant genetically engineered crops are disallowed from saving and reusing increasingly expensive seed, and non-genetically engineered producers are no longer physically able to produce uncontaminated conventional or organic canola crops.

Again, the courts have reaffirmed both types of enclosure in a most unsatisfactory and inconsistent manner. In *Schmeiser*, the courts confirmed Monsanto's ownership claims to its patented gene in a way that establishes control over the entire plant regardless of the means by which the gene enters a field. However, in the *Hoffman and Beaudoin* case, Monsanto's ownership and consequent tight restrictions on use claimed through the patent itself and its technology use agreements were no longer an issue because the courts construed Monsanto (and Bayer CropScience) as a mere marketer. Thus, if a patented, genetically engineered seed disperses into the environment, ownership interests attach in order to prosecute patent rights, but when issues of liability for contamination arise these same companies are absolved of responsibility through claims that they lack control by virtue of being mere marketers. Taken together, these two decisions provide a jurisprudential framework that supports the accumulation imperatives of agricultural biotechnology companies. Moreover, these examples demonstrate how contemporary processes of primitive accumulation are expanding the alienation Marx elaborated to include new strata of producers beyond the orthodox Marxist emphasis on the industrial proletariat and waged labour.

These cases are testament to the expanding range of actors caught up in practices of primitive accumulation and capitalist control of social production processes.

Considered in tandem with Chapter 2, what I hope to have demonstrated with the preceding discussion is that the capitalist appropriation of agricultural biotechnology represents a contemporary example of primitive accumulation that touches on all three elements of this process as articulated earlier. As biotechnology developed from the 1970s onward to a stage sufficient for capitalist valorization, capital began exerting a stranglehold over this technoscience in what can be interpreted as an instance of another area of social existence now brought under capitalist control, thus reinforcing the idea that primitive accumulation remains a continuous social process. As shown in this and earlier chapters, large swaths of the social knowledge in respect of agricultural production have been increasingly appropriated by capital, which represents intensified efforts to privatize once public goods.

Finally, efforts by capital to bring agricultural biotechnology profitably within its control have involved the same spatial ambitions outlined previously with respect to primitive accumulation. Given the informational characteristics of biotechnology as well as the embodied existence of genetic materials, both the global North and South have been confronted by the efforts of capital to bring this technoscience firmly within its ambit. The fact that capital continues to arrogate control over both material and intellectual means of production in our contemporary context is clearly evident in the realm of agricultural biotechnology, which is characterized by increasing attacks on both the biological and knowledge commons as part of broader capitalist accumulation strategies.

5
Regulatory Capture and Its Critics

Having elucidated how capital has strategically used genetic engineering, contract law, and intellectual property mechanisms to establish command over biotechnology in this country, this chapter will examine the ways in which the regulatory system for this technoscience is being subverted in service of capitalist accumulation imperatives. Technological change is presumed to be progressive unless proven regressive. The burden of proof therefore typically falls on those who would seek to regulate new technologies rather than those who profess its benefits. Although the environmental movement of the 1970s resulted in the introduction of some prospective technological assessments, the neo-liberal agenda adopted in many Western countries, particularly in North America, has resulted in a mood that is generally hostile to regulatory initiatives. Both industry and government are quick to emphasize the potential negative economic implications of stringent regulation on commodifiable genetic research and development. Indeed, the current federal science and technology policy developed by the Harper government contends that *"the most important role of the Government of Canada is to ensure a free and competitive marketplace*, and foster an investment climate that encourages the private sector to compete against the world on the basis of their innovative products, services, and technologies" (Government of Canada 2007, 19 [emphasis added]; 2009). Within this broader context, the emphasis in this chapter is placed on locating the major gaps in our regulatory approval processes that portend an array of critical implications for the environment, human and animal health, and the economy of agricultural producers. In terms of the theoretical framework informing this book, the following pages will illustrate the ways that Canadian regulatory policy and practice facilitate enclosures of the BioCommons.

After first briefly elaborating the departmental structure of the Canadian regulatory environment in respect of genetically engineered products, this chapter will discuss this regulatory regime's major defects as articulated by members of the Royal Society of Canada. The force of this critique assumes added weight when considered in tandem with the emerging scientific evidence of adverse environmental and health effects from genetically engineered crops, covered in the subsequent part of this section of the chapter. The next section of the chapter considers the formal critique submitted to the Government of Canada by a group of scientists critical of current agricultural biotechnology regulation. Although this critique reinforced several of the shortcomings highlighted in the report by the Royal Society of Canada, it failed to spur the government into addressing the continued weaknesses of this country's system of regulatory oversight, as reflected in the latest regulatory decision to approve SmartStax corn.

Regulatory Capture

The discussion begins with a review of the current state of the Canadian regulatory regime with respect to genetically engineered organisms. At the Canadian federal level, there are two main entities that have responsibility for the regulation of biotechnology: Health Canada and the Canadian Food Inspection Agency (CFIA). Health Canada, under the *Food and Drugs Act* and the *Food and Drug Regulations,* is mandated with assessing the safety of genetically engineered foods meant for human consumption as well as veterinary drugs, pharmaceuticals, medical devices, cosmetics, biologics, radiopharmaceuticals, genetic therapies, and pesticides.[1] The CFIA is responsible for assessing the environmental safety of genetically engineered organisms under the jurisdiction of the *Seeds Act.*[2] The CFIA approves field trials and commercial growing (that is, unconfined release) and is responsible for variety registration of most new seeds in this country.[3] The CFIA also maintains regulatory oversight of animal feed under the *Feeds Act* and its regulations and assesses veterinary biologics under the auspices of the *Health of Animals Act.*[4] Finally, the CFIA enforces food safety standards through a system based on inspection and monitoring activities (see Table 5.1 for a list of biotechnology products and the departments/agencies and regulatory legislation relevant to them). The various critiques levelled against Canada's agricultural biotechnology regulatory regime tend to be rooted in concern over the dual promotion/regulatory role assumed by government agencies, the application of "substantial equivalence" within the regulatory process, and the imprecision of genetic engineering techniques.

Promotion versus Regulation

As highlighted in Chapter 1, Canadian biotechnology policy suffers from an internal tension between the federal government's role in promoting economic growth and its responsibility to assess and regulate all products developed and marketed for human and animal consumption as well as environmental release. As might be expected, this inherent policy contradiction is operationalized in various line departments and agencies, representing

Table 5.1

Legislative responsibility for the regulation of biotechnology products

Products regulated	Department/ Agency	Act	Regulation
Foods, including novel foods derived through biotechnology	Health Canada	*Food and Drugs Act*	*Novel Foods Regulation**
Feeds, including novel feeds derived through biotechnology	Canadian Food Inspection Agency	*Feeds Act*	*Feeds Regulations*
Environmental release of plants with novel traits such as pest resistance or herbicide tolerance and, where required by the *Seeds Act*, variety registration	Canadian Food Inspection Agency	*Seeds Act*	*Seeds Regulations*
Plant pest risk assessment and permit to import plants, including plants with novel traits	Canadian Food Inspection Agency	*Plant Protection Act*	*Plant Protection Regulations*
Pest control products	Pest Management Regulatory Agency, Health Canada	*Pest Control Products Act*	*Pest Control Products Regulations*
All animate products of bio-technology for uses not covered under other federal legislation (as listed in *Canadian Environmental Protection Act*, Schedule 4	Environment Canada and Health Canada	*Canadian Environmental Protection Act*	*New Substances Notification Regulations*

*   Division 28 of the Food and Drug Regulations.

a conflict of interest that threatens to undermine the government's regulatory function. This danger is exacerbated by the fact that contemporary regulatory agencies are compelled to establish a solid working relationship with the industries they regulate in order to ensure an effective and enforceable system of regulation. However, such a "co-management" arrangement leads to problems involving industry capture of state regulators (Clark 2002; Leiss 2000, 2001; Salter 1993).

The Institute on Governance, a Canadian non-profit think-tank, opines that government credibility, which it sees as a major criterion in effectively dealing with biotechnology in Canada, is undermined by the government's dual role as promoter and regulator of this new technology (Boucher et al. 2002). Specific to agricultural biotechnology, this tension is particularly evident within the CFIA, where regulation and promotion warrants continue to be sources of significant criticism and concern about how well the agency discharges its regulatory duties. The Royal Society of Canada (2001, 213-14 [emphasis added]), which was commissioned by the federal government in 2000 to analyze Canada's regulatory system for genetically engineered food products, similarly critiques the promotion role assigned to regulatory agencies in its report, *Elements of Precaution: Recommendations for the Regulation of Food Biotechnology in Canada:*

> Such concern with industry development, though understandable, highlights another aspect of the regulatory conflict. The conflict of interest involved in both promoting and regulating an industry or technology ... is also a factor in the issue of *maintaining the transparency, and therefore the scientific integrity, of the regulatory process.* In effect, the public interest in a regulatory system that is "science based" – that meets scientific standards of objectivity, a major aspect of which is full openness to scientific peer review – is significantly compromised when that openness is negotiated away by regulators in exchange for cordial and supportive relationships with the industries being regulated.

These conflicting mandates at the CFIA have been a source of concern even among some federal government bureaucrats. In 1999, a number of Health Canada employees petitioned then Minister of Health Alan Rock to work to ensure that the control over genetically engineered foods remained within the purview of the health ministry. The petition pointed out the blatant conflict of interest that arises from the agricultural department being tasked with monitoring the same products – products that could pose human health risks – that it is in the business of promoting (Riley 1999). It is also worth recalling in this context the harassment and ultimate dismissal of Shiv Chopra and some of his Health Canada colleagues who resisted internal pressure to sign off on regulatory applications for products they considered dangerous to human and animal health. In addition to such inherent conflicts of interest among biotechnology regulators in this country, or perhaps as a direct result, the Canadian regulatory regime for agricultural biotechnology is also susceptible to scientific critique.

The Imprecision of Genetic Engineering Techniques
As early as 1864, George Perkins Marsh recognized that humanity's attempts to master nature filter through what Harvey (2000) refers to as the "web of life," a web of interconnections constitutive of the natural world that bring with them unintended consequences: "These intentional changes and substitutions constitute, indeed, great revolutions; but vast as is their magnitude and import, they are ... insignificant in comparison with the contingent and

unsought results which have flowed from them" (as cited in Harvey 2000, 219). Regulatory systems fail to adequately consider the secondary effects that accrue from manipulating an organism at its genetic level, despite the fact that genetic engineering affects an organism's metabolic pathways in ways that are often difficult to detect and determine (Ferrara and Dorsey 2001). As Friedrich Engels (1940 [1898], 289-90), Karl Marx's lifelong intellectual companion, pointed out long ago, "in nature nothing takes place in isolation. Everything affects every other thing and *vice versa*, and it is usually because this many-sided motion and interaction is forgotten that our natural scientists are prevented from clearly seeing the simplest things."[5]

Along similar lines, the more contemporary report from the Royal Society of Canada adopts a critical stance toward the notion that genetic engineering is a "precise" technique, asserting instead that inserting genes into new cellular environments carries the potential for unexpected and contingent consequences. In order to introduce exogenous deoxyribonucleic acid (DNA) sequences (transgene) into a plant genome, commercial firms typically avail themselves of either of two widespread processes – particle bombardment or *Agrobacterium*-mediated transformation. Referred to more colloquially as the "gene gun" method, particle bombardment (biolistic transformation) is typically employed when the host plant into which the transgene is to be inserted is not readily susceptible to *Agrobacterium*. In this method, miniscule particles of metal such as gold are coated with the plasmid DNA packet and fired directly into the cells of the host organism, thereby transferring the transgene into the plant's chromosome. The latter method relies on infecting plant cells with a disarmed pathogenic organism *(Agrobacterium tumefaciens)* that contains the transgene of interest.

Since scientists discovered methods for deleting its tumour-inducing genes while maintaining its capacity to transfer DNA, the self-replicating capacity of *Agrobacterium* can be harnessed to serve as a plasmid suitable for transferring part of its DNA into the plant that it infects.[6] Employing a variety of methods, including the use of restriction and ligation enzymes, scientists first insert the transgene into *Agrobacterium*.[7] In order to confirm that the transgene was transferred successfully, a second gene sequence for antibiotic resistance is incorporated into the plasmid. Once the host plant material has been infected with the *Agrobacterium* plasmid, an antibiotic is administered, thus ensuring that only the cells with the transgene plasmid survive. The process also requires what is known as a promoter, a third gene sequence capable of "switching on" the main transgene once it has been incorporated into the genome of the host plant. The most widely used promoter, which is also included in the plasmid containing the transgene of interest and the antibiotic gene sequence, is composed of a DNA sequence from the cauliflower mosaic virus (CaMV35S). The promoter ensures that the transgene will be expressed in all of the host plant cells.

One risk of employing viruses as vectors in genetic engineering is that they can enter their infectious phase, replicate, and escape from the target cell to introduce the gene insert into other cells or organisms (Wheale and McNally 1998).[8] In fact, the mosaic character of most vector constructs renders them structurally unstable and disposes them to recombination (Ho 1999). For these reasons, a number of researchers have made a plea to discontinue the use of the cauliflower mosaic viral promoter, which is employed in a substantial majority of transgenic crops released commercially or at the field trial stage (Ho, Ryan, and Cummins 1999). The British Medical Association (1999) and even the American Medical Association (2000), which otherwise is very favourable toward genetically engineered crops, have also issued appeals to stop using antibiotic resistance marker genes in genetic engineering out of concern that these genes, and antibiotic resistance, could be transmitted to bacteria in the environment, including the bacteria that reside in human and animal intestines. This has the potential to become a human health concern because as bacteria become more resistant to antibiotics, these drugs lose their effectiveness in treating human infections (Ferrara and Dorsey 2001; Kloppenburg 2004).

Precise placement of engineered genetic constructs is critical to proper gene expression. However, the use of "gene guns" and various vectors to transfer foreign genetic material into another cell is imprecise and random in terms of placing the transgene packet in the target plant chromosome. This randomness can result in physical disruption in the genome (insertional mutation at the site of the transgene insertion or at random locations throughout the genome) or in the regulation of the gene, which compromises the safety of the genetically engineered plant. Mutations that can occur as a result of the *Agrobacterium* method include deletions and rearrangements of host chromosomal DNA as well as the insertion of superfluous DNA, which can consist of additional whole or partial copies of the transgene, the vector DNA, or filler DNA (DNA newly created at DNA-DNA junctions). Even in cases where the inserted transgene achieves the desired target effect, research in actual field conditions has demonstrated that the insertion can impact significantly other phenotypic traits in ways that can affect the ecological behaviour of the species (Zeller et al. 2010).

While Jonathan Latham, Allison Wilson, and Ricarda Steinbrecher (2006, 3) point out a relative lack of scientific research describing insertion sites associated with particle bombardment methods, "it appears that transgene integration resulting from particle bombardment is usually or always accompanied by substantial disruption of plant DNA and insertion of superfluous DNA." Since transgenes are inserted into, or proximate to, functional gene sequences, insertion site mutations have the potential to cause inadvertent loss, acquisition, or misexpression of critical traits. Interaction effects might also occur between the inserted genes and the thousands of other genes in the organism being genetically altered. Endogenous genes might be inactivated or, conversely, silent genes may be activated. Other unintended effects could include modified metabolism, the production of new substances or novel fusion proteins, the induction of unanticipated changes in the manner in which the plant functions or interacts with other organisms or the environment, or alterations to the toxicity or nutritional value of the genetically engineered plant (Cheeke, Rosenstiel, and Cruzan 2012; Ho 1999; Ho, Ryan, and Cummins 1999; Jiao et al. 2010; Lotter 2009; Royal Society of Canada 2001; Wilson, Latham, and Steinbrecher 2006; Zeller et al. 2010; Zolla et al. 2008).

Latham, Wilson, and Steinbrecher (2006, 5) articulate nicely the overall message of the unfortunately scant scientific literature evaluating the implications of genetic engineering processes: "Even with the limited information currently available it is clear that plant transformation is rarely, if ever, precise and that this lack of precision may cause many of the frequent unexpected phenotypes that characterize plant transformation and that pose a significant biosafety risk."[9] Yet one is left wondering to just what extent regulatory decisions reflect these findings given that our regulatory regime is triggered by the presence of a novel trait in a plant and not by the process or method employed to introduce that trait.

Despite such dangers and a serious paucity of rigorous scientific study of the mutational effects of genetic engineering techniques, the Canadian government refuses to implement a post-market surveillance program for genetically engineered products, contending instead that "if a product gains market approval, it is the legal responsibility of the proponent to provide the Government of Canada with additional information regarding any untoward observations or effects. The Government of Canada *may* carry out post-market sampling, auditing and testing, either as routine post-market surveillance or on a case-by-case basis, or change its regulatory decisions, in response to additional information provided by the proponents, the public, or advances in scientific knowledge" (Office of the Auditor General of Canada 2002, para. 55 [emphasis added]).[10] Given the challenges and constraints the federal government currently faces in enforcing its post-market surveillance programs for food, drugs, and medical devices, it remains highly unlikely that the responsible regulatory bodies will engage in monitoring activities not legally mandated.

Substantial Equivalence as a Regulatory Criterion
Part of the rationale offered for the decision to regulate the genetically engineered product rather than the process by which transgenes are incorporated into host plants is found in the controversial concept of "substantial

equivalence," which is defined as a safety assessment criterion in the following manner: "A GM organism is 'substantially equivalent' if, on the basis of reasoning analogous to that used in the assessment of varieties derived through conventional breeding, it is *assumed* that no changes have been introduced into the organism other than those directly attributable to the novel gene. If the latter are demonstrated to be harmless, the GM organism is *predicted* to have no greater adverse impacts upon health or environment than its traditional counterpart" (Royal Society of Canada 2001, 182 [emphasis added]).[11]

However, comparing transgenic products, such as *Bacillus thuringiensis* (Bt) products that produce toxins, to other products, even an insecticide, is to draw a false analogy. Certainly, both can wreak havoc on the environment, but the latter does not pose the same threat of long-term genetic consequences as does the former (Critical Art Ensemble 2002). Similar to the way that genetic reductionism is employed in the service of intellectual property rights (as discussed in the previous chapter), the linear model employed to evaluate genetically engineered organisms, which supports the test of "substantial equivalence," incorrectly assumes that transferring genetic material into the DNA of another organism will only affect the genes transferred across species. In fact, a consortium of scientists who conducted a four-year US National Human Genome Research Institute study argued in a June 2007 report that, contrary to the principles upon which recombinant DNA technology has been developed since 1973, genomes are not collections of independent genes for which each sequence of DNA codes for a single function. The study, which included thirty-five groups from eighty different international organizations, found, instead, that genes function in a complex network environment that involves interactions, and they overlap with each other as well as with other components not yet fully understood (Caruso 2007).

The report compiled by the Royal Society of Canada, which contains fifty-three recommendations for overhauling Canada's regulatory system for genetically engineered food, also challenges the validity of a linear model of scientific assessment. It argues that equivalence claims cannot be made *a priori* but, instead, require an integrated approach of rigorous scientific evaluation to uncover how phenotypes are affected by genomes and their variants at multiple levels (DNA structure, gene expression, protein profiling, and metabolic profiling) (Royal Society of Canada 2001; Wills 2001). According to the Royal Society of Canada's (2001, ix) report, "[t]he Panel finds the use of 'substantial equivalence' as a decision threshold tool to exempt GM agricultural products from rigorous scientific assessment to be scientifically unjustifiable and inconsistent with precautionary regulation of the technology. The Panel recommends a four-stage diagnostic assessment of transgenic crops and foods that would replace current regulatory reliance upon 'substantial equivalence' as a decision threshold."[12] Other researchers express their opposition to substantial equivalence in even more critical terms, arguing that it

> is a pseudo-scientific concept because it is a commercial and political judg-ment masquerading as if it were scientific. It is, moreover, inherently anti-scientific because it was created primarily to provide an excuse for not requiring biochemical or toxicological tests. It therefore serves to discour-age and inhibit potentially informative scientific research ... the concept of substantial equivalence is being misapplied, even on its own terms, within the regulatory process. (Millstone, Brunner, and Mayer 1999, 526)

Rene Van Acker, an interview informant and professor and chair of the Department of Plant Agriculture at the University of Guelph, contends that genetic engineering is such a novel technology that scientists still do not fully understand whether "substantial equivalence" is true. The reality, ac-cording to Van Acker, is that, in countries such as Canada, the adoption of "substantial equivalence" is based on argument rather than on empirical testing. There are few studies around the world on the medical effects of feeding genetically engineered food to mammals. Yet, in practice, any reli-able evaluation of the toxicity of a genetically engineered food must rely on studies that test for toxicity on animals. As Van Acker laments, clearly more research needs to be conducted worldwide, but, without funding, nobody is doing it (Interview). Over the course of the last decade, José Domingo has conducted periodic literature reviews that document this paucity of rigorous research about the toxicological risks of genetically engineered foods. In addition to a scarcity of experimental data, this small amount of scientific literature tends to comprise short-term, mainly nutritional, studies that offer very limited toxicological information (Domingo 2007; Domingo-Roig and Arnáiz 2000). In their most recent review of the extant literature, Domingo and his colleague indicate that although the number of peer-reviewed studies has increased in the last few years, most of the work specific to edible genetic-ally engineered plants focuses on only three products: maize, soybeans, and rice. More disturbing is their finding that the bulk of studies claiming to find no difference in the nutritional content or safety of genetically engineered crops as compared to conventional varieties has been conducted by industry or scientists with close industry affiliations (Domingo and Bordonaba 2011; see also Diels et al. 2011).

Another recent literature review conducted by Greek scientists Artemis Dona and Ioannis Arvanitoyannis (2009) reaches similar conclusions. Al-though the number of studies into the health effects of genetically engineered foods has grown, the reviewers conclude that almost all of the studies conducted to date have not been sufficiently longitudinal to detect and evaluate possible signs of toxicity and pathology. They conclude that the research they have reviewed relies on samples that are too small and that genetically engineered foods should ultimately be subject to testing that is even more stringent than that demanded for drugs since genetically engineered foods potentially can be ingested by all humans rather than a particular population stratum affected by a certain malady. Echoing concerns articulated by others, Dona and Arvanitoyannis (2009, 164) point out that "in the absence of adequate safety studies, the lack of evidence that GM food is unsafe cannot be interpreted as proof that it is safe." Industry and state regulators ardently cling to the "substantial equivalence" concept as the basis for their claim that genetically engineered crops pose no dangers for human or animal health. Yet, as the work of this growing group of researchers clearly attests, the science behind this postulate, at best, lacks sufficient methodological rigour and, at worst, actually distorts research findings.

In its report on regulating genetically engineered foods, the Canadian Biotechnology Advisory Committee (CBAC) (2002a, 27) recommends that Canada continue employing the "substantial equivalence" criterion when assessing differences between conventional and genetically engineered food crops, asserting that in its review of various regulatory agencies it "found no evidence to indicate that substantial equivalence has been used as a decision threshold to exempt GM foods from appropriate regulatory oversight." This recommendation contrasts rather starkly with the concern voiced by the Royal Society of Canada that the CFIA is incorrectly applying the concepts of "familiarity" and "substantial equivalence" such that regulatory approvals are "*based upon unsubstantiated assumptions* about the equivalence of the organisms by analogy with conventional breeding" (Royal Society of Canada 2001, 182 [emphasis added]).[13]

In a report produced for the Polaris Institute, Lucy Sharratt (2002) writes that the Canadian government developed the categories "novel food"[14] and "plant with novel traits"[15] to camouflage genetic engineering among a range of other technologies in order to reduce public attention to, and knowledge of, various genetic engineering processes. Instead of examining the process, the product is evaluated in a manner that relies on comparisons through the use of concepts such as "familiarity" and "substantial equivalence." The result of this system of assessment is that the dangers of genetic engineering are not sufficiently scrutinized during the regulatory review system. As Sharratt (2002, 12) suggests, "[t]he industry and our government try to define genetic engineering as nothing new so that they do not have to develop new regulations and so that the public does not view genetic engineering as a new technology with new risks. This is a deliberately false representation that contradicts what many people think and know about genetic engineering."

It is only by asserting that genetically engineered products yield no risk beyond those associated with the products developed through conventional breeding and mutagenesis that the CBAC is able to make the substantial equivalence claim.[16] However, such claims miss the point of contention voiced by opponents of "substantial equivalence," including the Royal Society of Canada, that a system based on this principle is not sufficiently scientific-ally rigorous to assess genetically engineered foods. A later government publication mentions the work of the Royal Society of Canada but completely omits any discussion of how its findings were largely ignored by the recom-mendations made by the CBAC in 2002 (Government of Canada 2004).[17] Ultimately, the CBAC's position reinforces the status quo of the current regulatory regime, one that is biased in favour of industry (Magnan 2006).

Although the entire situation presented here in regard to genetically en-gineered crop regulation contradicts the Canadian Biotechnology Strategy (CBS), which envisions a review process that emphasizes science-based technical assessment methodologies when regulating biotechnology, it is not particularly surprising. After all, in the context of rapid technological development characterized by large economies of scale, a few multinational juggernauts are apt to play national economies against one another for fa-vourable corporate conditions. Naturally, a business-friendly regulatory regime will assume paramount importance (Jarvis 2000). According to some observers, science, at an aggregate level, has been overtly complicit in all of this: "It is willing and able to exclusively serve the needs of capital, not just by generating knowledge that can be applied for profit, but also by *not* gen-erating any knowledge or applications that could be detrimental to the maintenance and/or expansion of the system" (Critical Art Ensemble 2002, 41 [emphasis in original]).

In large part, the current situation is a result of government reliance on data supplied by the same companies seeking regulatory approval for their products – data that these companies and the government refuse to release publicly. As the Royal Society of Canada's (2001, 214) report critically points out, "the more regulatory agencies limit free access to the data upon which their decisions are based, the more compromised becomes the claim that the regulatory process is 'science-based.' This is due to a simple but well-understood requirement of the scientific method itself – that it be an open, completely transparent enterprise in which any and all aspects of scientific research are open to full review by scientific peers." In addition to citing a lack of scientific rigour, the report is also quite critical of the low levels of transparency inherent in the regulatory system for genetically engineered food:

[T]he Panel has concluded that there is no means of determining the extent to which these information requirements are actually met during the approval process, or of assessing the degree to which the approvals are founded on scientifically rigorous information. The Panel attributes this uncertainty to a lack of transparency in the process by which GMOs are approved within the present regulatory framework. (214)

The report goes on to argue that "there is no means for independent evaluation of either the quality of the data or the statistical validity of the experimental design used to collect those data. Furthermore, *it appears that a significant part of the decision-making process can be based on literature reviews alone*" (214 [emphasis added]). The panel reaches the conclusion that "the lack of transparency in the current approval process, leading as it does to an inability to evaluate the scientific rigor of the assessment process, *seriously compromises the confidence that society can place in the current regulatory framework used to assess potential risks to human, animal and environmental safety posed by GMOs*" (215 [emphasis added]).[18]

The fact that Canada's regulatory risk assessment framework is heavily influenced by economic and commercial imperatives must be conceded given that the entire product review process relies on publicly unavailable data supplied by crop developers, which, according to a number of critics and as we will see later in this chapter, serves to underplay the ecological and health uncertainties associated with the environmental release of genetically engineered crop varieties.[19] Unfortunately, this business/regulator nexus appears poised to become even more potent under the latest iteration of federal science and technology policy: "Through strong and clear environmental laws and regulations *that work with market forces, we will also create the conditions for businesses and people to respond to environmental challenges with entrepreneurial innovation*" (Government of Canada 2007, 13 [emphasis added]).

This same strategy makes a case for rendering Canada's regulatory and marketplace framework policies more conducive to private sector investment and commercialization. In fact, the discussion specific to Canada's regulatory system developed in this new science and technology policy concentrates exclusively on the perceived need to address intellectual property, information sharing, and confidentiality issues and on the purportedly important role that international regulatory co-operation has to play in this regard: "As announced in Budget 2007, the government will invest $9 million over two years to make Canada a best-in-class regulator by ensuring that *efficiency and effectiveness* are key considerations in the development and implementation of regulations through a new Cabinet directive on streamlining regulation" (Government of Canada 2007, 55 [emphasis added]).

Given that efficiency and effectiveness are two of the three "e"-words critical to the New Public Management paradigm, one can surmise with relative confidence that these considerations are constructed from the perspective of the businesses being regulated rather than from the broader population that is supposed to be protected by government regulations. With a regulatory system increasingly stacked in favour of industry interests, as implicitly admitted by federal policy, it takes little imagination to account for the conflicts of interest and informational gaps inherent in our contemporary regulatory regime.

In addition to dubious science, the current approval regime has proven to be susceptible to industry hijacking. In 1999, Monsanto refused to provide scientists at the CFIA and Health Canada with additional requested product data during a review of the company's application for two types of genetically engineered potatoes. These requests by government regulators came after CFIA audits determined that the field trials being carried out by Monsanto on 1,170 hectares of land in Prince Edward Island, New Brunswick, Ontario, Manitoba, and Alberta were woefully deficient. Monsanto was gearing up to expand its field trials to over 4,000 hectares across the country, or about 12 percent of total domestic potato cultivation. Internal CFIA documents reveal that its inspectors wanted the scope of the field trials scaled back, warning that succumbing to industry pressure would "compromise the integrity" of Canada's regulatory regime. In a memo from 19 February 1999, Morven McLean, a CFIA staff member who conducted the audit, wrote that "the production of 10,000 acres of transgenic potatoes, as proposed by the seed-potato industry, would put the CFIA, the minister and the industry at risk as such large-scale production cannot be grown under adequate conditions of confinement and the environmental, food and feed safety of these transgenic potatoes has yet to be determined" (as cited in Laidlaw 2001, B1). Monsanto, however, rejected these requests, contending that the data it supplied with its regulatory application adequately supported its conclusions that these products presented no significant environmental, feed, or food safety risk (Tam 1999).

Rather than simply allow Monsanto's applications to lay dormant until the company provided what the government scientists considered to be key scientific data, a deal was brokered by government officials that promised Monsanto a decision on its submissions within thirty days if it provided the requested information. Yet, even afforded such consideration, Monsanto responded by simply reformatting the data it had originally submitted instead of producing the requested new data (Laidlaw 2001). The genetically engineered potatoes were subsequently approved by both the CFIA and Health Canada and were on the market in time for spring planting in 1999. Michele Brill-Edwards, a former drug regulator at Health Canada, asserts that this deal offers further evidence of the weaknesses of Canada's regulatory review system caused by industry interference in the system (as cited in Tam 1999, A1). Given poor market performance, including rejection by major potato processors such as McCain Foods and McDonald's, Monsanto discontinued marketing its genetically engineered potatoes in 2001, although its official position was that the decision was part of a corporate strategy to streamline its genetically engineered crop lines (Spears 2001).[20]

According to one observer, "the bio-elites reside strategically as an influential network of allies – recombining behind the closed doors of increasingly complex and consistently stacked regulatory systems, of myopic and self-interested industry associations, and of government bureaucracies unrepresentative of the general public" (Hindmarsh 2001, 51). In fact, most companies have waged a relentless battle to ensure that the prevailing regulatory standard of substantial equivalence is not replaced by a more stringent system, such as the precautionary principle, which places the burden of proof on the proponent of a particular technology for which regulatory approval is being sought. The situation is quite similar in the United States, according to Henry Miller, a US Food and Drug Administration official heavily involved with biotechnology from 1979 to 1994: "In this area, the US government agencies have done exactly what agribusiness has asked them to do and told them to do" (as cited in Kloppenburg 2004, 301).[21]

Thus, not only do we have a situation in Canada in which the government-sanctioned knowledge commons for biotechnology remains unidimensional, but this putatively authoritative knowledge violates basic scientific tenets to render it insufficient for stringent regulation.[22] As a further indication of the degree to which capital and government shape and limit the knowledge commons, this is quite troubling in itself as an informational concern, but once again we note that biotechnological issues also extend into the biological commons. It is precisely as a result of a purposely circumscribed knowledge commons that genetically engineered organisms receive regulatory approval for unconfined release, which brings with it a plethora of environmental and health dangers that impact on our terrestrial commons.

Evidence of Adverse Environmental and Health Effects from Genetically Engineered Crops

While it is certainly true that gene exchange between different types of conventional crops and their wild relatives has been a natural phenomenon for millennia, genetically engineered plants pose novel and more formidable concerns. As the remainder of this section documents, there is a growing body of scientific literature that outlines the dangers inherent in the release of genetically engineered organisms into the environment – dangers that are increasingly coming to light as more and more hectares of farmland are planted with genetically engineered crops (Faure, Serieys, and Berville 2002; Hall et al. 2000; Lavigne, Klein, and Couvet 2002; Mikkelsen, Anderson, and Jørgensen 1996; Nottingham 2002).[23]

For example, a group of Pennsylvania State University researchers conducted a three-year study that reinforces the extent of our still limited comprehension of the important interaction effects between genetically engineered crops and the environment. These scientists determined that a squash variety genetically engineered to resist three of the top viral diseases to which cultivated squash are susceptible actually rendered the genetically engineered variety more vulnerable to a fatal bacterial infection transmitted by cucumber beetles. When a viral infection sweeps a field, the genetically engineered plants remain healthy and are thus the preferred food choice for the beetles that are exposed to the bacteria through their digestive tracts. When feeding, the beetles create open wounds on the plant's leaves into which the bugs' feces are deposited, thereby infecting the plant. So while the genetically engineered plant might offer a solution to the viral diseases, it simultaneously creates an additional susceptibility to other plant problems (Sasu et al. 2009).

One outcome of the environmental release of genetically engineered plants and subsequent gene flow has been the occurrence of genetic contamination, which, as a number of studies indicate, is a matter of "when" rather than "if" (Belcher, Nolan, and Phillips 2005; Knispel et al. 2009; Marvier and Van Acker 2005). For example, one research team has ascertained that genetically engineered canola volunteers can survive and produce progeny as long as a decade after they are first sown, thus providing further compelling evidence about the dangers of genetic pollution and contamination (D'Hertefeldt, Jørgensen, and Pettersson 2008). Another study confirms that not only are herbicide-resistance traits frequently detected among escaped canola populations in regions where these crops are widely cultivated, but progeny plants are also developing resistance to multiple herbicides (glyphosate, glufosinate, and imidazolinone) because gene flow occurs with high frequency and intensity among different types of canola (Knispel et al. 2009). The most recent discovery in North Dakota reveals feral canola populations resistant to either glyphosate or glufosinate growing at sites along roads, gas stations, and grocery stores, quite removed from agricultural production areas (the researchers, who took 604 samples at eight-kilometre intervals along 5,000 kilometres of highway in the state, also found two plants that contained both transgenes) (Gilbert 2010).

Not surprisingly, herbicide-tolerant volunteer canola, which was detected on the farms of 38 percent of those involved in another research project, is now a major concern that influences Canadian prairie farmers' risk assessment of this biotechnology (Mauro and McLachlan 2008). A corresponding consequence of gene flow among canola varieties is that the pedigreed canola seed production system in the prairie provinces has been contaminated to such an extent that conventional canola seedlots are unable to guarantee the absence of genetically engineered traits in the seed they sell (Friesen, Nelson, and Van Acker 2003). As we saw previously, the contamination of canola seedstocks in this country has effectively removed this crop from the roster of organic producers. According to the GM Contamination Register, there were 311 publicly documented cases of genetic contamination in sixty countries across the globe between 1996 and 2012, forty-six of which involved the illegal release of genetically engineered organisms (see Figure 5.1).[24]

Another of the most pressing agricultural and environmental problems associated with genetically engineered seeds is the mounting emergence of herbicide tolerance and resistance in various weed species. It is certainly true that gene flow via seed or pollen is not unique to genetically engineered plants and that evolved resistance is a universal problem that goes beyond any one herbicide. However, the effects of these problems are increasing in both range and velocity in precisely those countries where genetically engineered crops have been adopted extensively. That is, widespread adopters of genetically engineered cropping systems that rely overly on a particular herbicide are being invaded by resistant weed species with greater frequency and across more and more hectares of arable land. Figure 5.2 offers a dramatic graphic account of the sharp rise in documented cases of herbicide resistance among weeds that chronologically corresponds closely to the development and marketing of seeds genetically engineered to be herbicide resistant.[25]

A weed variety is considered tolerant when it is able to survive and reproduce if treated with a particular herbicide at the normal use rate. Weed management practices for tolerant varieties thus require farmers to apply increasingly higher doses of herbicide or to spray multiple applications of different herbicides – a growing problem elaborated in Chapter 6. Herbicide resistance confers upon the weed species the ability to withstand even higher pesticide application rates. Herbicide resistance can be acquired through outcrossing to wild crop relatives, through natural selection, and from genetically engineered remnant plants that emerge in subsequent crops. Outcrossing, or gene transfer, can occur between genetically engineered plants and their wild relatives as a result of wind-borne or insect-borne pollination.

Over the course of time, the regular use of a particular herbicide will invoke natural selection processes among individual plants present in the weed population that are resistant either naturally or because of random genetic mutations. These plants can then multiply and eventually achieve dominance in the weed population if the same chemical is applied long enough. Weeds that have developed resistance to a particular herbicide are referred to as "biotypes." A biotype might also develop cross resistance, meaning that the weed's ability to withstand a single herbicide and its associated mode of

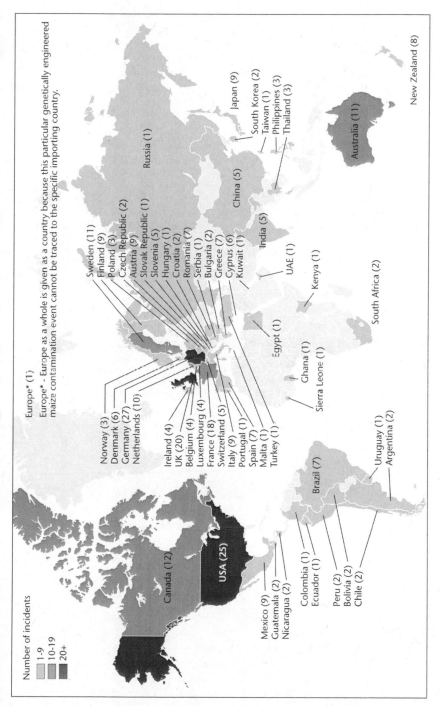

Figure 5.1 Worldwide incidents of GE contamination, illegal plantings, and negative agricultural side effects, 1996-2012. Data from GM Contamination Register.org.

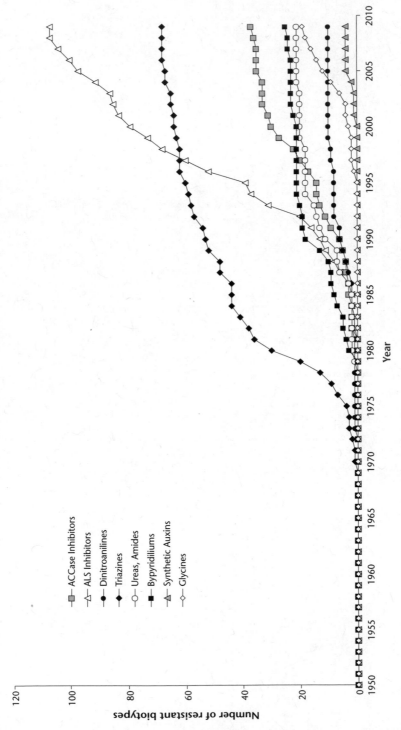

Figure 5.2 Documented cases of herbicide resistant biotypes. *Source:* Dr. Ian Heap from the International Survey of Herbicide Resistant Weeds.

action confers on it an additional imperviousness to all other herbicides within that particular herbicide group with the same mode of action. Multiple resistances may also occur among some weeds that develop resistance to several groups of herbicides with various modes of action. For example, a South Australian biotype of rigid ryegrass *(Lolium rigidum)*, the first identified glyphosate-resistant weed, has evolved multiple resistances to seven different herbicide modes of action.

The third type of resistance can develop when genetically engineered seeds spilled in fields either before or during harvest germinate in subsequent growing seasons. These genetically engineered remnant plants, referred to as "volunteers," typically have the capacity to crossbreed and thus pass on resistance genes to non-genetically engineered crops or acquire additional resistance genes from genetically engineered crops cultivated in following years. Volunteer plants are especially problematic in the case of seeds that possess long dormancy periods, such as canola. One recent Canadian study documented the introgression and persistence of an herbicide-resistant transgene into the gene pool of a weedy relative.[26] While the evidence does not indicate a substantial risk inherent in the particular plants studied, the authors make the point that such considerations (that is, gene escape and introgression among wild relatives) must figure more prominently in risk assessments of transgenic hybrids, particularly in the case of fitness-enhancing traits such as disease and pest resistance and tolerance to cold, drought, and salinity. Not only are these latter types of genetically engineered organisms still not well understood ecologically, but gene escape and subsequent introgression could also pose dire consequences for the surrounding environment beyond farmers' fields (Warwick et al. 2008).

According to the International Survey of Herbicide Resistant Weeds database, twenty-one species of weeds in eighteen different countries, including Canada, had developed resistance to glyphosate by early 2012.[27] Despite claims by Monsanto scientists in the late 1990s that the "evolution of glyphosate resistance ... would seem to be a low probability event," evolved resistance against glyphosate has, in fact, become especially problematic and prevalent given the popularity of this broad spectrum herbicide – in part due to the bundling practices discussed in Chapter 4 (Bradshaw et al. 1997, 195).[28] As Stephen Duke and Stephen Powles (2008, 323), who researched the agricultural applications of glyphosate, point out,

> most of the documented cases of evolved GR [glyphosate-resistant] weeds in the past 6 years have been in GR crops ... It is clear that in the USA, Argentina and Brazil (to a lesser but significant extent) the massive levels of adoption of GR crops means an overreliance on glyphosate for weed control across massive areas with insufficient diversity. Thus, there is a high selection pressure for resistance, and, consequently, glyphosate-resistant weeds

are evolving in these areas. Given the high popularity of GR crops, this process is likely to accelerate through the foreseeable future.

One of the major dilemmas in respect of evolved glyphosate resistance from an agronomical perspective is that scientists still do not understand clearly the resistance mechanisms in weeds or the genetics of the resistance traits (Johnson et al. 2009; Mortensen et al. 2012). Even the advocates of glyphosate point out that selection pressures and the emergence of glyphosate-resistant weed populations are serious and growing threats in areas of the world where agriculture relies on intensive planting of seeds genetically engineered to resist this herbicide (Gaines et al. 2010; Johnson et al. 2009; Powles 2008, 2010; Powles and Preston 2006).

Since resistant weeds are not absolutely impervious to glyphosate (or any other herbicide to which they might have evolved resistance), some advocate new weed management practices that include higher doses of glyphosate and tank-mixing glyphosate with a different mode of action in combination with tillage, crop rotation, and other diverse methods of weed management (Sammons et al. 2009). Increased tillage would belie one of the original benefits claimed of glyphosate-resistant crops, namely the ability to engage in no-tilling production. Tank mixing poses its own problems, such as evolved resistance against the other herbicides used and adverse interaction effects. The clear favourite among the putative solutions offered by Douglas Sammons and his colleagues (2009), who happen to be Monsanto employees, is to rely on glyphosate at full dose and in sequential applications.

However, this practice had, until only very recently, become increasingly problematic for agricultural producers given the escalating price of Roundup over the last few years. It also completely ignored the environmental impact of heavier herbicide loads on farmlands. As more and more weed species develop resistance as a result of increased applications of glyphosate, Monsanto's main competitors, BASF, Syngenta, Bayer CropScience, DuPont, and Dow AgroSciences, are now pouring large sums of research and development dollars into developing new herbicides to address the "glyphosate gap." Remarking on the business opportunities offered by herbicide resistance, the chief executive officer of Syngenta's Crop Science has been quoted as contending that "resistance is actually quite healthy for our market, because we have to innovate" (as cited in ETC Group 2008, 16). This admission helps explain the most recent and ongoing efforts by several of the biogopolists to develop new genetically engineered cultivars designed to resist other herbicide chemistries such as dicamba (another Monsanto product) and 2,4-D (2,4-dichlorophenoxyacetic acid, which is owned by Dow AgroSciences). Yet as several scientists point out, and despite claims by industry, it will undoubtedly only be a matter of time before weed species begin evolving resistance to these herbicides should their rates of application increase to the same precipitous levels as experienced with glyphostate over the last fifteen years (Mortensen et al. 2012).

More problematic than increased costs, at least from an environmental perspective, are the negative impacts on soil associated with glyphosate, as documented by a number of soil scientists. For example, Robert Kremer, a microbiologist housed at the University of Missouri who works for the US Department of Agriculture's Agricultural Research Service, and his colleagues have determined that, although crops sprayed with glyphosate appear normal on the surface, this herbicide impinges negatively on root growth and root-associated microbes (Kremer and Means 2009; Kremer, Means, and Kim 2005; Yamada et al. 2009). Other studies have found that glyphosate can reduce the uptake and transport of calcium, magnesium, iron, and manganese in both genetically engineered and non-target (that is, non-glyphosate resistant) plants (Cakmak et al. 2009; Eker et al. 2006; Neumann et al. 2006; Tesfamariam et al. 2009). As a result, these plants contain reduced levels of micronutrients beneficial to human health and which play important roles in the proper functioning of a plant's disease resistance mechanisms. Other scientific studies indicate that, under certain conditions, the application of glyphosate exerts negative pressures on plant growth and micronutrient status, even among crops (soybeans) genetically engineered to be resistant to the herbicide (Bott et al. 2008; Zobiole et al. 2010).

Others have documented not only reduced nutrient efficiency but also an exacerbation of disease severity as a result of glyphosate-induced weakening of plant defences and increased pathogen population and virulence (Johal and Huber 2009). In particular, four primary types of soil fungi *(Fusarium, Phythium, Rhizoctonia,* and *Phytophthora)* have been found to become more active when glyphosate is present in the soil. Specific to Canada, studies have documented consistent associations between previous glyphosate applications in fields and increased presence of *Fusarium* pathogens responsible for diseases such as root/crown rot and *Fusarium* head blight, although determining the relative contribution of glyphosate to increased *Fusarium* infection requires additional research (Fernandez et al. 2005, 2009).

For the first time ever, scientists at the University of Guelph have documented the persistence of glyphosate-tolerant transgenic maize DNA residues in animals within a soil food web (Hart et al. 2009). Although the study has left open questions about whether the transgenic DNA remained within plant residues; was present as free, extracellular DNA; or had been transformed into native soil bacteria, it has made clear that the *cp4 epsps* gene genetically engineered into Roundup Ready corn does not degrade significantly. As almost all of these scientists point out, their findings beg further research dedicated to answering the question of what might be the long-term impact of glyphosate on soil health and, thus, on future crop production systems.

Yet it is not only glyphosate-resistance genes that are causing problems in farmers' fields. As several studies have documented, a growing subset of pest populations is also beginning to develop resistance to particular Bt toxins genetically engineered into cotton and maize seeds. For example, researchers have discovered field-evolved resistance in the southeastern United States among cotton bollworm (*Helicoverpa zea* – a major lepidopteran pest)[29] to a Bt toxin produced by transgenic cotton (Tabashnik et al. 2008) as well as cross-resistance between Cry1Ac and Cry2Ab in various important cotton pests (Tabashnik et al. 2009). In early 2010, Monsanto disclosed to India's Genetic Engineering Approval Committee that pink bollworm has developed resistance to its genetically engineered Bollgard I cotton variety in the province of Gujarat (Bagla 2010; see also Dhurua and Gujar 2011).

In the case of genetically engineered traits against pests, plant-produced pesticides tend to induce selection pressure more vigorously than synthetic biocides, resulting in the "genetic treadmill" discussed previously, in which continued dependence on proprietary genetic technology substitutes for sustainable agricultural management practices. This is clearly the case in India, where, in response to evolving resistance against its Bollgard I cotton variety, Monsanto has recommended that farmers migrate to Bollgard II seed varieties. However, drawing on five years of research data, scientists in Australia have already discovered field-evolved resistance to the Cry2Ab protein genetically engineered into Monsanto's stacked Bollgard II cotton seed (Downes, Parker, and Mahon 2010).

Similarly, field-evolved resistance to Cry1F in Bt corn has been documented in Puerto Rico, and resistance to Cry1Ab in Bt corn has been found in South Africa (Storer et al. 2010; Tabashnik, Van Rensburg, and Carrière 2009; van Rensburg 2007). Other studies indicate that the Cry1Ab protein, which is toxic to the European corn borer (*Ostrinia nubilalis)* and genetically engineered into Monsanto's Bt corn MON810, also exerts a negative impact on the corn earworm *(Helicoverpa zea)*. However, the corn earworm, aside from feeding on corn, is also a predator of other pest insects such as the western bean cutworm *(Striacosta albicosta),* which is not susceptible to the Cry1Ab protein. As a result, western bean cutworm is losing a natural competitor in areas where substantial amounts of Monsanto's YieldGard corn are grown. A number of scientists, who have studied the extensive spread of the western bean cutworm across corn-growing areas in the United States over the past few years, hypothesize that genetically engineered corn, including YieldGard, has affected the competitive success of the cutworm, making it the more damaging and geographically extensive pest (Catangui and Berg 2009; Dorhout and Rice 2010; Then 2010).

In Iowa, researchers recently detected field-evolved resistance to Cry3Bb1 maize among the western corn rootworm *(Diabrotica virgifera virgifera)*, which is one of the most serious maize pests in the United States. This is also the first time that field-evolved Bt resistance has been observed within a coleopteran species, as all previous resistance has involved Lepidoptera (Gassmann et al. 2011). Similarly, scientists at agricultural extension offices at the University of Illinois and the University of Minnesota have described substantial performance problems with Bt corn rootworm traits that growers have detected over the last couple of years (Gray 2011a, 2011b; Ostlie and Potter 2012). According to an internal memorandum dated 22 November 2011 from a staff ecologist to a senior regulatory action leader at the US Environmental Protection Agency, scientists at the agency suspected evolved resistance to Cry3Bb1 among western corn rootworm in four different states (Iowa, Illinois, Minnesota, and Nebraska). This same team of scientists was critical of Monsanto's overall resistance monitoring strategy, concluding that it was inadequate and likely to miss early resistance events. Based on Cry3Bb1 performance inquiries reported to Monsanto by growers, these scientists recommended that future monitoring be expanded geographically to include Colorado, South Dakota, and western Wisconsin (Martinez 2011).

Additional environmental dangers being determined include the ill effects of herbicide-tolerant crops on the surrounding bird, mammal, fish, and amphibian populations. Studies in the United Kingdom have tracked significant declines in bird and insect populations around farmlands planted with genetically engineered canola and sugar beets as compared to conventional farm plots (as discussed in Borromeo and Deb 2006). A recent laboratory and field study conducted by environmental scientists at Indiana University indicates that the toxin contained in genetically engineered Bt corn increases mortality and reduces the growth of caddisflies, aquatic insects related to the pest targets of the Bt toxin. These results raise a concern because caddisflies are a source of nourishment for fish and amphibians, and the pollen and other plant remnants from fields planted with Bt corn are being washed into neighbouring streams and other waterways (Rosi-Marshall et al. 2007; Tank et al. 2010).[30]

Other studies demonstrate that genetically engineered maize containing the Bt toxin Cry1Ab increases mortality and reduces female sexual maturation and overall egg production levels among the water flea *Daphnia magna*, a crustacean anthropod typically used as a model organism in ecotoxicological studies (Bøhn et al. 2008; Bøhn, Traavik, and Primicerio 2010). Although not specific to genetically engineered crops, other research has found that Roundup herbicide is highly lethal to amphibians and, thus, a potential danger to the vitality of amphibian populations (Relyea 2005). Bt cotton, which has been planted for a number of years already, is also beginning to demonstrate problems. Not only have pest and natural enemy populations been affected by the toxins genetically engineered into the cotton, but so too have the beneficial parasites and populations of secondary pests (Hisano 2005). Indeed, as systematic reviews of the scientific literature on Bt crops demonstrate, scientists still do not have a complete understanding of the effects of transgenic insecticidal proteins on all of their natural enemies, including some Dipteran species (Lövei and Arpaia 2005; Lövei, Andow, and Arpaia 2009). As Lövei (2009, 301-2 [emphasis added]) and his colleagues point out,

> it is not yet possible to infer toxin specificity from toxin structure, and thus toxin specificity of a Cry toxin is a scientific hypothesis, *not a scientific fact* ... Based on our review of the literature, it is clear that *conclusions that Bt and PI [proteinase inhibitors]*[32] *transgene products have "no harm" to natural enemies are currently overgeneralized and premature.*

Other studies have documented the persistence of Bt toxins in soils, which could threaten the long-term health of the latter (Desser 2000; McAfee 2003). The Navdanya organization in India recently completed a study of the effects of Bt cotton on soil fertility, in which researchers compared twenty-five fields planted with genetically engineered cotton for the past three years to adjoining fields where traditional cotton varieties or other non-genetically engineered crops were grown. The selected test plots, which were based in regions that account for the highest concentration of Bt cotton harvests in India, showed significant reductions in soil fertility and naturally occurring organisms and other enzymes vital to the maintenance of soil fertility: actinomycetes levels were reduced by 17 percent, bacteria by 14 percent, microbial biomass by almost 9 percent, acid phosphatase by over 26 percent, and nitrogenase by almost 23 percent (Navdanya 2009). As the authors of the study point out, the lack of research on the soil impacts of genetically engineered crops, including within the regulatory process, represents a significant and very dangerous lacuna in debates about agricultural biotechnology.

A recent policy paper in the journal *Science* that was drafted by seven environmental scientists in the United States levels similar criticisms, contending that the data available are too vague to permit an accurate assessment of the benefits and drawbacks of genetically engineered crops (Marvier et al. 2008). As the editors of *Scientific American* ("A Seedy Practice" 2009, 28) bemoan,

> [u]nder the threat of litigation, scientists cannot test a seed to explore the different conditions under which it thrives or fails. They cannot compare seeds from one company against those from another company. And perhaps

most important, they cannot examine whether the genetically modified crops lead to unintended environmental side effects. Research on genetically modified seeds is still published, of course. But only studies that the seed companies have approved ever see the light of a peer-reviewed journal. In a number of cases, experiments that had the implicit go-ahead from the seed company were later blocked from publication because the results were not flattering ... It would be chilling enough if any other type of company were able to prevent independent researchers from testing its wares and reporting what they find ... But when scientists are prevented from examining the raw ingredients in our nation's food supply or from testing the plant material that covers a large portion of the country's agricultural land, the restrictions on free inquiry become dangerous.

Indeed, the findings documented by these various studies belie industry and government regulators' claims that genetically engineered organisms receive marketing approval only after being subject to comprehensive and stringent safety assessments. Instead, such new research underlines the logic of applying the precautionary principle to biotechnological products in the absence of a clearer understanding of the consequences of gene flow; the effects on waterways and the aquatic organisms living within or around them caused by runoff material from fields planted with transgenic seeds; and the interaction effects between genetically engineered crop systems and the environment, including soil fertility.

This level of disquietude regarding genetically engineered crops assumes added significance when we consider that even sanguine proponents of this technology point out that "it seems inevitable that [genetically engineered] volunteer plants and cross-pollinated varieties will be co-mingled in the commodity food system" (Smyth, Khachatourians, and Phillips 2002, 539). Given this seeming inevitability, a number of researchers anticipate a variety of major health and safety concerns specific to human ingestion of genetically engineered foods, including the following: higher levels of food toxicity, masked allergens,[32] a nutritional decline in food as a result of an increasingly genetically homogenous food supply, potential food shortages due to reduced biodiversity, and an increase in antibiotic resistance (Ferrara and Dorsey 2001; Ho 1999).

A widely referenced study conducted by the internationally recognized lectin specialist, Arpad Pusztai, contradicts the notion of "substantial equivalence" for transgenic foods. This study, which was carried out at the Rowett Research Institute in Scotland, investigated whether potatoes that had been engineered to express an insecticidal compound present human health effects. Using rats, Pusztai and his team compared the nutritional composition and effects of transgenic potatoes to non-altered potatoes. His research determined that protein, starch, and sugar levels in the genetically engineered potatoes varied by as much as 20 percent compared to conventional potatoes. More importantly, the rats fed the altered potatoes experienced weight reductions in a number of vital organs, including the intestine, pancreas, kidneys, liver, lungs, and brain. These same rats also suffered a significantly depressed immune system. Evidence of intestinal inflammation and infection was also detected. Pusztai, having found that potatoes sprayed directly with lectin did not produce the same effects in lab rats, concluded that the negative responses found stemmed from some other part of the DNA construct employed to facilitate the genetic engineering, such as the viral promoter from cauliflower mosaic virus that is commonly used to transport lectin genes into the recipient (Ewen and Pusztai 1999).[34]

Findings determined by a different group of scientists suggest that diets composed of significant amounts of genetically engineered soybeans can influence pancreatic metabolism in mice (Malatesta et al. 2003). A more recent study that examined the effects on mice fed diets of genetically engineered soybeans demonstrates the effect this crop can exert on liver features during the ageing process. As the scientists who conducted the research point out, because the influence mechanisms remain unknown, it is vitally important that additional investigations be conducted in order to ascertain the long-term consequences of genetically engineered diets as well as any potential synergistic effects among ageing, xenobiotics, and/or stress conditions (Malatesta et al. 2008).

David Schubert (2008) examined some of the human health implications of consuming food products that are genetically engineered to produce elevated levels of various molecules such as vitamins, omega-3 fatty acids, and amino acids.[35] The major findings presented in this study demonstrate that enhancing metabolic pathways through gene overexpression can result in unexpected and unpredictable consequences that potentially impact human health negatively. On the basis of his findings, Schubert (2008) concluded that the genetically engineered product and the genetic engineering process, itself, should be subject to scientific assessment as part of any regulatory review. Other researchers, who have detected unintended compositional changes in genetically engineered rice, have reached similar conclusions (Jiao et al. 2010). These studies critically disarm the typical contention advanced by biotechnology proponents that the risk of genetically engineered crops is no larger than the risk inherent in conventional breeding methods. As interview respondent and agricultural scientist Ann Clark points out, equating a lack of evidence that genetically engineered foods do not pose health risks with the contention that they are thus safe is false, given that we have almost no data and certainly no valid assays that compare the food safety profiles of genetically engineered products against those bred conventionally. It will only be through rigorously developed and executed epidemiological studies that scientists will be able to detect any human health implications posed by these genetically engineered crops. This is why labelling is so important. Without it, post-market surveillance of adverse health reactions to genetically engineered food is rendered impossible.

In March 2003, Monsanto received regulatory approval for unconfined release of its MON 863, an insect-resistant corn variety that produces the Cry3Bb1 protein. Two weeks subsequent to the approval from the CFIA, Health Canada approved MON 863 for human consumption. This transgenic corn, which is engineered to produce an insecticide that renders the plant resistant to western corn rootworm *(Diabrotica virgifera)* and northern corn rootworm *(Diabrotica barberi),* has been the subject of health concerns since at least 2002, when experts at the French Genetic Engineering Commission began articulating critical reservations about the test data provided by Monsanto in support of its application for European regulatory approval of MON 863. Simultaneous studies were emerging in Germany that linked the Cry3Bb1 protein to other comparable proteins that exhibit high relevance for human health.

After three years of lobbying by Greenpeace and other activists in Europe, and despite intensive pressure by Monsanto to maintain the confidentiality of the results included in its application for regulatory approval, an order from a German Court of Appeal (Münster) compelled authorities in 2005 to release the original data provided by Monsanto for independent testing (Brandon 2007). At around the same time, a story broke in the United Kingdom when a secret Monsanto study was made public. According to the 1,139-page confidential report, Monsanto's own scientists had determined that rats fed its genetically engineered corn MON 863, compared to control group rats fed a non-genetically engineered diet, had developed smaller kidneys and abnormalities in their blood composition. According to doctors and researchers, the changes in blood composition could be indicative of damage to the rat's immune system or perhaps tumour growth that the immune system was fighting against (Lean 2005).

An ensuing study conducted by the Comité de Recherche et d'Information Indépendantes sur le Génie Génétique (Crii-Gen) (Committee for Independent Research and Information on Genetic Engineering), the results of which appear in a 2007 peer-reviewed article in the *Archives of Environmental Contamination and Toxicology,* disputes the claim made by Monsanto that this genetically engineered variety exhibits no significant biological differences from conventional corn. In addition to criticizing the methodology of the study conducted by Monsanto researchers, these independent scientists determined that "the two main organs of detoxification, liver and kidney, have been disturbed [by MON 863] in this study. It appears that the statistical methods used by Monsanto were not detailed enough to see disruptions in biochemical parameters ... we strongly recommend a new assessment and longer exposure of mammals to these diets, with cautious clinical observations, before concluding that MON 863 is safe to eat" (Séralini, Cellier, and de Vendômois 2007a, 601; see also Séralini et al. 2009). The particle bombardment method used by Monsanto to inject engineered genes into the seed was also problematic, according to these scientists. Beyond the inherent imprecision of this process, it also "may cause insertional mutagenesis effects, which may not be directly visible by compositional analysis" (Séralini, Cellier, and de Vendômois 2007a, 597).

As these researchers point out in a later article that responds to critiques levelled by biotechnology proponents, even with the methodological weaknesses of industry-provided data, there are clear signs of toxicity associated with genetically engineered diets. On the basis of such findings, and given the pronounced limitations of past research, dangerous lacunae exist regarding the possible health implications of genetically engineered foods that are begging to be addressed by additional methodologically and analytically rigorous research (Séralini et al. 2009). One suggestion for detecting possible insertional mutagenic effects is to compare genetically engineered plants not only to their parent plants but also to parent plants genetically engineered to express an empty construct – that is, a vector that is similar in all ways to the one used in the genetically engineered plant except that it does not contain the transgene. Thus, any effect observed in the plant engineered with the empty construct, but not in the parent plant, would likely be caused by the genetic engineering process itself (Pryme and Lembcke 2003). Given the growing consensus among the scientific community that placement within the genome has important implications for phenotypic expression, this method of genetic engineering alone should be enough to warrant *sui generis* regulations for genetically engineered products and, at a minimum, more stringent and independent review of the scientific data being produced and submitted by these biogopolists to regulators.

Greenpeace has provided the results of this independent testing to the CFIA and to all Canadian federal, provincial, and territorial ministers of agriculture. Not only has the CFIA not issued a response, but its website continues to display Supplemental Decision Document DD2003-43 from March 2006, which outlines the agency's decision to permit the indefinite and unconfined release of MON 863. The agency has also yet to state whether the data provided by Monsanto in support of its application in Canada – data that, no doubt, are very similar to those provided to European regulators – will be made publicly available (Brandon 2007). Given the track record of the Government of Canada in protecting confidentiality in what it broadly considers proprietary information, it seems very unlikely that this information will be forthcoming without some form of judicial compulsion.[37]

In the course of further study, the French Crii-Gen research group, in a preliminary report issued in June 2007, voices deep concerns about the methodological design and resulting interpretation of test data for the studies conducted by Monsanto and provided in support of its application for approval in Europe of NK 603, another Roundup-tolerant corn variety (NK 603 was approved in Canada in 2001) (Séralini, Cellier, and de Vendômois 2007b). In their latest follow-up study that examined the effects of three genetically engineered corn varieties (MON 810, MON 860, and NK 603) on mammalian health, these researchers again found evidence of hepatorenal toxicity among rats fed these three types of genetically engineered maize. They noted negative impact on kidney and liver function as well as some effects on heart, adrenal, spleen, and blood cells. Yet, as the study's authors point out, they were unable to determine whether the toxicity they discovered was a result of the different pesticide residues specific to each genetically engineered event (Cry1Ab in MON 810, Cry3Bb1 in MON 863, and glyphosate and AMPA in NK 603) or of the unintended metabolic effects induced through the genetic engineering process. For these reasons, the researchers strongly recommended that additional long-term studies, preferably multigenerational, be conducted using multiple mammalian species. In the absence of such research, the scientific community lacks valid data about the acute and chronic toxicity effects of genetically engineered crop varieties (de Vendômois et al. 2009; Séralini et al. 2009).[36]

Heeding their own call, Séralini and colleagues recently completed the first long-term study of the health impacts of Monsanto's genetically engineered maize (NK 603) on rats. The two-year study, which tested for health effects at three different doses of NK 603, also investigated the health consequences among rats of direct exposure to Roundup herbicide (Séralini et al. 2012). Overall, Séralini and his research team determined that the rats fed diets of Monsanto's genetically engineered maize developed tumours larger than, and at rates two to three times greater than, the rats fed non-genetically engineered diets. Pathological symptoms were detected in rats at all three dose levels (11, 22, or 33 percent NK 603 maize added to the dry rat feed), which suggests a threshold effect at low levels of genetically engineered maize consumption. Although some tumours began appearing at four to seven months, the majority were observed after eighteen months. Similarly, the rats exposed to Roundup herbicide, even at doses below currently acceptable European safety limits, experienced severe hormone-dependent mammary, hepatic, and kidney disturbances. These findings underscore the complete inadequacy of the standard ninety-day feeding trials currently employed by industry and regulators to evaluate the potential toxicity of genetically engineered crops (ibid.).[37]

In a study of the effects of genetically engineered corn on Atlantic salmon, a Norwegian research group determined that Monsanto's Bt maize MON 810 (which contains the transgenic protein Cry1Ab that provides resistance against the European corn borer *[Ostrinia nubilalis]*) affects white blood cell populations, which has implications for immune response (Sagstad et al. 2007). The study's authors further point out that the insertion of transgenic DNA could result in the production of unknown proteins during the engineering process, which is particularly problematic given that transgenic DNA sequences of varying sizes have proven to survive feed processing and remain in the digestive tract of Atlantic salmon. In fact, this could be an even more widespread problem if Jack Heinemann (2009) is correct in his assessment that genetically engineered plant material can be transferred to animals exposed to such crops in their diets or environment and that there are residual differences in animals or products produced from such animals as a result of exposure to genetically engineered feed. Finally, the study, similar to previously published findings, demonstrates that MON 810 has effects on liver and intestinal activity, which are indicative of a mild stress response (Hall et al. 2000; Lavigne, Klein, and Couvet 2002; Sagstad et al. 2007). Another study conducted under the auspices of the Italian government's National Institute of Research on Food and Nutrition has documented substantial negative impacts on the immune system of young and old mice fed Monsanto's genetically engineered maize MON 810 (Finamore et al. 2008).

The willingness of some scientists to proffer answers about the dangers, or lack thereof, of releasing genetically engineered organisms into the ecosystem notwithstanding, science is currently unable to predict accurately the long-term effects that would accrue from such action. Yet despite this limitation, policy-makers continue to appeal to science as the neutral arbiter when deciding questions about biotechnology (von Schomberg 1998). As Philip Davies (2004, 75) writes, "[t]he hypothesis that GE crops will have no significant environmental impact may be a good method for the advancement of scientific knowledge. However, as a tool for making decisions about whether GE crops should be released into the environment, it lacks the caution necessary for responsible, long-term, environmental stewardship." Similar to Canadian critics, Hans Bergmans, former secretary of the Commission on Genetic Modification in the Netherlands, argues that existing field experiments with genetically engineered organisms have only proven that the experiments were carefully planned and have not, as proponents of genetically engineered organisms maintain, proven their safety for widespread release (as discussed in von Schomberg 1998).

Jeffrey Smith (2003) outlines a number of cases in the United States where biotechnology companies have designed their studies in such a manner as to produce results that would not indicate problems with genetically engineered foods. For example, Aventis heated its StarLink corn four times longer than standard before it tested for intact protein and also substituted protein derived from bacteria for protein from StarLink corn for testing purposes.

Monsanto fed adult animals a diet that contained only 10 percent of its protein derived from genetically engineered soy, and in another study Monsanto used stronger acid and more than 1,250 times the amount of a digestive enzyme recommended by international standards in order to garner evidence of how quickly a particular protein degraded. In tests looking at recombinant bovine growth hormone (rBGH), Monsanto researchers injected cows with only one-forty-seventh of the dosage of the growth hormone before testing for hormone levels in the milk and then pasteurized the milk 120 times longer than normal to see if the hormone was destroyed; Monsanto removed sick cows from rBGH tests and included cows that conceived prior to the testing to support the claim that the hormone does not affect fertility; and the US Food and Drug Administration ignored evidence of antibody reactions found in rats fed rBGH. These are just a few examples of dubious scientific data offered by industry in support of its various regulatory applications (Smith 2003, 2007).[38]

The overarching and compelling message of this emerging scientific research is that it is necessary that we investigate the long-term consequences of ingesting genetically engineered crops, including the potential for interaction effects with ageing, xenobiotics, and stress conditions (Malatesta et al. 2008). Claims by agricultural biotechnology proponents notwithstanding, Domingo (2007, 731) poses the very relevant question: "[W]here is the scientific evidence showing that GM plants/food are toxicologically safe?" Indeed, according to a position paper of the American Academy of Environmental Medicine, "[t]here is more than a casual association between GM foods and adverse health effects ... The strength of association and consistency between GM foods and disease is confirmed in several animal studies ... GM foods pose a serious health risk in the areas of toxicology, allergy and immune function, reproductive health, and metabolic, physiologic and genetic health" (Dean and Armstrong 2009, paras. 4, 9).

Accordingly, on 8 May 2009, the Executive Committee of the American Academy of Environmental Medicine approved a statement that asks:

- Physicians to educate their patients, the medical community, and the public to avoid GM foods when possible and provide educational materials concerning GM foods and health risks.
- Physicians to consider the possible role of GM foods in the disease processes of the patients they treat and to document any changes in patient health when changing from GM food to non-GM food.
- Our members, the medical community, and the independent scientific community to gather case studies potentially related to GM food consumption and health effects, begin epidemiological research to investigate the role of GM foods on human health, and conduct safe methods of determining the effect of GM foods on human health.

- For a moratorium on GM food, implementation of immediate long term independent safety testing, and labeling of GM foods, which is necessary for the health and safety of consumers. (Dean and Armstrong 2009, paras. 12-15)

In June 2009, the Irish Doctors' Environmental Association followed suit and issued a call for an immediate moratorium on genetically engineered foods and crops because of what it believes is insufficient testing for possible adverse health effects.

Unfortunately, the negative health implications of genetic-engineering production systems extend beyond concerns about the food crops themselves. Robert Bellé, who is associated with the National Center for Scientific Research in the United States and the Pierre and Marie Curie Institute in France, contends that there is a link between cancer and Monsanto's Roundup herbicide.[39] According to Bellé, "Roundup provokes the first stages that lead to cancer ... The tested doses were well below those which people normally use" (as interviewed in *The World According to Monsanto* 2008). A study conducted by Swedish oncologists similarly raises concerns about the linkages between exposure to glyphosate and non-Hodgkins lymphoma:

Gene mutations and chromosomal aberrations have been reported in mouse lymphoma cells exposed to glyphosate. Furthermore, the incidence of hepatocellular carcinoma, leukemia, and lymphoma was somewhat increased in one study on mice. In cultures of human lymphocytes, glyphosate increased the number of sister chromatid exchanges. Recently, we published an increased risk for hairy cell leukemia, a rare type of NHL *[non-Hodgkins lymphoma]*, for subjects exposed to glyphosate as well as for subjects exposed to other pesticides. For these reasons, glyphosate deserves further epidemiologic studies. (Hardell and Eriksson 1999, 1359)

Another study that evaluated the effects of four glyphosate-based herbicides on three different types of human cells provides evidence that "the proprietary mixtures available on the market could cause cell damage and even death around residual levels to be expected, especially in food and feed derived from R formulation-treated crops" (Benachour and Séralini 2009, 104). Moreover, these effects are most likely amplified by vesicles formed by adjuvants. Adjuvants, which are added to glyphosate-based herbicide formulations, allow the product to stick to leaves or other parts of the plant and aid in the absorption of the active ingredient by the plant's cells (Gasnier et al. 2009). Indeed, a number of other studies indicate that exposure to the Roundup formulation (not just the active ingredient glyphosate) interferes with estrogen synthesis at levels lower than those found in the recommended agricultural use (Richard et al. 2005), impinges on transcription (Marc et al.

2005), depresses respiratory activity (Peixoto 2005), disrupts steroidogenic acute regulatory protein expression that, in turn, inhibits steroidogenesis (the synthesis of steroids within the adrenal cortex, testes, and ovaries) (Walsh et al. 2000), and induces alterations in the mitochondria and the internal membrane of mitochondria in carp (Szarek et al. 2000).

One of the latest studies, which emerges from Argentina (a country ranked third globally in terms of the highest amount of land planted with genetically engineered crops), has found that both the active ingredient glyphosate and the glyphosate-based herbicide formulation (that is, glyphosate and adjuvants) interfere with critical molecular mechanisms responsible for regulating early development in both frog and chicken embryos (Paganelli et al. 2010). These findings are particularly troublesome since the mature human placenta has been determined to be pervious to glyphosate. Given the increasing evidence of growing weed resistance to Roundup and the consequent increases in its application, such findings portend serious future health risks for the people involved in growing these crops as well as those who live around such areas. This troublesome connection assumes an added scale of magnitude if we recall that Monsanto is the same company to bring us a variety of carcinogenic products such as polychlorinated biphenyls, dioxin, and the defoliant Agent Orange – all originally claimed to be benign for both the environment and human health. Moreover, the company is suspected by the Environmental Protection Agency of being responsible for over fifty Superfund sites in the United States (the American terminology for an uncontrolled or abandoned environmental site contaminated by hazardous waste) (Barlett and Steele 2008).

Given this burgeoning corpus of scientific data, governments need to take the implications of genetically engineered crops and food for the environment and human and animal health more seriously. The consequences of gene flow from pesticide-resistant crops must be subjected to more stringent scrutiny by regulators when developing policies and rules that police the environmental release of seeds with traits that could impact the ecosystem. As we have also seen, the inherent instability and incomplete understanding of the effects of various genetic engineering methods can result in the silencing of genes, altered levels of expression, or possibly the expression of extant genes that were not previously expressed. Inserted genetically engineered genes might similarly interact with existing genes to cause an unpredictable disruption of metabolic pathways or result in the creation of new toxic compounds or an increase of already existing ones.

Indeed, emerging epigenetic research refutes the essentialism inherent in the genetic reductionism paradigm and, instead, demonstrates that particular genes are only partially responsible for various biochemical processes within an organism. Epigenetic research contemplates and seeks to elaborate the levels of control that organisms have beyond a specific gene as well as the interaction effects between genes and the organism as a whole. This more holistic type of research is providing critical insight into the unpredictability and instability of genetic engineering techniques.

Overall, the studies highlighted earlier provide independent evidence not only that agricultural biotechnology companies, particularly Monsanto, conduct methodologically weak scientific assessments of their products but also that, in fact, genetically engineered crops pose both environmental dangers and health risks. Although the Canadian federal government claims that its scientists and regulators are keeping abreast of new developments in the field, it continually fails to act on any of this mounting negative evidence in respect of genetically engineered crops. Instead, it uncritically parrots the rhetoric employed by agricultural biotechnology producers about increased yields and nutritional content (Government of Canada 2007).

Formal Critique of Canada's Regulatory System and the Government's Response

These concerns about the efficacy of Canada's regulatory system for genetically engineered organisms have been raised formally by a group that includes the Council of Canadians, the Canadian Institute for Environmental Law and Policy, and interview respondents and agricultural scientists Ann Clark and Bert Christie. In a petition submitted on 9 May 2000 to the federal government under the *Auditor General Act*, these petitioners asserted that Canadian regulations and policies concerning genetically engineered organisms fail to ally with the principles adopted by the Government of Canada for sustainable social, economic, and environmental development (Office of the Auditor General of Canada 2000a).[40] The petitioners outlined the potentially adverse and irreversible environmental effects of genetically engineered organisms that might include: the evolution of new pests; an exacerbation of dangers posed by existing pests; ancillary harm to beneficial species; species extinction; and broader negative impacts on ecosystem processes and functions such as horizontal gene transfer and the accumulation of dangerous levels of residual endotoxins in soil.[41]

The Ecological Society of America released a similar list of possible ecological risks related to genetically engineered seeds in its 2004 position paper on genetically engineered organisms and the environment (Snow et al. 2005; see also Steinbrecher 2001). This potential for broader effects on overall ecosystems is particularly troubling because it means that unconfined release into the environment can pose dangers that were untested or unanticipated during controlled trial releases since the testing stage only collects and measures data that were actually contemplated during the development of the study's design. Yet large-scale commercial release of genetically engineered organisms into the environment can produce widely divergent effects on different ecosystems, in different years, with different crops, and at different scales of introduction. The petitioners cite international research studies that have documented negative environmental impacts stemming from genetically engineered organisms as well as other evidence demonstrating that the pesticide reduction and increased crop yield arguments touted as benefits of genetically engineered crops have been seriously overstated.

Studies began emerging as early as the late 1980s to suggest that weeds were evolving to become resistant to some new herbicides, thus calling into question agrochemical company claims that genetically altered seed strains reduce aggregate levels of herbicide use (Levidow and Tait 1995). The petitioners also reminded the Canadian government that the data it receives from companies seeking regulatory approval for their products, and upon which the approval process relies, fail miserably in addressing human health issues of toxicity and allergenicity. Seventy percent of the available crops approved in Canada as of 2000 did not test or measure for human toxicity, and no measure of allergenicity was provided for any of the forty then approved genetically engineered events. A further human health concern raised by the petitioners involved the potential development of antibiotic resistant pathogens as a consequence of inserting marker genes into genetically engineered organisms (Office of the Auditor General of Canada 2000a).

With regard to economic impact, the petitioners pointed out that, while the agricultural supply industry has been subject to significant consolidation that has served to increase profits for large multinational firms, little investment has been devoted to the development and promotion of more sustainable forms of agricultural production such as organic farming and integrated pest management. In fact, certain biotechnologies pose dangers to the viability of such alternative types of agriculture, as we saw in Chapter 4 with regard to organic canola. The social impacts of biotechnology, according to the petitioners, include concerns about interspecies genetic transfer, questions about the appropriateness of patents on genetic materials, and the pressure toward increased capital intensity of agriculture that flows from biotechnology and its attendant applications, which threatens the economic and social viability of rural communities (Office of the Auditor General of Canada 2000a).

In a set of internal CFIA memos obtained under the *Access to Information Act*, Bradford Duplisea determined that the Wheat, Rye, and Triticale Subcommittee of the Prairie Registration Recommending Committee for Grain (an advisory committee to the CFIA) had a "definition of merit" clause in its operating procedures since 1990 that permitted an assessment of new crop varieties according to their potential to pose production or marketing risks.[42] According to sections 3.2.3 and 3.3 of the committee's operating procedures, "candidates that introduce production or marketing risks for their own or for other wheat classes may be rejected regardless of merit in other traits." Once this operating procedure was discovered, the CFIA instructed members not to employ the clause, which was subsequently removed in 2002 (as discussed in Wilson 2003). As we saw previously, a legislative attempt to make economic impact a relevant and required regulatory assessment criterion (Bill C-474) was defeated by the Conservatives and the Liberals in the House of Commons.

In addition to illustrating the inherent conflict of interest that emerges from the dual regulatory and promotional roles of the CFIA, the petitioners were extremely critical of the fact that the government relies on industry-provided data and does not engage in any independent testing of products for which approval is sought. This fact, combined with the basic assumption driving the regulatory system that genetically engineered products do not differ in any substantial way from their natural correlates (that is, acceptance of the concept of "substantial equivalence"), establishes a relatively low safety and data requirement threshold for industry when seeking approval for its biotechnology products. These regulatory weaknesses are further aggravated by the lack of a labelling system for genetically engineered products in this country, which curtails post-market epidemiological monitoring and study.

The ultimate effect of such an anti-precautionary system is that the onus is shifted from business having to demonstrate adequate safety in the first instance to citizens having to demonstrate harm after the fact. In the case of genetically engineered organisms released into the environment, this situation can have far-reaching and devastating consequences. The petitioners therefore recommended that the Government of Canada implement a mandatory labelling scheme and overhaul the product review process to ensure comprehensive safety assessments based on independent testing and broad and inclusive safety standards (Office of the Auditor General of Canada 2000a). In its response, which was drafted jointly by the Departments of Agriculture and Agri-Food, Fisheries and Oceans, Industry, Environment, Health, and Natural Resources, the Government of Canada failed to engage in any depth with the multiple issues raised by the petitioners and, instead, rather dismissively parroted what appears to be a ministerial talking point that the existing regulatory system assesses the risk of biotechnology products from a sustainable development perspective (Office of the Auditor General of Canada 2000b).

Genetically Engineered Corn: The Most Recent Critique of Canada's Regulatory Regime

On 20 July 2009, Monsanto announced that the CFIA had granted regulatory authorization for its new SmartStax corn, which would be marketed under its new trait brand "Genuity." Developed through research collaboration and cross-licencing agreements with Dow AgroSciences, this genetically engineered corn is a stacked variety that contains eight different traits to render it both herbicide (glyphosate and glufosinate) and insect (root-worm and corn borer) resistant.[43] A fact sheet developed by the Canadian Biotechnology Action Network (CBAN) contends that the European corn borer and the corn rootworm are not serious pests in most Canadian corn fields, meaning that a major roll-out of these new seeds will force farmers to incur increased input costs to purchase patented technology of little effective value in their production processes. This expensive, partly superfluous technology no doubt helps explain some of the most recent producer backlash against Monsanto highlighted in Chapter 2. Nonetheless, given past experience in seed markets, whereby the reigning seed oligopolists increasingly removed non-genetically engineered varieties from the marketplace, a reasonable expectation is that Monsanto will begin limiting the supply of its current single, double, and possibly even triple-stacked products on the market to compel farmers to purchase its SmartStax technology. In fact, according to Trish Jordan, spokesperson for Monsanto Canada, SmartStax will be the basic platform for all of the company's future versions of the crop (as discussed in Mittelstaedt 2009a).

Within a week of this announcement, CBAN determined that the CFIA had failed to publish any "decision documents" for this new genetically engineered seed. The lack of a public "decision document," which summarizes the regulatory approval decision, led CBAN to conclude that the agency had failed to conduct an environmental risk assessment for the product. This failure is perhaps not completely surprising given that CFIA Directive 94-08 on the Assessment Criteria for Determining Environmental Safety of Plants with Novel Traits allows for the unconfined environmental release of all progeny and sister lines derived from the original transformation event. More specifically, section 2.4 of the directive states that proponents of a plant variety derived through intentional trait stacking – resulting from either intra-specific or inter-specific crosses between genetically engineered varieties previously approved for unconfined environmental release – need *merely notify* the CFIA of their intent to release the new stacked variety. The Plant Biosafety Office at the CFIA may request and review data to support the safe use of the genetically engineered plant in the environment.

The CFIA also reduced refuge area requirements around this new variety by 75 percent, from 20 percent to 5 percent.[44] These legally mandated refuge areas, which are planted with a non-Bt variety of the same crop, are designed to provide a buffer zone within 0.4 kilometres of any Bt crop in order to hinder the evolution of insect resistance to the toxin introgressed into the genetically engineered plants. According to the CBAN website, this decision by the CFIA, the rationale for which has not been articulated by the agency, seriously undermines current environmental stewardship rules for genetically engineered crops. The only public statement issued by the CFIA about its rationale for reducing refuge requirements for SmartStax can be found on its website:

> The CFIA has evaluated the potential impact on and risk to the environment of using a 5 per cent non-Bt refuge strategy for this product, and has concluded that a conditional authorization until December 31, 2012, of the use of this refuge poses minimal risk to the environment. During the interim authorization, in addition to applicable terms and conditions applied to the individual events in the stacked product, the proponents are required to conduct field evaluation of corn rootworm adaptation to this product, including monitoring of actual emergence of susceptible and resistant corn rootworm adults in the field.[45]

The utter folly of leaving post-market surveillance to the companies marketing these products requires no further elaboration.[46] Moreover, as entomologist, and recognized expert in insect resistance to Bt toxins, Bruce Tabashnik points out, "no one knows how much shrinking the refuge will speed up resistance. If you kept the refuge size the same, it would delay resistance, [because of the stacked traits] but shrinking the refuge will accelerate resistance. You have got a plus on one side and minus on the other" (as cited in Little 2009, paras. 14-15). The reduced refuge requirements might thus offset any potential added benefits conferred by the stacked seeds. Indeed, as scientists at the US Environmental Protection Agency point out, evolving resistance among western corn rootworm to Cry3Bb1 undermines the rationale for reduced refuge requirements of 5 percent for SmartStax and might ultimately compromise the other unrelated toxin genetically engineered into this seed to control the pest (that is, Cry34/35) (Martinez 2011). This emerging evidence, coupled with the environmental dangers of genetically engineered crops discussed earlier, renders astonishingly inexplicable (aside, of course, from presumed industry pressure) the CFIA's decision to reduce one of the few environmental requirements currently demanded by the Canadian regulatory system, a system judged insufficient by the Royal Society of Canada's report.

This charge is bolstered by the latest published refuge compliance data released by the Canadian Corn Pest Coalition, which conducts a biennial survey of corn growers. It appears that compliance levels with refuge requirements have slipped from a high of 85 percent in 2003 to a dismal 61 percent in 2009 (Dunlop 2010). The CFIA has also reported declining levels of compliance with the 20 percent refuge requirement, although its numbers are higher than those determined by the Canadian Corn Pest Coalition. The most recent field audits conducted by the CFIA in 2008 revealed refuge compliance levels of 76 percent, which are down from 94 percent compliance in 2004 and 2005 (Canadian Food Inspection Agency 2009).[47] However, the compliance figures cited in this latest report are much higher for 2004 than those reported in the actual 2004 audit document, which revealed that only 80 percent of interviewed Bt corn producers complied with the 20 percent refuge requirement. Why the figures were subsequently adjusted upward by 14 percent remains unclear.

In any event, actual compliance with refuge requirements is lower than the rates claimed by producers during field audits. Laboratory analyses conducted by CFIA scientists determined that 29 percent of growers audited by CFIA inspectors in 2008 had at least one refuge sample test positive for Cry protein (62 percent of these growers had all three refuge samples test positive). Effective compliance with refuge requirements among those who actually planted a refuge was thus actually as low as 71 percent in 2008, which represents a further decline from 2004, when the level of effective compliance was 77 percent (Canadian Food Inspection Agency 2004, 2009). Yet even these poor numbers may not be completely accurate since the CFIA audits only 1 percent of Bt corn growers and participation is completely voluntary, thus making it impossible to develop a proper random sample. As a result of the low levels of compliance found among growers during the 2008 field audits, on 4 March 2010 the CFIA sent letters of "non-compliance with the conditions of authorization of Bt corn products" to Dow AgroSciences Canada Inc., Syngenta Seeds Canada Inc., Pioneer Hi-Bred Production Ltd., and Monsanto Canada Inc. requesting each company to provide a written Corrective Action Plan to the Plant Biosafety Office of the CFIA within three months. According to a plant biosafety management analyst at the CFIA, all four companies furnished the agency with plans deemed satisfactory.[48]

Even more astonishing than poor audit practices and compliance levels, the CFIA does not actually monitor for the development of evolved insect resistance to genetically engineered crops. Instead, the agency leaves this task to public sector actors such as the University of Guelph, the Canadian Corn Pest Coalition, the Ontario Ministry of Agriculture, Food, and Rural Affairs, and private sector industry stakeholders.[50] According to Heather Shearer, a plant biosafety management analyst at the CFIA, since the agency does not engage directly in post-market surveillance for evolving resistance, it has not collected leaf samples from farmers' fields as part of its audit practices since 2008 (Shearer also pointed out that the agency attempts to make the most effective use of the resources available to it, which might be interpreted as bureaucratese for "we do not have enough resources to properly execute our oversight functions").[50]

Instead, in 2010 the survey focused on corn seed retailer awareness of insect resistance management (IRM) requirements and targeted the communications between developer and retailer, and between retailer and

grower.[51] In 2012, the survey will focus on grower awareness of IRM requirements, but, again, no leaf samples will be collected or analyzed to determine
actual producer compliance with refuge requirements. Based on records I
obtained from the CFIA under an access to information request, in addition
to low levels of willingness among growers to partake in audits conducted
by the agency to monitor compliance with refuge requirements attached to
seed authorizations, it appears that the companies producing these genetically engineered seeds are very slow to produce dealer lists to facilitate the
audit process and even more resistant to providing the biological reference
material necessary for laboratory analysis of the leaf tissue samples collected
during field inspections.

CBAN also maintains that Health Canada ignored recommendations set
out in the Codex Alimentarius Commission's *Guidelines for the Conduct of
Food Safety Assessment of Foods Derived from Recombinant-DNA Plants* by failing to conduct a food safety assessment for SmartStax. Although not specific
to stacked traits, the Codex Alimentarius guidelines nonetheless speak to
the potential for unintended effects when inserting foreign DNA sequences
into plants: "Unintended effects in recombinant-DNA plants may also arise
through the insertion of DNA sequences and/or they may arise through
subsequent conventional breeding of the recombinant-DNA plant. Safety
assessment should include data and information to reduce the possibility
that a food derived from a recombinant-DNA plant would have an unexpected, adverse effect on human health" (Codex Alimentarius 2003, para.
14). CBAN considers this failure to adhere to the Codex Alimentarius recommendation all the more surprising given the active role played by the
Canadian government in negotiating the guidelines. As Lucy Sharratt argues
in a CBAN press release, "our standards should be at least as high as Codex,
if not higher" (Canadian Biotechnology Action Network 2009b, para. 7).[52]

According to a statement issued by Health Canada, "when a company
chooses to breed or cross approved genetically modified plants with other
approved GM or non-GM plants, the company must inform Health Canada
only if there is a change in the safety of the product" (Mittelstaedt 2009a,
A6). Yet herein lies the rub – Monsanto's position is that, because each individual trait has been assessed and gone through prior regulatory review,
the subsequent combining of these particular traits does not trigger any
additional safety assessment requirements. As a further indication of the lax
Canadian regulatory regime for genetically engineered products, we thus
have a situation in which the regulator relies on the proponent of the technology to determine potential safety concerns, while that same proponent
simply contends that no additional safety testing is required.[53] According
to Trish Jordan, spokesperson at Monsanto Canada, and Brenda Harris, regu-

latory and government affairs manager for Dow AgroSciences Canada, the two companies provided only insect resistance modelling data and a proposed insect resistance monitoring plan as part of their notification to the CFIA (Jordan and Harris 2009). Yet, as one group of scientists studying stacked seeds argues, "it is recommended to at least carry out agronomic, morphological and compositional studies on the GM stacked event in order to identify potential adverse effects that might result from interbreeding of GM cultivars. These studies will reveal if the phenotype and the composition, including the amount of naturally occurring allergens and toxins, of the GM stacked event will be equivalent to its comparators" (De Schrijver et al. 2007, 108; see also Wilson, Latham, and Steinbrecher 2006).

Of course, when it comes to patenting, Monsanto's forked tongue once again takes centre stage. The company is seeking intellectual property protection for its SmartStax seed on the basis that, by grouping these previously patented individual traits into one seed, it has developed an innovative new product that meets the threshold for patentability. Similar to the way in which the company employs a shifting definition of "natural" depending upon whether it is dealing with government regulators or patent examiners (as discussed in Chapter 4), here too we see the subjugation of the "novelty" definition to similar explicatory circumvolution.

In response to these glaring instances of failed regulatory oversight, CBAN sent a letter to Health Canada on 28 July 2009, asking that the department immediately rescind its authorization of SmartStax and initiate a full food safety assessment of the product according to the Codex Alimentarius guidelines. Noting the regulatory system's weaknesses in identifying potential risks of stacked-trait products, the letter also requested that the department direct the CFIA to place a moratorium on the approval of any further stacked-trait seed varieties until Health Canada has reviewed both its Novel Food category under the *Food and Drug Regulations* and the entire regulatory system for genetically engineered foods and crops.[54]

In addition to this formal letter, CBAN has issued a number of press releases to alert the public about the lax safety and environmental assessments of SmartStax corn, developed fact sheets about this new genetically engineered seed, placed all of this material on its website, and organized an electronic letter writing campaign for Canadians to register their disapproval of this product with the ministers of health and agriculture and agri-food; the president of the CFIA; and the health critics of the federal Liberal, Bloc Québécois, and New Democratic Parties. As part of its expert lecture series, on 1 December 2009, CBAN hosted a public event in Ottawa titled "Unsafe genetically modified corn? Canada's Lack of Assessment for Monsanto's 'SmartStax' GM Corn."

Conceptualizing the State through the Lens of Primitive Accumulation

> The sum total of these relations of production constitutes the
> economic structure of society, the real foundation, on which arises
> a legal and political superstructure and to which correspond
> definite forms of social consciousness.
>
> .   – Karl Marx, "Preface to a Contribution to the Critique
>   of Political Economy"

Despite the almost hackneyed reliance by more orthodox Marxists and critics alike on that infamous passage from the "Preface to a Contribution to the Critique of Political Economy" that privileges the economic base over the politics of the superstructure, Marx's broader oeuvre suggests a more nuanced conception of the relationship between the state and the economy through which the political cannot simply be read as a reflection of the economic but, rather, must be understood in terms of the development of the capital relation.[55] That is, the economic and the political are both forms of social relations, which, in capitalist society, are based upon the antagonism and exploitation inherent in the basic capitalist production relationship. Put more explicitly, the capitalist mode of production and the bourgeois state form are organic to one another rather than merely conjuncturally related (Bologna 1979; Harvey 1976; Tronti 1979). Any analysis of capitalist state forms must therefore find its basis in the class character of capitalist relations of production (Holloway and Picciotto 1978).

An understanding of state actions demands a materialist theory of the state through which the political form can be derived through an analysis of the social relations of production.[56] Rather than conceptualize the state as a captured institution that functions merely as capital's handmaiden, a more fruitful line of investigation seeks to understand the state's relationship to capitalist accumulation and crisis. These, in themselves, are premised on an underlying antagonistic set of social relations driven by private property as well as the imposition of work and consequent extraction of surplus value through the commodity form (Holloway and Picciotto 1978). Put another way, behind the obvious ensemble of institutional arrangements typically attributed to the state, a deeper understanding conceives of the state as a facet of the antagonistic class relation inherent in capitalist forms of social reproduction. As Simon Clarke (1983, 118) contends, "the state does not constitute the social relations of production, it is essentially a regulative agency, whose analysis, therefore, presupposes the analysis of the social relations of which the state is regulative. The analysis of the capitalist state conceptually presupposes the analysis of capital and of the reproduction of capitalist relations of production, despite the fact that in reality, of course, the state is itself a moment of the process of reproduction."

The advantage of this approach is that it avoids the functionalist economic reductionism inherent in some Marxist theory in which state activities are considered to emerge from the requirements of capital, while also rejecting those approaches that maintain a strict binary division between the political and the economic (in which theorists concerned with the former realm of activity omit any scrutiny of capitalist accumulation processes). Instead, and in a dialectical fashion consonant with Marx's own method and conceptualization, social relations of production are given existence through forms of economic, legal, and political relations. The fundamental and antagonistic class relations that emerge from practices of primitive accumulation – through which the actual producers are separated from their products and processes of production – directly impact the political, economic, and ideological forms through which these analytically prior class relations express themselves (Burnham 2006). By construing the relationship between the state and the market as both internal and necessary, rather than external and contingent, we are able to conceive of the state as an element of the social relations of production (albeit with different institutional forms based upon the historical disposition of the class struggle). It similarly permits us to encompass the complexity of the relations between the economic and the political within our analysis and assess them as complementary forms of the fundamental class relation based upon the prevailing social relations of production.

As we have seen, the Canadian government has long proclaimed the importance of biotechnology as a motor for economic growth and development. By similarly accepting and promoting capitalist control of this technoscience, the government has been compelled to operate in ways that ensure the success of this industry in contributing to capitalist accumulation. Viewed from this perspective, we can thus construe the state as the political form, through whose actions and policies the circuit of capital might be safeguarded – keeping in mind, of course, that struggle is always contingent and open (Burnham 2006). Yet we need to remember that the state, embodied in this particular (capitalist) form of social relations, depends on capitalist reproduction for its existence. This dependence represents a structural constraint on the state in terms of its room to manoeuvre. Since it occupies a space beyond the immediate production process, the state is typically only able to react to outcomes of capitalist processes of production and reproduction. The overall result is that the state's actions and specific functions emerge in a reflexive manner that reacts to the ways in which the processes of accumulation unfold (Hirsch 1978; Holloway and Picciotto 1978). We, therefore, note that the power of the state thus derives from its capacity to safeguard the capitalist law of property and contract in ways that organize labour-capital relations most optimally for purposes of capital accumulation.

Harkening back to the dominant conceptual theme of primitive accumulation that informs the analysis in this book, we note that, at the most basic level of the divorce between the producers and their means of production, Marx provides a conceptual lens through which to view the entirety of social relations, including their changing political form. Indeed, in his account of primitive accumulation, Marx takes great pains in outlining the historical applications of state power through such things as law and brute force, which helped to create the social conditions necessary for capitalist relations of production and exchange. Thus, despite the oft-attributed, and perhaps misunderstood, quote about base and superstructure in "Preface to a Contribution to the Critique of Political Economy," there is a dialectical interaction between the two rather than the base conjuring the existence of the superstructure sequentially.

Recalling the earlier discussion about the forms of alienation consequent to processes of primitive accumulation, we note similar outcomes in respect of the state. In tandem with the balance of class forces, the power of the state to vouchsafe private property and guarantee and enforce the law of contract legitimizes the expropriation of the direct producer's product, the bestowal of the consequent private property right to that product on the capitalist rather than the worker, and the former's control over the production process. As Werner Bonefeld (1992, 116) maintains, "[t]he formal safeguarding of rights inverts into the substantive guarantee of exploitation ... and specifies the state as a moment within the 'context of the valorisation process.'" This same possibility of guaranteeing the general social conditions of capitalist reproduction gives rise to a symbiotic relationship between the bourgeois state and capitalist reproduction processes, since the former can only maintain its form by guaranteeing the latter and thus its own material basis. As Joachim Hirsch (1978, 66) writes, "[t]his will necessarily manifest itself as the specifically political and bureaucratic interest of the direct holders of state power and their agents in the safeguarding of capital reproduction and capital relations. This is why the bourgeois state must function as a class state even when the ruling class or a section of it does not exert direct influence over it."

More specific to the empirical content offered in the current work, as science and technological development become increasingly vital to capitalist accumulation so too do the external, material conditions of production and reproduction that the state must produce in service of the basic function of maintaining processes of accumulation. Despite the rhetoric of neo-liberal apologists, there is a clear relationship between economic markets and political regulation and control. The question is not *whether* the state will intervene but, rather, *how*. As the evidence presented in both the first and current chapters attest, science and technology policies in Canada have become part of the general and external guarantees of the social conditions of production, even if, in practice, such policies are directed toward specific capitals or sectors of capital.

Considered in tandem, Canadian biotechnology policy and its consequent regulatory regime demonstrate the degree to which the economic is embedded in political structures of power as well as the reflexive nature of those political forms that depend, in part, on the economy for their continued existence. That is, the CBS and its emphasis on harnessing the commercial potential of biotechnology to ensure Canada's economic growth and international competitiveness has manifested itself in a regulatory regime that, at its heart, is crippled by a critical conflict of interest, promoting the science of biotechnology as a new means of facilitating capital accumulation, on the one hand, and protecting the health and safety of Canadians and their environment, on the other. This internally conflicted regulatory regime evinces a number of critical weaknesses that, by calling into question its scientific rigour and the veracity of its claims about genetically engineered organisms, challenge its profession to serve the public interest. Instead, the Canadian regulatory regime uncritically equates the public interest with the private interests of capital.

Moreover, by restricting the regulatory approval system for genetically engineered products to issues of what the Canadian government and industry deem to be appropriate science, the Canadian regulatory regime circumscribes the biological knowledge commons. Through their construction of a particular "black box" of biotechnology, to borrow from Latour, government regulators regularly omit from discussion a plethora of social, ethical, environmental, and economic issues that attach to the technological developments derived from this science. Instead, capitalist science and the capitalist state's regulatory system continue to adhere to the Cartesian binary that separates natural science and philosophy in ways that deny the label of "science" to skills and knowledge that cannot be readily integrated into capitalist relations of production (Gorz 1976). This enduring separation inherent to capitalist social relations thus serves to mystify the political economic interests that promote and maintain the hegemony of a particular type of science that, in its subservience to capital, also serves to foreclose discussions that challenge the dominant agricultural production models. While there is certainly a vocal opposition in this country that is battling to inject such considerations into debates about biotechnology, the government nonetheless occupies a potent gatekeeper function that translates into significant power to shape broader public discourses and influence what knowledge is considered legitimate.

6
Capture and Control of Biotechnology Discourse in Canada

Members of the public have been exposed to a rhetorical whirl-
wind, battering them from all sides. They are told that GM is good
for them by a host of authoritative sources: the White House, the
Vatican, Downing Street, other political and religious leaders,
learned societies, university scientists, government commissions,
international corporations, commercial leaders and some of the
press. This onslaught has deployed every persuasive rhetorical
strategy imaginable: from august oratory to the chattiest synthe-
sised egalitarianism, from broadsheet bombast to tabloid humour,
from complex philosophy to advertising and PR. It has compared
GM with the greatest of human achievements, and its opponents
to Nazis and terrorists. One might expect the combination of such
power and persuasiveness to have succeeded ... A cause for optimism is that those without vested interests in GM technology
remain critical of both GM and the language used to promote it.

– Guy Cook, *Genetically Modified Language:*
The Discourse of Arguments for GM Crops and Food

Beyond rehearsing some of the mounting evidence of deficiencies in a linear
model of scientific assessment, the previous chapter also briefly highlighted
the obstinate refusal of both regulators and regulated companies to expand
the regulatory system to account for citizens' and some scientists' additional
concerns in respect of biotechnology. Taking a cue from Guy Cook (2005),
this chapter will expand on this theme, elaborate upon some of the ways
that biotechnology and biotechnological information are culturally constructed, and describe how capital and government disseminate an ideology
of biotechnological information and communication. The chapter will attempt to prise apart the ideological limitations imposed on the concepts,
vocabulary, and tools necessary to interrogate the nature of biotechnology,

which is touted by capital and government alike as one of the latest drivers of economic growth and prosperity. It similarly seeks to reveal the contradictions and exclusions imposed (or that could be potentially imposed) by biotechnology that are masked by prevailing capitalist-directed discourses.

In pursuing this task, I am less interested in rendering a normative assessment of the ontology of biotechnological information than in outlining how certain discursive constructs of biotechnological information are privileged over others. This privileging brings with it a plethora of social, cultural, political, economic, and environmental implications as certain renditions and their attending values are elevated over others. After setting the general context in which information about genetically engineered foods is typically disseminated to the people of this country, the chapter will turn our attention toward refuting some of the marketing myths propagated by the biotechnology industry on their own scientific terms. The tactic employed here should not be construed as acceptance of the circumscribed parameters within which biotechnology proponents seek to confine debate but, rather, as proof that many of the supposed benefits of this technoscience cannot be legitimated even on such limited terms. The following two sections of the chapter will elaborate, respectively, on some activists' struggles to expand the scope of the debate concerning agricultural biotechnology and the rejection of the Canadian Biotechnology Advisory Committee's (CBAC) public consultation exercises by many of these same activists.

Corporate/Government Construction of Agricultural Biotechnology Discourse

Though perhaps susceptible to dismissal as a trivial concern, I agree with Cook (2005) that we need to be critical of the substitution in much of the debate around biotechnology of the term "genetic modification" for the initial term "genetic engineering." Although both terms refer to the same set of activities, the former invokes the notion of non-intrusive and minor adjustment of something already in existence, while the latter connotes mechanical and impersonal processes, perhaps even evoking memories of now discredited ideas about "social engineering." The subsequent abbreviation of "genetic modification" to GM succeeds in further concealing what is at stake with the technical applications of this science, something George Orwell (1987, 263-64) pointed out in *Nineteen Eighty-Four*: "It was perceived [in Newspeak] that in ... abbreviating a name one narrowed and subtly altered its meaning, by cutting out the associations that would otherwise cling to it."[1]

Commenting more broadly on capital's propensity for exaggerated hype about new technology, the Critical Art Ensemble (2002, 53) reminds us of the negative repercussions that historically have attended capitalist technological innovation: "As always, capital makes techno-revolutions sound good, and to the extent that the interests of individuals and of capital overlap, the revolution will be good. Unfortunately, we do not know how big this overlap will be, and if we are to judge from past experience, we can expect much more to be worse than better. Further, while the utopian promises have yet to really manifest themselves, the numerous problems ... are already manifesting themselves."

Claims by multinational agrochemical corporations (the self-styled "life sciences" companies) that their products are designed to solve ecological and food shortage problems provide ready discourses grounded in ethics.[2] Such discursive constructions help to obscure these same companies' efforts to secure state subsidies and a relaxed regulatory environment in which genetically engineered organisms are assessed by governments in the same fashion as traditional health and food products. Biotechnology companies actively invoke information dissemination practices that predominantly emphasize the favourable and sanitized aspects of this new technology and its applications (Hindmarsh 2001; Murdock 2004). Monsanto engages in deliberate attempts to shape public debate around biotechnology in a manner propitious to its own product lines and designed to deflate opposition. It does so through a combination of tactics that draws on technological determinism, the apparent infallibility of science and scientists, and the portrayal of biotechnology as completely natural.[3] Such corporate strategies provide companies with a window of opportunity to sell as many of their genetically engineered products as possible in an attempt to integrate them so deeply into markets that regulated withdrawal would result in such economic upheaval that it is no longer considered a viable policy option (Levidow 1995). For example, in an effort to overcome largely negative European public sentiment toward genetically engineered food, Monsanto engaged in an aggressive 1998 public relations campaign in major European newspapers that was designed to try and convince readers that the only way to feed the world was through genetic engineering.[4] Monsanto's public relations efforts were condemned almost unanimously by African delegates to the Food and Agriculture Organization's negotiations on the International Undertaking for Plant Genetic Resources, who

> strongly object that the image of the poor and hungry from our countries is being used by giant multinational corporations to push a technology that is neither safe, environmentally friendly, nor economically beneficial to us ... We do not believe that such companies or gene technologies will help our farmers to produce the food that is needed in the 21st century. On the contrary, we think it will destroy the diversity, the local knowledge and the sustainable agricultural systems that our farmers have developed for millennia and that it will thus undermine our capacity to feed ourselves. (*Let Nature's Harvest Continue* 1998)

Indeed, the rhetoric of feeding the poor needs to be revealed as the smoke-screen that it is, employed by business as an insincere discursive construct designed to disarm opposition to agricultural biotechnology in a way that promotes agricultural biotechnology as the only solution, completely failing to address the deeper underlying structural causes of poverty and their link-ages to hunger and the plight of small landholders. As a number of com-mentators point out, the paradox of hunger in an era of high growth in food production and general overproduction (relative to markets and income distribution) is the result of a capitalist system driven by the quest for profits rather than the goal of achieving and sustaining human well-being (Magdoff, Foster, and Buttel 2000b).

Part of the way government and industry confuse debate is through their use of terms such as "sound science" and "natural" versus "unnatural." The term "sound science" is not a mere tautology – after all, is science not required to be sound in order to be considered science – but actually pro-vides a rhetorical strategy for disarming biotechnology opponents, implying that, in addition to "sound science," there is also "unsound science," a charge usually directed toward opponents of genetic engineering. Of course, scien-tific research can be methodologically flawed and thus unsound in its find-ings, which, as we saw in the previous chapter, characterizes much of the science being conducted by industry. Similarly, the terms "natural" and "unnatural" represent ideal types that are best considered as existing on a continuum comprising degrees of naturalness, along which various phe-nomena can be situated. For example, a plant growing in the wild represents our prototypical understanding of natural, while a genetically engineered variety developed in a laboratory would be located at an opposite and fairly distant position on the continuum.

An exact degree of "scientific" distinction between the two terms is ren-dered problematic because, aside from describing an attributed ontological status, they can also express an oftentimes normative judgment (Cook 2005). Recalling our discussion of intellectual property, we recognize that move-ment of corporate and government discourse along the continuum is invari-ably driven by the profit-seeking goals of industry. For example, as Thacker (2005, xviii) points out, "the claim that a genetic sequence or GMO is arti-ficial underscores the 'tech' part of the biotech: it is in some minimal way the result of human intervention, industry, and technology." Yet the same companies that employ such arguments when justifying their intellectual property claims also market their products as being "natural" and therefore safe for the environment, humans, and animals. The rhetoric disseminated by these multinational firms to market their genetically engineered products relies on a discourse that endeavours to situate biotechnology as a comple-ment to the natural processes and rhythms of human health, nutrition, and the environment. The biotechnology industry has thus manoeuvered itself into an internally inconsistent logic, stressing the "tech" part of biotechnology when seeking intellectual property protection and the "bio" side of the biotechnology concept when promoting the range of applications derived from this science.

In Canada, it is not only industry that has actively engaged in concerted efforts to deliberately manipulate public opinion of agricultural biotechnology. As early as 1999, at a time when consumers and advocacy groups grew increasingly apprehensive about the burgeoning number of genetically engineered food products appearing on grocery shelves, Ottawa began developing public relations campaigns designed to win public approval for agricultural biotechnology. In April 1999, then federal minister for agriculture and agri-food Canada, Lyle Vanclief, called together a secret meeting of government bureaucrats, industry insiders including the head of Novartis, Byron Beeler, and then Monsanto Canada president, Ray Mowling, and communication experts Joyce Groote of the industry lobby group BIOTECanada and Diane Weatherall of the Food Biotechnology Communications Network (FBCN).[5] Anna Hobbs, an associate editor of *Canadian Living* magazine, was also invited to participate. Thanks to documents obtained by researcher Bradford Duplisea under the *Access to Information Act*, we also know that Hobbs wrote to the minister the day following the meeting to thank him for the invitation and to advise him that "based on my experience with the food-safety concerns surrounding Alar and the apple industry, communication is most effective when government and industry partner with a credible, independent third party that, in consumers' perception, does not have the vested interest of a stakeholder" (as cited in Stewart 2002b, B3).[6] The chief executive officer of Foragen Technology Ventures, Murray McLaughlin, would reiterate the same sentiments to the minister in a letter dated 28 April 1999: "We need a champion to convey the information to the public while at the same time bring the industry together. A possible vehicle for the communications component is the Food Biotechnology Communications Network" (as cited in Stewart 2002b, B3). In any event, the minister listened.

For example, in late March 2000, the Canadian Food Inspection Agency (CFIA) produced and distributed to all Canadian households a pamphlet titled *Food Safety and You* at a cost of $2.5 million. Although a purported source of politically neutral information designed to inform Canadians about the ostensible stringency of our food regulatory system, it omits any discussion of the fact that the data used to "assess" applications for regulatory approval derive entirely from the applicant. As Ann Clark points out, the pamphlet was "very carefully worded. The phrase that they 'conduct assessments' is probably true, but it's not research" (as cited in Stewart 2000, B3). As part of its public relations/promotion role, the CFIA also paid $300,000 in that same year to *Canadian Living* in a non-tendered contract for a pro-biotechnology supplement titled *A Growing Appetite for Information,* which was to be placed in one of the magazine's editions (Abley 2000b). Hobbs' participation in the secret meeting and subsequent letter to the agriculture minister clearly paid dividends for the magazine.

According to a segment on the Canadian Broadcast Corporation's (CBC) radio program *World Report,* from 11 October 2000, the original text of the supplement included comments from groups skeptical about the safety of genetically engineered foods, including David Suzuki, the Council of Canadians, and Greenpeace. However, all critical commentary was removed from the final edition included in the magazine. While CFIA official Bart Bilmer maintained that the order did not come from his agency, a public relations consultant hired by the magazine told CBC News that comments made by biotechnology opponents were contrary to the intentions of the CFIA (McKie 2000). Given its predominantly female demographic, it is not particularly surprising that the CFIA chose *Canadian Living.* After all, women continue to be in the majority position when it comes to family health care and food purchasing decisions.

Between 1997 and 2003, Ottawa spent more than $13 million to fund a variety of communication strategies designed to foster public acceptance of agricultural biotechnology. Of this amount, $1.3 million went to the Consumers' Association of Canada (which only recently reversed its opposition to mandatory labelling, despite overwhelming evidence that Canadians want genetically engineered food labelled), $5.7 million to BIOTECanada, and $1.5 million to Pollara to conduct polls and report its findings to the CBAC (Abley 2000a, 2000b; Aubrey 2003; Freeman 2001, 2003; Freeze 2002). According to a passage on the BIOTECanada website, long since removed, it executed an Industry Canada-funded study "to examine the biotechnology communications strategies and outreach activities undertaken by the Canadian biotechnology community since 1992. The goal was to provide recommendations for the improvement of public awareness about biotechnology" (Abley 2000a, A7).

Based on the evidence and discussion presented thus far in this book, it is fairly easy to surmise what type of public awareness the federal government had in mind. As Brewster Kneen (2000, para. 7) writes, "[i]t is vitally important not to underestimate the manipulation and misrepresentation that is taking place in a frantic, not to say hysterical, campaign to convince the public that we really love GE and will starve in a polluted environment without biotech. We ignore the deceitful, highly centralized and extremely well-funded character of the campaign at our peril." This blatant use of public money to promote agricultural biotechnology and reduce public concerns about genetically engineered foods would be a source of later criticism articulated in the report from the Royal Society of Canada (2001,

212): "The more the regulatory agencies are, or are perceived to be, promoters of the technology the more they undermine public trust in their ability to regulate the technology in the public interest."

The CFIA also provided at least $750,000 to the now defunct FBCN over the course of its lifetime, which, in 2001, claimed to be a leading "information source for balanced, science-based facts about food biotechnology and its impact on our food system" (as cited in Stewart 2001, B3). This organization's website further maintained that because it includes a range of stakeholder perspectives from farmers to consumers, it "brings both neutrality and strong credibility to the information we provide" (as cited in Stewart 2001, B3; 2002c). However, the 150 corporate partners from the biotechnology and pharmaceutical industries who were members of the FBCN put the lie to claims of neutrality.

The FBCN made a variety of information resources available to the public, including information kits, resource sheets on a variety of topics, a referral network of experts, and a toll-free information telephone number. Callers to this number could receive a gratis copy of a booklet titled *A Growing Appetite for Information,* which was a slick promotional piece co-produced with the Consumers' Association of Canada that claimed to offer a bias-free introduction to food biotechnology in Canada. As Duplisea determined through federal access to information requests, this publication, as well as the FBCN's resource sheets and toll-free information line, were all funded by the CFIA. Yet the CFIA never made public its involvement in funding the FBCN (as discussed in Stewart 2001). As Duplisea comments in a CBC interview, "[w]hen I read this booklet, I thought I was reading government and biotech industry propaganda because it parroted their message on every issue from labelling to regulation" (as cited in Nunn 2002).

Although the FBCN was disbanded by 2002, there were other bodies eager to step into the breach. For example, the Dieticians of Canada, which describes itself as "the nationwide voice of over 5,000 dieticians, bringing trusted information on food and nutrition to Canadians," published a pamphlet in May 2002 titled *Modern Food Biotechnology: Principles and Perspectives* (Stewart 2002d). Although the Dieticians of Canada claims that the information brochure was a neutral educational resource to help its members comprehend the issues involved with genetically engineered foods, the project was funded by the Council for Biotechnology Information, an industry group dedicated to persuading the public about the purported benefits of biotechnology.[7] The Dieticians of Canada chose not to disclose this funding source when announcing the release of the pamphlet, which was written by Milly Ryan-Harshman, a former Monsanto employee. Ryan-Harshman had also been involved in the promotion of genetically engineered foods, including efforts to lobby the Chilean government (on an initiative funded by the Canadian

International Development Agency) to loosen its regulation of genetically engineered foods (Stewart 2002d).

Prominent biotechnology opponent Jeremy Rifkin (1998) is extremely critical of the often one-sided nature of contemporary debates over biotechnology. His analysis of reports about biotechnology that have emanated from the trade and business press and the general media reveals a picture of a communication landscape heavily influenced by the messages circulated by geneticists and the biotechnology industry. Critical assessments of this new technology have received a dearth of coverage, a trend in both Canada and the United States that has been documented by other writers (Hornig Priest 2006; Hornig Priest and Ten Eyck 2004).[8] Instead, media hype around biotechnology has tended to promote investor confidence and public support for these new technologies. As Steven Best and Douglas Kellner (2004, 206) write, "[i]ndeed in the highly competitive cloning marketplace, where companies are scrambling to patent the first major breakthrough in stem cell research, PR and manipulation of media are lab tools as basic as a microscope."

Those who do voice concerns about the trajectory of biotechnological development risk being marginalized by the "scientific establishment" and biotechnology industry as heretics attacking the conventional wisdom developing around mainstream biotechnology (Rifkin 1998). Such dismissals are underscored by mainstream media coverage of agricultural biotechnology issues.[9] Rather than offering critical or even alternate perspectives, the bulk of news reports uncritically reproduce the assertions and assumptions made by scientists and the biotechnology industry, in part because of what Oscar Gandy (1982) calls "information subsidies" provided by these actors – specific information that portrays new technologies in a positive manner. Some observers propose that this lack of critical media coverage has lulled policy-makers, at least in the United States, into a false sense that biotechnology has not engendered any popular resistance (Hornig Priest and Ten Eyck 2004).

Specific to Canada, one researcher has determined through a content analysis of Canadian newspaper coverage of biotechnology issues for 2004 that these same media trends and biases exist in this country (Knezevic 2005). As we saw in Chapter 3, 2004 was an important year for biotechnology given the substantial opposition being generated against Monsanto's application for regulatory approval of its genetically engineered Roundup Ready wheat. Yet Irena Knezevic (2005) ascertained that a mere 279 articles (of which eighty were duplicate articles that appeared in multiple newspapers) were published in seventeen newspapers. More troubling than this minimal coverage was the inherent industry slant found in the articles, which accepted the biotech industry's claims about genetically engineered organisms representing progress, aiding farmers, producing healthier foods, and serving environmental remediation purposes. Conversely, opponents of biotechnology and genetic engineering were typically portrayed as self-serving and fundamentally uninformed activists promoting their own personal political agendas. For the most part, industry and government scientists were referred to as experts, while scientists voicing critical assessments of this technoscience were often discredited as engaging in "junk science" (Knezevic 2005, 108). The conclusion Knezevic (2005, 115) draws from her study is that lack of public knowledge around biotechnology issues "is not a case of voluntary ignorance; it is a systemic problem where those who are supposed to inform us [media] continue to obscure the truth not only about GMOs, but also about the wider context in which this industry has managed to flourish."

This dismal picture is confirmed by almost all of the agricultural bioactivists in Canada, who tend to hold out little hope that the messages they are struggling to disseminate will be picked up and transmitted by the dominant media in this country. According to Lucy Sharratt, co-ordinator of the Canadian Biotechnology Action Network (CBAN) and former co-ordinator for the Sierra Club of Canada's Safe Food, Sustainable Agriculture Campaign, the media attention given to the latter organization's first movement against genetic engineering in Canada was more a function of the novelty of the issue and the kind of inherent controversy that the media could envision developing around biotechnology. Today, however, she believes the media have largely lost interest in biotechnology issues. For example, according to Sharratt, the media never reported on two incidents concerning experimental genetically engineered pigs that had not been approved for human consumption but that, nonetheless, accidentally infiltrated the animal food chain and, subsequently, the human food system. These two incidents occurred on separate occasions, by two companies conducting two different experimental procedures.[10]

Sharratt goes on to bemoan the fact that, when media coverage is forthcoming, it is heavily biased in favour of industry. For example, the 2002 counter conference to the industry Biotech conference in Toronto that same year organized a picnic that fed 2,000 people an organic lunch. Yet the reporter from the *Globe and Mail* who covered the event, and with whom the organizers of the counter conference spoke, elected to pick up only on the press releases posted by participants at the industry conference. In one small article, the accompanying picture was of people at the industry conference with their industry-filled bags and a caption along the lines of "people pick up swag at industry conference." Meanwhile, David Suzuki, Vandana Shiva, and other prominent activists were speaking at a nearby local park to some 2,000 people in a festival-like atmosphere, of which the paper made no mention and offered no picture. For Sharratt, this disregard represented a blatant example of the almost total blind eye the media turn toward the issues brought forward by critics of biotechnology.

Kneen contends that it is arduous to entice the media into covering stories about the loss of biodiversity or the expansion of corporate control over agriculture and the food system, given the ownership structure and resulting bias of the media. In fact, Kneen believes that it has been an unwritten policy among both the corporate and government sectors to restrict public discussion of biotechnology issues to a minimum, and this policy is further facilitated by a media environment that devotes little critical coverage to the topic. Nonetheless, he also observes a quiet, but much broader, discussion about biotechnology, "one that the corporations continually try to counter with all their hype about feeding the world because they know that public opinion is becoming more and more suspect about the confluence of corporate control and pesticides, agrotoxins, genetic engineering, etc." According to Kneen, people are becoming more suspicious and reading the labels much more, and genetic engineering is one of the things that elicits growing mistrust. People are also beginning to ask whether genetically engineered products are even necessary. Kneen sums up the feelings of many critics of the biotechnology sector:

> Well I'd like to see an example and so would a lot of other people. What is it you're talking about? What have you delivered? Because people know there hasn't been any quality and this is where it gets interesting because people's realization about fresh foods and nutrition and long distance travel and then you get the energy equation put into it, I think the public know, and that's reflected in Loblaws and other stores in terms of their organics and their refusal to push GE. You know what they're doing is responding to the public ... I think the challenge is and it's difficult with this stupid government we have now that doesn't care about what the public thinks, what possibilities are there to affect policy. It may well be that the courts are one way of doing this, particularly given some of the interesting decisions they're making. But that's a long and expensive way to go. Myself I wish we had the population of India and we did do some direct action. Put 5,000 people in the field. (Interview)

Perhaps more importantly, even when there is wider media attention to biotechnology issues, coverage tends to assume an alarmist tone, distorting the critical message that opposition movements are attempting to publicize. In a conversation with David Suzuki, Sharratt asked him why we have this problem with critical media coverage in Canada. His response was that the climate for science in Canada is not such that we can effectively look for allies, and, if we should happen to find them, they usually tend to be marginalized. In terms of moving forward, Sharratt is unsure how relevant media campaigns will be to the work of CBAN. It depends on the strategy:

> If the goal is to attempt to move an issue politically so that Members of Parliament pay attention, CBAN will be compelled to try and attract media attention. If the objective is to communicate to people about the benefits of purchasing local food in support of farmers and to convince them to write letters to their Member of Parliament, CBAN can accomplish such things by employing a variety of public channels outside of the media, who, in any event, are normally not particularly interested in taking up such stories. (Interview)

From its creation in 1971, Greenpeace has tasked itself with trying to be effective in terms of communicating to the wider public. As Josh Brandon (2007) laments, media coverage is not at all balanced: "[I]ndustry proponents receive ten times the coverage that Greenpeace manages to get in the papers." Brandon recounts an incident with a journalist from the *Vancouver Sun*, who told him that the newspaper would not cover protests against genetic engineering nor would it publish information or messages from Greenpeace unless the organization could provide scientific data. Some time later, Greenpeace did come out with a scientific report about the toxicity of genetically engineered corn that was peer reviewed and that appeared in a scientific journal. Greenpeace sent out a press release, and Brandon contacted that same reporter from the *Vancouver Sun*, only to be told that, while the article might have been peer reviewed, the newspaper wanted information from BC scientists.

Given the nature of the commercial media and the way they operate, Greenpeace has long recognized a strategic necessity: to engage in highly visible tactics that ensure its underlying message is carried through those media widely viewed by the public. As Eric Darier points out, it is not by accident that Greenpeace undertakes spectacular actions. It is not just because Greenpeace likes doing them but, rather, that they fulfil a purpose. Such actions, many of which are outlined in detail on its website, help ensure that Greenpeace and its message make their way into the news in order to inform the general public and to move the issue politically (Interview).[11] Thus, unlike some other groups that discount the value of the media, Greenpeace uses the media and the existing means of communication as a major component of its strategy to reach people on a large scale. Yet Greenpeace is well aware that once an opposition movement does something, it loses control of the message. For this reason, Greenpeace develops and implements a clear communication strategy to prevent side issues from emerging. As Darier points out, it is impossible to communicate everything, so Greenpeace must be clear, it must have one message, it must have one target, and it must adhere to the basic rules of communication that apply in the context of a current media environment dominated, and often constrained, by heavy corporate control. As Darier readily admits, it is not an exact science, and it changes over time, but he believes that Greenpeace's messages are reflected fairly accurately in the media for the most part (Interview).

Although the messages the National Farmers Union (NFU) tries to communicate through the media appear relatively intact, attracting media attention to its issues is often challenging. Moreover, the coverage given to its positions, relative to those held by industry or science, is typically quite small, particularly in the mainstream media. Aside from very sparse coverage, according to Terry Boehm, the national papers tend to lack thoroughness and depth in their reporting. The NFU has more success in convincing local media, such as little community weeklies, to take its press releases and print them verbatim. More pernicious is the latest trend in the media environment of press release services, which Boehm likens to marketing food products in the supermarket where you pay for shelf space. Now organizations can pay a not insignificant fee to ensure that the media will pick up their releases before other random releases come in. This practice highly limits the media coverage the NFU can achieve. As Boehm states, "whether media is going to admit it or not that this is taking place, these fee for dissemination services are there. Freedom of the press for a price. I guess it wouldn't really be freedom of the press, it's probably more freedom of public access to the press for a price" (Interview). Overall, Boehm believes that the pressures in the media are creating a situation that impinges on the level and quality of journalistic research being conducted for stories. Given the fairly substantial resources government and, particularly, industry can direct toward influencing public debate, the following section seeks to debunk some of the myths being propagated in respect of biotechnology.

Genetically Engineered Seed Myths

Ann Clark (2003), previously introduced in Chapter 5, forcefully asserts that proprietary genetic technologies address only the symptoms, rather than the underlying management problems, in contemporary agriculture. Ecologically dysfunctional crop rotation practices exacerbate selection pressure and contribute to the development of vigorous weeds, which creates opportunities for herbicide-tolerant crop varieties. However, as Clark points out and a number of scientific studies discussed in the previous chapter attest, the genetic traits inserted into such seeds to render them resistant to particular herbicides can transfer to other species and promote additional weed resistance, thus exacerbating the initial problem even further.[12] Emerging studies show that Roundup Ready soybeans require more of the active ingredient glyphosate than competing herbicides. Even more damaging, the increased weed resistance that emerges after several years of planting Roundup Ready soybeans compels agricultural producers to expand their herbicide spraying regimens by either using additional herbicides or applying Roundup multiple times (Clark 2003).

One early study noted that farmers who planted Roundup Ready soybeans were applying approximately 0.56 kilograms per hectare more herbicide than those growing conventional soybeans – or about nine million kilograms of additional herbicide per year – in the United States alone (Benbrook 2001). In his most recent findings based on an analysis of United States Department of Agriculture data, Benbrook (2012a, 2012b) has determined that although *Bacillus thuringiensis* (Bt) corn and cotton have allowed producers to reduce their insecticide applications by some 56 million kilograms, this reduction was more than offset by increased herbicide usage, which totaled 239 million kilograms over this sixteen-year period. As a net result, during their first sixteen years of cultivation (1996-2011), genetically engineered crops in the United States have required the application of 183 million more kilograms of pesticide than would likely have been the case had the same acreage been sown with conventional seeds.[13] On average, this net increase represents an additional 0.21 kilograms of pesticide active ingredient for every genetically engineered-trait hectare planted. These results are even more startling when we disaggregate the numbers. For example, in 2011, soy, corn, and cotton seeds genetically engineered to be herbicide tolerant required 0.82, 0.46, and 0.97 more kilograms of herbicide per hectare, respectively, than their conventional counterparts (Benbrook 2012a, 2012b). Consistent with his previous findings, Benbrook (2001, 2009a, 2012a) determined that around 70 percent of this total increase in herbicide use (167 million kilograms) is attributable to herbicide-tolerant soybeans.

As already pointed out, there has been a significant increase in herbicide applications since 2007 due to the emergence of glyphosate resistance among a variety of weed species. One industry solution has been to genetically engineer crops to tolerate even heavier doses of glyphosate, as is the case with Monsanto's Roundup Ready Flex cotton, which was introduced in 2006. The label on this new variety recommends an application rate of glyphosate that is almost 1.5 times higher than for original Roundup Ready cotton (2.3 litres per hectare for Roundup Ready Flex cotton versus 1.6 litres per hectare for original Roundup Ready cotton). The environmentally disastrous result of such solutions has been a dramatic upsurge – 65 million kilograms – in herbicide usage over the last two years in the United States (Benbrook 2012b).

Such industry solutions to the increasingly severe weed resistance problem no doubt explain the widening differentials in herbicide applications between conventional and genetically engineered crops. Between 1996 and 2011, soybean producers increased their application of herbicides on herbicide-tolerant varieties from approximately 1 kilogram to 1.89 kilograms per hectare. Herbicide usage for conventional soybeans decreased in the same time period from 1.34 kilograms to 1.08 kilograms per hectare. Producers of herbicide-tolerant corn expanded their use of herbicides from 2.34 kilograms to 2.72 kilograms per hectare. Their conventional corn-growing counterparts reduced herbicide applications from 2.99 kilograms to 2.25 kilograms per hectare. Although conventional cotton farmers increased their herbicide applications slightly from 2.11 kilograms to 2.15 kilograms per hectare, those producers who planted herbicide-tolerant cotton varieties had to spray 3.11 kilograms per hectare in 2011, which was up markedly from 1.27 kilograms per hectare in 1996 (Benbrook 2012a). This precipitous increase is no doubt the result of the severity of the glyphosate-resistant weed problem that plagues large areas of cotton-growing regions in the southern United States. Despite herbicide reductions in their early years on the market, these statistics demonstrate clearly the increasing non-sustainability of herbicide-tolerant crops, particularly soybeans and cotton.

In addition to the environmental consequences, these escalating herbicide application requirements are becoming increasingly onerous for farmers as the price of Roundup herbicide has increased from about US$32 per gallon in 2006 to US$45 a year later and to as high as US$75 per gallon by June 2008. These steep price increases had abated substantially by late 2009, when Monsanto announced that, due to increased competition from generic glyphosate herbicides, especially those produced in China, it would reduce the price of its Roundup brand herbicides by up to 50 percent. In addition to slashed prices of around US$22 per gallon, Monsanto established financial incentive programs totalling more than US$100 million to woo back farmers who have switched to less expensive generics.[14] Although, as mentioned earlier, by 2011 Monsanto had achieved some success in stabilizing and even increasing the average net selling price of Roundup and other glyphosate-based herbicides. In any event, factoring in the original cost of purchasing Roundup Ready seed each year, as required by both intellectual property law and the Monsanto technology use agreements imposed on farmers, as well as the loss of the price premium offered for non-genetically engineered soybeans, conventional soybean varieties are beginning to experience a resurgence of interest (Shannon 2008).

Clark (2003) argues that, in the case of Bt corn, the genetically engineered crop holds out very little promise of reducing insecticide use since relatively little insecticide is applied to control the target pests. Most of the insecticides sprayed on corn are designed to control root pests, which are not affected by the current Bt corn breeds (SmartStax has been engineered to combat certain root pests, but the effectiveness in the fields of the latest stacked variety remains unclear). With regard to Bt cotton, a long-term study conducted by researchers in China has documented that, after seven years of cultivating this crop, populations of other insects have increased so much that farmers now have to spray their fields up to twenty times per season (Wang, Just, and Pinstrup-Andersen 2006). Rather than reducing pesticide applications, one of the vaunted benefits widely publicized by proponents of agricultural biotechnology, these genetically engineered plants require additional insecticide. The result has been that initial increases in income achieved by farmers planting Bt cotton have now been offset by the expansion in pesticide applications. Overall, farmers who planted genetically engineered cotton experienced net declines in their income by an average of 8 percent as compared to conventional farmers, given the price premium for genetically engineered cotton seeds (ibid.). More recent research specific to China, which confirms such findings, outlines a twelve-fold increase since 1997 in populations of the mirid bug, another cotton pest forcing Chinese farmers to spray more insecticides (Lu et al. 2010).

Clark (2003) also refutes the myth propagated by the seed industry that genetically engineered crops produce higher yields. As the international group of over 400 researchers and scientists who collaborated on the International Assessment of Agricultural Science and Technology for Development point out, data on genetically engineered crops indicate a great deal of yield variability (McIntyre et al. 2009). According to Clark, almost 99 percent of all genetically engineered crops are designed to be either herbicide tolerant or insect resistant. That is, only one active transgene is inserted. In some cases, a seed will be stacked, meaning that it contains both herbicide tolerance and Bt traits. Yield, however, depends upon the interaction of multiple genetic traits. The only way that these current genetically engineered seeds augment yield is by decreasing production losses that might occur as a result of weeds, as in the case of herbicide-tolerant crops or the insect targets of Bt crops.

To determine how much of any increased yield is the result of the inserted genetic trait requires a comparison to what could have been achieved using other tools (Clark 2003). Indeed, according to a recent report from the Union of Concerned Scientists in the United States, intrinsic yield increases in corn and soybeans during the twentieth century can be traced to breakthroughs in traditional plant breeding and not to genetic engineering. Moreover, overall operational yield gains in herbicide-tolerant soybeans and corn have been non-existent as compared to yields for conventional planting methods.[15] Improved yields for insect-resistant varieties have been achieved only when pest infestations have been high (Gurian-Sherman 2009).

In fact, several recent research projects support Clark's claim that genetically engineered seeds do not produce significantly increased yield loads. A Canadian study based on surveys and interviews of agricultural producers in Alberta, Saskatchewan, and Manitoba demonstrates that Canadian farmers have not experienced significant yield improvements associated with genetically engineered canola (Mauro and McLachlan 2008). A group of scientists who compared transgenic and non-transgenic cotton production systems in the US state of Georgia determined that Roundup Ready cotton seeds tend to produce lower yields than cultivars from other systems. Overall, these researchers found no significant differences in financial returns between the two types of production systems, although they do allow for the possibility that hidden labour savings often inherent in transgenic crop systems could offer an important benefit (though presumably not for the agricultural workers rendered redundant!) (Jost et al. 2008).

In the case of Roundup Ready soybeans, several studies have shown that these seeds produce fewer beans than conventional seeds (Benbrook 1999, 2001; Elmore et al. 2001; Gordon 2007). Along similar lines, researchers who reviewed nine years of US Department of Agriculture data conclude that Monsanto's Roundup Ready soybeans not only produce lower yields than conventional soybeans but also that they require more chemicals at higher doses (Gunian-Sherman 2009; see also Benbrook 2009a). Other studies have been able to show definitively that the reduced yields from Roundup Ready soybeans are the result of either the gene inserted or the insertion process (Clark 2003; Elmore et al. 2001; Gordon 2007). That is, pleiotropic effects (the phenotypic expressions of gene insertion through which one gene or protein influences multiple phenotypic traits) may occur as a result of altered metabolic pathways in plants that could silence the expression of pre-existing genes in an organism.

Monsanto recently announced the release of a new herbicide-tolerant soybean called Roundup Ready 2 Yield, which the company claims can achieve yield increases of between 7 and 11 percent over previous Roundup Ready varieties. One initial report based on a survey of farm managers and seed distributors in five American states indicates that this new variety is failing to meet expectations (Kaskey 2009). Interestingly, the purported yield increase is the result of a change in gene insertion meant to avoid the yield drag inherent in the original Roundup Ready soybean. The genetic treadmill clearly is gaining velocity!

German federal scientists who report to the national Parliament's *(Bundestag)* Committee for Education, Research, and Technology Assessment maintain that there is insufficient scientific data to support the claim made by industry that genetically engineered seeds provide higher yields and, therefore, higher revenues for farmers, despite the fact that such genetically engineered crops have been planted for over twelve years now. This report further contends that industry claims are based on faulty methodological studies (Maurin 2008).

An economic analysis of genetically engineered crops produced for the German Federation of Organic Food Producers goes further, demonstrating not only that these technologies produce negligible macro-economic benefits for producers but also that, when considered across the entire food chain, they actually result in increased overall costs. Over the last thirty years, yields of maize and soybeans have increased by a factor of 1.7, while seed prices have quintupled. By way of comparison, seed prices for rice and wheat, products for which there are no genetically engineered varieties on the market, have risen roughly in line with yield increases over the same period (Then and Lorch 2009). Similar to other studies, this report indicates that, overall, genetically engineered seed varieties have failed to deliver significantly higher yields than conventional varieties. When one factors in the additional costs associated with evolving weed and insect resistance, segregation measures and analyses, and contamination events (for example, StarLink corn,[16] LibertyLink rice,[17] and most recently flax[18]), it quickly becomes clear that the short-term marginal benefits produced by genetically engineered crops are outstripped by the aggregate economic costs to food and agricultural markets (ibid.).

As one recent commentator points out, industry is quick to invoke the ingenuity involved in its introduction of genetically engineered traits into seed varieties when accounting for years of increased yields, but it quickly resorts to rationalizations about the complexity involved in genetic engineering and the interaction effects between plants and the broader ecosystem for years of reduced yield performance (Kuruganti 2009). This contrary and growing evidence, considered in tandem with the multiple environmental and health ramifications of genetically engineered crops outlined in the previous chapter, represents a multifaceted corpus of research that seriously undermines the veracity of even the constrained industry/regulatory discourse limited to questions of purported yield improvements and safety. However, people's apprehensions around biotechnology encompass wider normative questions of social justice and economics, conceptions of nature, and cultural values.

The Struggle to Expand the Biotechnology Debate beyond the Confines of Science

As considered briefly in Chapter 5, restricting the biotechnology debate to issues of science only is a function of Canada's regulatory process, which simply does not permit the injection of ethical, political, or socio-economic considerations into the discussion. Indeed, proponents of biotechnology from both industry and science steadfastly refuse to entertain such concerns.[19] For example, the Conference Board of Canada opposes any efforts to introduce non-scientific, ethical, or socio-economic issues into the regulatory review process out of fear that this would lengthen review and approval time frames (Munn-Venn and Mitchell 2005). In order to avoid any expansion of regulatory scope, industry has become adept at deploying discursive strategies designed to characterize all opposition to biotechnology as uninformed, irresponsible, and even hysterical. For example, biotechnology advocates have advanced the claim that opposition groups are "interweaving political, societal and emotional issues ... to delay commercialization and increase costs by supporting political, non-science-based regulation, unnecessary testing, and labelling of foods" (as cited in Nestle 2003, 140).

Former Monsanto chief executive officer, Robert Shapiro, has asserted that "those of us in industry can take comfort ... After all, we're the technical experts. We know we're right. The 'antis' obviously don't understand the science, and are just as obviously pushing a hidden agenda – probably to destroy capitalism" (as cited in Smith 2003, 252). The Canadian federal government has voiced similar sentiments: "A complicating factor [to gaining public acceptance of biotechnology] is that people often do not know about or do not understand the benefits to them of various biotechnology applications" (Industry Canada 1998, 3). Completely disregarding the possibility that citizens might have legitimate concerns about biotechnology, the federal government prefers to retain a paternalistic perception of Canadians, who, in its view, are merely ignorant of the advantages this purportedly beneficent technology will reap for the country. No Canadian policy document raises concerns about biotechnology and its applications, at least not in any substantial depth. The consistent government line is that biotechnology is largely safe and the Canadian regulatory system will protect Canadians. What such statements and positions miss is that social movements are interested in emphasizing and interrogating those issues that professional science has excluded, rather than merely debating scientific knowledge *per se* (Cozzens and Woodhouse 1995).

For example, and as discussed earlier, in the last few years farmers and commodity groups – newly informed by the personal financial implications of increasing global rejection of genetically engineered crops over the previous years – marshalled economic arguments in support of their objections to the introduction of Monsanto's Roundup Ready wheat in Canada. Similarly, the NFU has long recognized that the control issues stemming from legal constructs around genetics and biotechnology, such as gene patents, plant breeders' rights, and regulatory and legislative initiatives, have been detrimental to the economy of ordinary farmers and will likely continue to be so in the future. The NFU further opines that there has not been adequate research in terms of human health and broader environmental concerns, nor have the biotechnology issues been debated in the public forum, in part due to a lack of a reasonable supply of knowledge and research on either side of the debate. Instead, the biotechnology agenda, which has been greatly facilitated by the Canadian federal government, continues to be promoted and dominated largely by industry. So, despite an almost complete failure to sufficiently debate things such as market harm and health and social concerns involving biotechnology, Canadians are forced to live with the results of this technology. That being said, the NFU does have members who use biotech crops – as an organization, it does not seek to control what its membership does. The important point for the NFU is that people be afforded the opportunity to participate in informed debate about whether or not to use these new biotechnologies. While such deliberation is certainly occurring within the NFU, Boehm laments that it has yet to really emerge at a broader societal level.

Boehm further suggests that a new "religion" has developed around science, which has marginalized people. Where they might previously have been comfortable debating ethical and moral issues without having a particular education in that field, this is no longer the case because the scientific community tends to present itself as the "high priest of the truths who should be almost blindly trusted." Boehm maintains that people have witnessed and experienced enough examples in the past of detrimental results that can flow from science that "we should no longer be prepared to adhere to blind faith" (Interview). We need to overcome what Humphrey Jennings, a British documentary filmmaker and cultural critic, referred to as the "fatalism among the mass about present and possible future effects of science, and ... [the] tendency to leave them alone as beyond the scope of the intervention of the common man" (as cited in Robins and Webster 1999, 36).

According to Boehm, the promoters of biotechnology have attempted assiduously to create a private good out of what was once a public good through legislative and other legal and technical constructs. The Canadian government has been actively promoting the development of biotechnology as an industrial sector for the last twenty years, so it, too, has a vested interest in this project. The multinational corporations that dominate the biotechnology sector are interested in marketing and pushing this technoscience as fast as possible. Boehm believes that if court decisions in both Canada and the United States had been decided differently in terms of gene patents (and, by extension, control of the seed and the plant that results from that seed), there would be much less interest in advancing and marketing biotechnology as fast as possible. Boehm similarly contends that biotechnology development and promotion would no doubt slow significantly if health and environmental liability issues were dealt with in a manner that protected agricultural producers and the environment.

In its encounters with government regulators, the NFU finds that the various departments and agencies myopically focus on regulatory definitions, which has the effect of containing the debate on approval decisions. Any attempts to insert broader contextual questions are met with obdurate resistance on the part of government bureaucrats. Even when regulators acknowledge that more extensive social and economic concerns might be quite valid, they still refuse to consider them. As Sharratt argues, the government and industry have worked together closely in constructing the regulatory process, which, in turn, has implications for what type of information about genetic engineering is required by the public regulatory review system. Both industry and government regulators remain steadfast in their attempts to ensure that the discourse around biotechnology is limited either to the science of these new technological innovations or to the claims that biotechnology and genetic engineering will contribute to economic growth.

Darier, like others, suggests that the Canadian government and its regulators deliberately focus on the issue of safety in order to silence debate on all other biotechnology concerns. He is suspicious of institutions that concentrate inordinately on one aspect of an issue, arguing that such tactics are usually done to compensate for something being neglected in another important area. So, according to Darier, if one delves below the surface of government and industry rhetoric about safety or science-based regulation, it quickly becomes apparent that the discussion is, in fact, not science based and that the issue often has relatively little to do with science (Interview). Darier's suspicion seems accurate if we recall that the safety studies used by the government for product approval rely on data produced and designed by industry – data that both industry and government refuse to make public, thus precluding independent assessment.

As a consequence of the emphasis on this type of informational content that tends not to be particularly relevant to citizens, farmers, and consumers, the public is left largely uninformed. In order to extend the terms of the debate in regard to biotechnology in this country, CBAN is trying to construct a type of dialogue about farming that makes sense to urban people, a discourse that gives people some sense of empowerment, not just as consumers but, rather, as citizens and as participants in democracy. CBAN views the urban rural connection as an important issue, but one that is not facilitated by government given the lack of political power enjoyed by farmers and the relatively low priority accorded the agricultural sector by the government. According to Sharratt, although federal agricultural policy in this country is weak, policy and investment in biotechnology is quite strong because it is perceived as a new form of high technology that will secure Canada's future economic prosperity. This inherent conflict is also part of the message that CBAN seeks to disseminate among the public (Interview).

In order to expand the terms of public debate about biotechnology, Sharratt believes an essential first step is to legitimize the instincts of people who oppose a particular technology based on socio-economic factors that go beyond science-based discourse. By providing comprehensive information, opposition movements can empower people to develop their own positions on biotechnology and genetic engineering. By providing critical information, CBAN and other groups can help validate lay knowledge and legitimate the particular stance toward biotechnology adopted by the average person. While CBAN readily accepts the responsibility of equipping the public with sufficient and credible information, it also recognizes the substantial challenge posed by such a task. This is so because genetic engineering affects people's lives differently and touches many diverse sectors of society.

Thus, there emerges the demanding and strategic question of how to communicate with such an assorted array of people. In part, this desideratum contributed to the choice made by CBAN to position Terminator technology as the focus for its first major campaign. CBAN members believed that this issue provided a ready vehicle for explaining the impacts of genetic engineering on farming as well as furnishing a conduit that facilitates international solidarity among a variety of social groups and movements. Sharratt contends that CBAN's discourse on Terminator technology is far stronger than that of the government, arguing that the current Canadian regulatory system, based almost exclusively on science, is obviously not adequate to deal with the full ambit of challenges posed by this new technology (Interview). The regulatory regime was not adequate to deal with either recombinant bovine growth hormone or genetically engineered wheat, which compelled the federal government to develop exit strategies for these particular biotechnologies. According to Sharratt, to the extent that opposition movements such as CBAN succeed in injecting broader social, economic, political, health, and environmental dimensions into the biotechnology debate, this type of conflict will continue to confront and potentially stymie industry and government.

Although Greenpeace initially focused its debate in terms of biodiversity and genetic engineering, it very quickly realized that it was not by accident that Monsanto, which controls the overwhelming majority of all genetically engineered seeds in the world, is a chemical company. In a way, this recognition forced Greenpeace to enlarge its own debate and open up a space around broader agricultural concerns that admits issues such as the kind of agriculture we want, the kind of food we want, who should control it, and so forth. Thus, in Canada, genetic engineering has become an issue that has permitted Greenpeace to carve out a dialogical terrain that encompasses wider agricultural concerns, particularly in Quebec and British Columbia, in which there are strong organic sectors. The strength of these sectors, according to Darier, has helped promote a greater awareness of biotechnology and genetic engineering over the last five years.

Darier similarly believes that Canadian society must expand the discourse around biotechnology to avoid technological determinism and the technological imperative, which provides capital and government a convenient strategy to circumscribe social debate about the ways biotechnology should be developed and deployed (Interview). Technology is not neutral. The perception of science and technology as asocial catalysts for progress independent of purposive human agency serves to obscure the social sphere from the design and development stages of technological innovation. Such an ostensibly objective ideal of scientific and technological development not only relegates the social implications of new technologies to the instances of their application but also casts the social effects of such science and technology on society as secondary and contingent.

Such a conception of scientific and technological development suggests that scientists and technologists are the discoverers of laws and processes immanent within an external and autonomous natural realm – the social is subsumed under the natural. Progress putatively rooted in the natural order of a world that triumphs over historical and social particularities thus comes to be viewed as unassailable (Robins and Webster 1999). Moreover, this discursive framing easily explains away negative social effects as unavoidable by-products of history's teleological march of progress that can be mitigated through the perspicacious application of the new technologies. Overall, this perspective, while admitting the presence of social priorities at the application stage, elides such concerns at the developmental stage, thus refusing alternative visions of science and technological development informed by broader social imperatives.

Part of the problem is the way that biotech proponents, who hail from industry, government, and science, frame biotechnology issues. Les Levidow (1999, 64) writes: "Biotechnology offers putative solutions which predefine the problems to be solved. Its reified problem-definition in turn influences forms of public participation and safety regulation." Risk assessment is embedded in a foundation that contemplates the potential dangers of agricultural biotechnology from the perspective of industrialized monoculture. It refuses to admit consideration of the broader environmental impact of this type of intensive agriculture. It excludes alternative definitions of the problem. Evaluating harm entails an assessment of the effects that might impact the agrochemical control of weeds. The result of this approach is that potential detrimental effects of new genetically engineered plants are considered tolerable if they can be dealt with by subsequent technological advances.

Of course, the problem is deeper and older than the advent of biotechnology. From the earliest phases of industrialization, the social, political, and environmental effects of technological development were interpreted as secondary effects to be resolved through additional technological innovation (Moser 1995). However, biotechnology and genetic engineering bring their own set of spiralling demands for ever-greater technological applications, leading many to suggest that the "chemical treadmill" of industrial agriculture will be replaced by Levidow's "genetic treadmill," in which corporations

respond to the negative repercussions of biotechnology through new generations of the technology that created the problems in the first place. The
upside for biotechnology firms in such a circular system is that dependence
on technological solutions ensures a continued generation of new markets
(Levidow 1999; Levidow and Tait 1995; Schmitz 2001).[20]

Reacting to such one-sided representations of technological development,
Darier articulates the need for Canadians to make collective decisions about
what kinds of technology they want, at what cost, and for what purposes.
Yet he also laments the lack of a suitable political space to conduct such
debate within the existing political structure, aside from the formal ones
such as the House of Commons. Unfortunately, even there, this broader
type of debate rarely happens, and certainly not in any depth, according to
Darier. In part, this democratic deficit is one of the reasons that Greenpeace
does not follow everything in Canada, and nothing in Ottawa, because even
in those few instances in which the public is consulted such events are insubstantial and insincere staged public relations exercises designed to ratify
what has long since been agreed upon behind closed doors. Greenpeace is
therefore very selective, choosing to participate only when it sees an opportunity to expose the issue and the consultation process in mainstream
media and to the wider public. That, as Darier comments, is the sad reality
in Canada today (Interview).

As is no doubt clear from the evidence thus far presented, a key theme
that emerges among a number of the groups struggling against particular
aspects of agricultural biotechnology is the need for greater citizen participation in scientific and technological decision-making and policy development.
Similar to the admonitions advanced by critical commentators going back
two decades when biotechnological research and development began exhibiting the scale and scope we know today, contemporary critics recognize
that attempts to insert their voices hold the most potential for success at
the outset of the research and development stage, when change is much
easier to effect than is the case once a technology or scientific development
has been deployed (Young 1985; Yoxen 1983). Cognizant of the predominance of the business-government relationship as the dominant power relationship driving Canadian science and technology policy, the people and
groups involved in critical debates over biotechnology development insist
that the reflexive manner in which science is socio-politically constructed
necessitates the injection of far more democratic mechanisms for steering
policy debates around this science.

Critical Dismissal of the CBAC Consultations

In part, the intransigence within the government against expanding the
terms of the biotechnology debate beyond safety issues helps explain the
unwillingness of a number of groups to engage in public consultations,

particularly those organized by the CBAC. Although these findings might be considered moot given that the CBAC has been disbanded, they certainly can be extrapolated to government consultation processes and thus warrant some discussion. As briefly outlined in Chapter 1, the CBAC originally identified and engaged in five special research projects around issues of the regulation of genetically engineered food; intellectual property concerns related to biotechnology; novel uses of biotechnology, such as stem cells; consequences for privacy that emerge around biotechnology; and how to integrate ethical and social issues into biotechnology policy (Government of Canada 2004). Between 2001 and 2007, it published a number of reports that it either commissioned or authored in response to these research areas (Canadian Biotechnology Advisory Committee 2002a, 2002b, 2004a, 2004b, 2006a; Expert Working Party on Human Genetic Materials Intellectual Property and the Health Sector 2005).

Although the CBAC contracted Decima Research to conduct telephone interviews and focus groups in order to elicit Canadian attitudes toward personal privacy and medical biotechnology, the only other research stream that attempted to engage a variety of stakeholders around the country related to the regulation of genetically engineered foods. The remaining projects were executed either in house or through the use of expert advisory bodies commissioned by the CBAC. This pronounced lack of enthusiasm for public consultation, aside from traditional hesitance among policy elites to avoid anything but perfunctory lip service to transparency, might also be explained by the experience concerning the genetically engineered food file.

Rather than hold consultations open to the general public, the CBAC engaged in what it termed multi-stakeholder meetings on an invitation basis only. Yet, from the outset, almost all of the groups organizing around agricultural biotechnology boycotted the public consultations organized by the CBAC on food biotechnology issues, accusing this governmental advisory body of not being interested in a true public consultation process. As Sharratt (2002, 41) charged, the CBAC consultations were "designed to control public input and legitimate government decisions." In general, the CBAC project on the regulation of genetically engineered food was criticized for the way it framed the issue – how to regulate genetically engineered foods rather than a more fundamental discussion and debate about whether such food should be grown in the first place (Abergel and Barrett 2002; Magnan 2006). In a petition sent to the Government of Canada, fifty groups justified their decision not to actively participate in public consultations in the following manner: "We believe the [CBAC process] is fundamentally and importantly flawed and that NGO participation in the consultation could legitimate CBAC's wholly inadequate mandate and process and undermine demands for true democratic processes and widespread public consultation" (as cited in Tansey 2003, 62).

Herb Barbolet, who participated in early CBAC meetings, contends that, in its initial constellation, the CBAC contained a group of open-minded industry members who indicated that they would be willing to accept constraints on the biotechnology sector so long as they were clear and reasonable. Interestingly, it was government representatives who, according to Barbolet, dug in and asserted that industry constraints were infeasible. Very soon thereafter, new, staunchly pro-industry representatives were brought to the table, as well as a new chair, whose impartiality was suspect, given her work as a business consultant. Additional personnel changes resulted in the influx of a number of people who clearly recognized that their career interests would be ill served if they were to adopt a stance too critical of industry. Barbolet adds that a new and somewhat confrontational chair, Arnold Naimark, exacerbated the increasingly poor working environment within the CBAC (Interview).

The effect of such changes was a membership iteration through which, as a number of other commentators point out, the CBAC became an unabashed promoter of the biotechnology industry on behalf of the government and lost any real interest in balanced dialogue and debate. Sharratt, who also participated in some of the initial consultations held by the CBAC, points out that this body committed some major errors in judgment from the outset of its consultation process. She accuses it of showing a "real ignorance and a real arrogance in its attempts to exclude certain critical perspectives from becoming members," instead preferring for membership those people who had previously proven to be vehement biotechnology industry proponents (Interview). GE Alert, which included interview participant Ann Clark, submitted position papers to the first few CBAC calls for public consultation, and Clark was nominated to be on the CBAC.[21] However, Clark and others from GE Alert eventually came to the conclusion that this advisory body was paying no attention to what they were writing and that membership in it was for the handpicked few, after which they declined to respond to further calls for proposals. Interview informant Devlin Kuyek (2002, 75) has been even more critical in his assessment of the advisory functions of the CBAC in Canada: "It is understood from the outset that the government and the advisory bodies share a common agenda. The advisory bodies, and the government itself, are only there to act out the roles of and make a few adjustments to a script that, in many ways, has already been decided upon behind closed doors."

Given such practices, it is not surprising that the groups organizing against particular aspects of Canada's Biotechnology Strategy dismissed public consultations with the CBAC. The NFU, following the strategic decision among many other groups, similarly chose not to participate in the CBAC consultations. At the time of these consultations, the NFU actually lacked a well-developed policy on biotechnology. With regard to possible participation in future public consultations, Boehm expresses the concern that it is always a very difficult internal debate with two different schools of thought – one that you have to be at the table and the other that you are being co-opted when you are there. For the NFU, the decision depends on the specific issue and where it thinks it can do the best work to advance its fundamental goal of creating social and economic justice for family farmers. As mentioned previously, Boehm believes that, not only have the terms of the debate been completely confined to what is called "sound science" but also that the government and industry committee members in such fora tend to lay claim to being the only ones to possess this "sound science" (Interview). Thus, from the perspective of the NFU, the CBAC and similar bodies would not be very successful places to participate.

When asked about the CBAC and possible participation in future consultations, Cathy Holtslander is unequivocal in her response: "CBAC is a completely illegitimate organization. It's set up as an apologist for the industry, so we wouldn't participate in that." She adds that "even participating in CFIA is a bit problematic because you know, are they sincere? Is this just a matter of going through the motions so they can say they did consultations before they bring in their plans anyways? And it's always a problem. It's something NGOs are always struggling with. Inside versus outside strategy" (Interview).

Given Kuyek's overall dismissal of the CBAC, it is not surprising that he rejects participation in any of its consultations. He does not, however, completely dismiss consultation processes in general and has, in fact, encouraged involvement in the more recent seed variety registration public consultations conducted by the CFIA. Nonetheless, he points out that engaging with government processes can be very difficult and time-consuming. Involvement in these exercises demands a lot of energy and resources, and, oftentimes, little of what an organization actually advocates is ultimately included in the policy. As Kuyek points out, the government offers no resources to support people's participation. Since civil society organizations typically have limited time and resources available, what a group can achieve is usually pretty narrow, especially since things often are decided elsewhere, irrespective of opinions put forward during the consultation process (Interview).

Accounting for Capture and Control of Biotechnology Discourse in Canada

The material offered in this chapter suggests a need to probe beyond the discourse constructed by governmental and corporate actors that positions biotechnology as a panacea for problems such as hunger, disease, and low crop yield due to environmental conditions. As Antonio Negri (2005, 204) tells us, "capital, having itself become social" is optimally positioned to obscure "the contours of the totality" in order "to disguise its hegemony over society and its interest in exploitation, and thus to pass its conquest off as being in the general interest." By manipulating the scope and tenor of the discourse concerning biotechnology in this country, capital is actively engaging in tactics designed to control the discursive norms and institutional contexts that surround this technoscience. Yet, as was made clear in the discussion about the myths of genetically engineered seeds and the growing body of scientific evidence outlining some of the adverse impacts of genetically engineered food on health and the environment, a number of industry claims about the benefits of biotechnologies are increasingly being called into question. That is, the benefits ascribed to agricultural biotechnology are susceptible to challenge precisely on scientific grounds, the one and only domain that industry and government concede as being legitimate for assessing biotechnologies.

I suggest that the intransigence of industry and government regulators to expand the terms of the biotechnology debate is designed to serve capitalist accumulation purposes. Recall our previous characterization of primitive accumulation as comprising those strategies designed to establish and maintain the social conditions conducive to capitalist valorization. According to Marx, precisely because class struggles are inherent to capitalist relations of production, capital is compelled, in order to safeguard its existence, to engage in strategies of primitive accumulation that provide the basis for accumulation proper. Any object that threatens the historically contingent balance of power between classes represents an impediment to capitalist accumulation and is thus susceptible to capitalist strategies of primitive accumulation. Viewed from this perspective, strategies of enclosure aimed at eliminating obstacles to capitalist processes of accumulation legitimately can be categorized as instances of primitive accumulation.

I posit that the efforts by multinational biotech companies, and, to a lesser degree, by Canadian regulators, to limit debate surrounding biotechnology to issues of safety and science should be construed as a discursive strategy of enclosure in service of primitive accumulation. By marshalling the considerable resources at its command, including a sympathetic government willing to contribute its own scientific and financial capacities, capital is able to engage in public relations campaigns, vitriolic attacks on opponents, intense lobbying, and efforts to stifle the dissemination of information unsympathetic to industry-controlled biotechnology. All of these activities are designed to shape and constrain the discourse around biotechnology in a manner that limits the knowledge commons in respect of this science and its applications.

The message being disseminated by industry and government, which often uncritically equates technology to progress, is that biotechnology is a neutral science and technology that will deliver economic prosperity for the country. Yet, as we have seen, many of the interview respondents reject the neutrality

thesis and its corresponding specious and teleological claims about the putative capacity of science and technology to guarantee social progress. Again,
we need only refer back to Marx and his chapters in volume I of *Capital*
(1992) on "Co-operation," "The Division of Labour and Manufacture," and
"Machinery and Large-Scale Industry" to note his own critical dismissal of
the apparent neutrality of technology as the inexorable outcome of modern
science. He instead demonstrated how science and technology are more
properly construed as being heavily influenced by the capitalist valorization
imperatives under which their development occurs. Put more explicitly,
Marx (1975b [1844]) recognized that the design, development, and deployment of machinery and technology emerge under particular socio-historical
conditions.

More contemporary observers are similarly resolute in their assessment:
"Scientific technology is not, in fact, 'neutral' in its meanings or value, but
is a vast global system of moving parts that materially reproduces the transnational corporate order as a totalizing mechanism to serve an absolutist
value-set of turning money into more money for investors" (McMurtry 2002,
97). Herbert Marcuse (1968a, 1968b, 120, 130) elaborated these themes in
his analyses of the ways that technological rationality infiltrated the social
realm, which works to the oppressive detriment of humanity: "[S]cientific-
technical rationality and manipulation are welded together into new forms
of social control ... Today, domination perpetuates and extends itself not
only through technology but as technology, and the latter provides the great
legitimation of the expanding political power." Followed to its logical conclusion, political questions risk being rendered as technical issues to which
only experts might respond, thus replacing political rationality with technological rationality in a manner that serves to denude critical opposition of
its legitimacy.

Notions of technological neutrality, necessity, or rationality serve as ideolo-
gies to choke off debate and opportunities for active citizen participation in
determining the types of human needs technology should be designed to
promote. The obvious benefit, to industry at least, of framing any discussions
about biotechnology in such a manner is that these frames immediately
foreclose admittance to broader social, environmental, ethical, and political
concerns. In eliding these broader issues that attach to this technoscience,
both the Canadian government and industry have been carefully crafting
messages designed to disarm opposition in order to facilitate continuing
processes of primitive accumulation of seeds in this country. Put more explicitly, we are witnessing discursive battles in Canada that portend very
real material consequences for the developmental trajectory of agricultural
biotechnology and our seed and food production systems more generally.

In order to re-assert some semblance of democratic control over technological development and implementation, a critical approach to technology

must reject its neutrality and instead admit the constituting influence of the values and interests of the dominant social system and its ruling classes over the design and deployment of new technologies. That is, we must recognize that technological development contains an inherent bias toward the status quo of a given social and ideological constellation. As Marcuse (1968a, 223-24) pointed out,

> [n]ot only the application of technology but technology itself is domination (of nature and men) – methodical, scientific, calculated, calculating control. Specific purposes and interests of domination are not foisted upon technology "subsequently" and from the outside; they enter the very construction of the technical apparatus. Technology is always a historical-social project: in it is projected what a society and its ruling interests intend to do with men and things.

Yet springing from this admittedly pessimistic account, we might also conceive technological development to be an ambivalent and socially contested process in which different alternatives compete, though certainly not always from an equal position of power.

Technological design and development might thus be construed to be ontological decisions with substantial political, social, and economic implications (Feenberg 1999, 2002). A full appreciation of the opportunities for shaping the trajectory of biotechnological development therefore presupposes a firm grasp of the social and economic conditions that prevail during its developmental stages. As the evidence presented in this book confirms, biotechnology in Canada contributes to capitalist accumulation in a manner that both presupposes and reinforces dominant social relations. Part of the struggle being waged by opponents of agricultural biotechnology is finding ways to open these social relations to critical scrutiny and contestation. By conceptualizing technological development in this recursive manner, these groups hope to glimpse and exploit opportunities to resist or even alter a particular path that might otherwise appear beyond control – that is, they seek to open pathways for resistance.

Conclusion

As I hope to have demonstrated convincingly in the preceding pages, biotechnology affords capital a scientific instrument to enclose genetic resources that have been cultivated in common for millennia. It seems to me the situation illustrated in the earlier pages in respect of agricultural biotechnology confirms the basic ontological connection between primitive accumulation and expanded reproduction, as originally articulated by Karl Marx. In an instance of primitive accumulation facilitated by technological development and state policy, capital is striving ardently to subvert the product of countless generations of common labour for its own valorization purposes. Some of the enclosure strategies of which capital avails itself include infiltration of the regulatory system and deliberate misinformation campaigns that emphasize only the purported benefits of such technologies. Capital's goal is to garner public acceptance and curtail the development of other more sustainable approaches to the problems that industry claims biotechnology can solve, since these approaches are not readily subverted in service of capital accumulation imperatives. Through state-sanctioned mechanisms, such as intellectual property protection, capital avails itself of an effective instrument to exploit the products that derive from, and might otherwise promote, the commons.

Intellectual property claims over genes and other forms of life thus represent private title over social and historical processes of knowledge creation. The contradictions between increasingly social forms of production and the private appropriation of the products of such production are amplified by the increasing pressure for technological innovation and consequent applications designed to facilitate surplus value extraction. This pressure results from the inherent contradictions of capitalist accumulation imperatives, including struggles against labour. As Werner Bonefeld (1992, 107) writes, "[c]apitalist social reproduction is social reproduction in inverted form: private production in a social context." Enclosing seeds and the knowledge they embody through bundled technological packages and intellectual property mechanisms have helped the biogopolists secure enormous profits and consolidate their control over the biotechnology industry. Consequently, this technoscience is being rendered susceptible to the exigencies of atomized market exchange relationships leading to an inequitable distribution of the social wealth generated by human activity. The private ownership of the so-called "code of life" by "life science" multinationals should be recognized for what it is: an affront to the public character of our species-being (Barber 2001). In order to exert some control over our own humanity, it is thus vital that we reclaim political and economic control over the developmental trajectory of such technologies in a manner that opens up a space for informed and reflexive debate.

By employing concepts such as primitive accumulation, enclosures, and commons, the theoretical schema developed in this book is an attempt to account for efforts by capital and the state to render agricultural biotechnology subservient to capitalist social and production relations as well as opponents' struggles against these efforts. This schema is meant to do so in a way that apprehends the informational and terrestrial commons implicated in current engagement around this technoscience. In addition to admitting broader issues of social change, this approach to analyzing the place of biotechnology in society today politicizes its development by conceptualizing it as a science and technology that has implications for access to, and control over, what I have termed the BioCommons. The people opposing agricultural biotechnology in this country are united not by particular class interests but, rather, by a general rejection of the property regimes that capital has invoked around the products produced by this technoscience. They are struggling against capital's attempts to appropriate and exploit agricultural biotechnology in order to inject some degree of democratic control over its developmental trajectory.

In rejecting the universalisms promoted by capital based on commodification and the "free" market, these people are mobilizing based on practices of co-operation, communication, social solidarity, and justice across town and country with a pronounced sensitivity to the environmental issues thrown up by biotechnology that hold serious implications for humanity and humanity's relationship to the natural world. The people involved in such biopolitical activism are interested in recovering the scientific knowledge and social co-operation in agricultural biotechnology from the control of capital. They are seeking to appropriate and re-deploy it in ways that might help us realize precisely those universalisms that the capitalist market promises but will never be able to deliver. By no means complete and always subject to appropriation by capital, these biopolitical movements nonetheless are developing alternative pathways, connections, and networks designed to build oppositional capacity from below that is resilient enough to contest capital's control over biotechnology.

In responding to the research questions set out in the introduction of this volume about the counterstruggles engaging capitalist enclosures of the BioCommons, and the information issues implicated in these struggles, I tend to agree with Michael Hardt and Antonio Negri (2004, 102 [emphasis in original]) that mapping such resistance is a useful exercise in promoting the autonomy of oppositional subjects:

> The places of exploitation ... are always determinate and concrete, and therefore we need to understand exploitation on the basis of the specific sites where it is located and specific forms in which it is organized. This will allow us to articulate both a *typology* of the different figures of exploited labor and a *topography* of their spatial distribution across the globe. Such an analysis is useful because the place of exploitation is one important site where acts of refusal and exodus, resistance and struggle arise.

Along these lines, and by way of conclusion, I thus offer the following summary overview of the dominant issues articulated by various groups in their struggles against biotechnology in this country.

As we have seen, biotechnology continues to elicit a range of resistance in Canada around the following issues: multinational corporations' control of seeds, the lack of rigorous scientific study of some of the claims made by industry and other biotechnology proponents, uncritical media championing of biotechnology, the heavy industry slant of government consultation bodies (particularly the now defunct Canadian Biotechnology Advisory Committee [CBAC]), the lack of funding for public good agricultural research, and the enclosing effect that high levels of patentability are having on biological resources and information. Opposition is organized predominantly around the genetic engineering of agricultural crops and food and has involved a number of prominent organizations, including the Canadian Biotechnology Action Network, the National Farmers Union (NFU), Greenpeace, the ETC Group, and the Council of Canadians. While the subjects mobilizing against agricultural biotechnology described in this book can hardly be categorized as anticapitalist, many do share the underlying goal of responding to, and resisting, the destructive repercussions that accompany the expansion of capitalist enclosures enveloping increasing swaths of biological existence (terrestrial commons).

Those critical of capitalist-controlled agricultural biotechnology recognize that the capitalist appropriation of biotechnology in service of broader accumulation imperatives impedes the free development of human capacities that rely on mutual co-operation and respect among people and between people and their natural environment. This point is made not to reduce the diversity of the various struggles emerging in response to these new forms of primitive accumulation – indeed, the manner in which such struggles define themselves is both historically and politically contingent – but, rather, the intent is to establish a basis of commonality through which we can identify and map the potentialities for transversal interlinkages among groups that might give rise to a broad-based and global attack on capitalist social relations. While this volume is limited to the Canadian context, some of the lessons might provide a starting point for future projects to respond to David Harvey's (2006) admonition that multiple groups resisting capitalist-ordered social relations identify international connections that could foster collaboration on a global basis.

All of the interview informants resisting biotechnology in Canada voiced concern about the corporate control of seeds and agriculture as well as the power such corporations are able to wield in political and economic forums. Many of the interview respondents articulated a strong desire to establish some form of civil society control over genetically engineered seeds. There is a belief among those struggling against biotechnology that the commodification of life in the wake of this technoscience is altering our relationship to life at the expense of people and community. Indeed, talk of the "biotechnology revolution" found in much of the celebratory discourse surrounding this technoscience masks the fact that its development and application are being driven largely by accumulation imperatives to expand capital's reach into deeper areas of social existence. At an even more fundamental level, we need to recognize that science is inherently political because the concepts and theories about nature that it develops are firmly rooted in the social and political ethos prevailing at the time of their development. Such a conceptualization should not be construed as an appeal to relativism but, rather, as an admonition to bear in mind that politics do impact the way science is practised in a particular historical epoch (Diamond 1981). Thus, while a case might be made that the science behind biotechnology is revolutionary, the neo-liberal context in which it is applied is really business as usual in terms of developing new revenue streams in what can be considered novel instances of primitive accumulation.

Such laudatory and often misleading discourse, as I have endeavoured to demonstrate throughout this book, must be subjected to systematic critique. Behind the apparent revolution is a more fundamental issue about access to, and control over, biological and knowledge resources. In order to facilitate contemporary enclosures, capital and government have sought to confine debate about biotechnology to aspects of health and science. Since such discourses, by their nature, are technical and abstract, they typically appear opaque and foreign to the majority of the population, which helps render them beyond "popular" critique. This type of discursive construction tends to discount the public as ill informed, uncritical, and susceptible to manipulation by what champions of biotechnology tend to disparage as "self-serving" non-governmental organizations (NGOs) and, to a lesser extent, the media.

However, in terms of educating the public, many scientists seem interested only in monologic communication. That is, scientists often adopt a "deficit model" when considering resistance to biotechnology, believing that opposition stems from a lack of knowledge about biotechnology. Yet gunpowder, DDT, thalidomide, and napalm, among others, are all examples of technology that biotechnology exponents conveniently forget to mention when naming litanies of technologies alleged to be beneficial and opposed at the initial stages of their development.

Industry proponents speak of the radical and putative beneficial transformations that biotechnology will have for humanity, yet they continue to lobby hard for keeping the regulatory status quo, arguing that these new genetically engineered products are substantially equivalent to what has come before. The inherent paradoxical nature of such corporate promotion appears lost on both industry and Canadian regulators. Moreover, the way regulators define genetic technology as a science issue restricts debate by excluding a range of other points of view and interests. It also affects the way science (at least corporate science) itself is defined and practised. Veils of secrecy invoked by capital through trade secrets, proprietary information, and patents tend to place important data beyond the purview of government reviewers and certainly beyond the reach of the public and other scientists, thus violating the basic tenet of openness in the scientific enterprise.

Yet government continues to reject the demands articulated by several interview informants that the scientific data provided by companies seeking regulatory approval be made public for independent testing and review. Indeed, the lack of transparency in the regulatory review process remains a source of criticism, including from the drafters of the Royal Society of Canada's report. The dual regulatory/promoter role of the Canadian Food Inspection Agency continues to evoke critical concern among all of those groups critical of biotechnology in this country. Some interviewees did, however, suggest that regulators are waking up to the presence of mobilized opposition groups and the broader issues and critiques they are levelling, even if this is still not being admitted out loud in Ottawa.

All of this similarly has decisive implications for stakeholder consultations, since such bodies are struck based on an underlying agenda that uncritically accepts the purported benefits of biotechnology and promotes industry development as a powerful economic goal. The result is a consultation process that limits the type of information solicited and, thus, the scope of ideas developed, which ultimately serves to constrict policy options. There is unanimous agreement among all of the activists that public consultations in respect of biotechnology in this country have been staged public relations exercises designed to ratify decisions previously reached behind closed doors. The CBAC, in particular, was criticized by all groups for being populated with industry cheerleaders and for functioning according to a pre-ordained mandate designed to champion this technoscience as a new motor for Canadian economic growth. Those individuals more sanguine about possible future consultation processes nonetheless expressed concern about the difficulty of manoeuvring a fine line between participation and co-optation.

Many citizens contesting capitalist-controlled biotechnology believe that their success in responding to enclosures on the BioCommons hinges upon the ability to inject broader knowledge issues into the public debate surrounding this science and its attendant technologies. Those interviewed recognize that biotechnology is a complex topic and that the science that surrounds it has been purposely mystified to discourage average citizens from debating the political, economic, environmental, ethical, and moral issues involved. Capital is exploiting the complexity of the science around biotechnology in a deliberate attempt to circumvent a deeper understanding of its inherent issues and implications, thus mitigating the rise of autonomous possibilities for transformative re-appropriation. By asserting a stranglehold on the nature and flow of information about biotechnology, capital and the Canadian government are trying to circumscribe the breadth of issues that might legitimately be debated.

In response, many biotech activists are engaging in research and rigorous analysis to ensure that those most affected by this technoscience are adequately informed so that they may make their own reasoned decisions. That is, almost all of the interview informants perceive the need to respond to a substantial information gap in this country with regard to biotechnology. Many groups are therefore engaging in information dissemination campaigns designed to correct misinformation and to offer balanced accounts of biotech issues that advance beyond the one-sided, laudatory content propagated by biotechnology proponents. Some activists sense that biotechnology issues must be rendered relevant to the urban population. All acknowledge that communicating to an assorted array of people who are affected by biotechnology in different ways is complex and difficult.

People struggling against biotechnology value networks based on practices of co-operation on issues of mutual interest, information sharing, and grassroots activism. Information dissemination relies heavily on relationships with like-minded organizations, public meetings and other fora, membership discussions, witness testimony at parliamentary committees (although only the NFU seems to possess enough clout to appear regularly before committees), e-mail and listservs, websites (although there are difficulties associated with keeping websites current given the speed of developments and resource constraints), one-on-one grassroots conversations (in this respect, mainly Greenpeace), op-ed pieces, research papers and briefs, and letter-writing campaigns. All of the activists recognize the critical necessity of disseminating information through common networks to alert the broader public about what is happening in respect of multinational agribusiness.

A fundamental element of the information production and dissemination efforts these groups have tasked themselves with is an analysis of the corporate control of biotechnology and, more broadly, agriculture. A variety of respondents believe that the legal constructs around genetics and biotechnology have been detrimental to the economy of ordinary farmers, such that previously common biological resources and information are being privatized and commodified. There is a perceived need to determine and make public who owns and controls what and to analyze the implications of such ownership patterns for the types of products being developed as well as their social implications once they are introduced into society. Respondents are also struggling for a mandatory labelling regime for genetically engineered products, which is viewed as a vital instrument in removing genetically engineered organisms from the food chain. At a basic level, the lack of a mandatory labelling system is an information issue that, from a purely market perspective, creates information asymmetry and thus violates the assumptions of neo-classical economics. The irony, however, remains lost on industry and government, both of whom continue to oppose mandatory standards.

While one informant made the case that some success can be had if you know how to plot your message into the media, most other informants were very skeptical about media coverage of biotechnology issues – reasons include superficial and/or biased reportage, the concentrated structure of the media industry, and fees for newswire service that most groups are unable to afford. According to various interviewees, part of the difficulty in expanding the debate around biotechnology is that the media tend to take their cues from industry and government. They frame coverage about this science and its attendant technologies almost solely in terms of business potential or food safety implications, which contributes to an individualized consumption discourse. In the words of Ann Clark,

> the biotech stuff, the real biotech stuff appears on the business pages as stock promotion so it's quite clear that the corporate concentration includes interlocking directorships between the banks, the media, the pharmaceuticals, the petrochemical companies, the food companies. So it's not a surprise that the media are not covering it or if they are covering it they are covering it as page one news that there may possibly be a breakthrough that in the future could possibly help 0.1 percent of the population with a hangnail, but the side effects might be death and then the stock price goes up so everybody's happy.

Speaking to the situation in Saskatchewan, which is considered to be big biotech country, Cathy Holtslander, an interview participant who has been working on issues of agriculture and genetic engineering for over a decade, believes that most media tend to avoid stories that might have a negative connotation for the biotechnology sector and, instead, focus on the more glamorous breakthrough-type stories. Since most industry hype is based on innovations that have yet to appear, a lot of early activist work has been trying to shatter such myths. This task, however, is complicated by weak media coverage that fails to engage in any real analysis or investigative work, which is particularly troublesome given the emphasis the Canadian government places on positioning this country at the forefront of biotechnology and genetic engineering. A number of people interviewed contended that the media fail to question the prevailing economic model and, instead, have uncritically bought into the idea about the dominance and importance of the "knowledge economy," including Canada's need to become a leader in technological innovation. The superficial central mantra about progress through innovation invoked by the media and government has circumvented any critical public dialogue and debate about what all of it means, including the broader implications that this science and these technologies pose for society and humanity. Instead, the Canadian media, or at least most of the Canadian media, have blindly accepted, and repeated, the message that the biotech industry puts out. This type of "sound bite" reporting, as Devlin Kuyek terms it, fails utterly in enhancing people's perspectives.

However, Kuyek believes it is not lack of popular interest that is motivating media silence. Journalists who want to take up this issue can draw on a wide pool of knowledgeable people working on biotechnology issues, so it is not difficult to put together a balanced story. The few journalists who have done this have produced good reportage, but it is very sparse. As outlined previously, some of the very minimal critical reporting of biotechnology issues in this country has tried to alert the public to the oblique links between the biotechnology industry, government, and a number of ostensibly independent NGOs. Particularly with respect to the latter groups, Lyle Stewart (2000, 2001, 2002a, 2002b, 2002c, 2002d) shows that a number of them that are dependent upon government funding tend to adopt positions that mirror, and conform to, the communication strategies pursued by the government-industry nexus.

Indeed, industry and government, as we have seen, make adept use of public relations campaigns designed to garner public acceptance of biotechnology, usually through various lobby groups that promote themselves as neutral arbiters in debates around this technoscience. This type of almost covert biotechnology advocacy is supplemented by more aggressive lobbying carried out by industry groups, such as the Biotechnology Industry Organization, CropLife International (formerly known as Global Crop Protection Federation), CropLife Canada (formerly the Crop Protection Institute), and BIOTECanada (Hisano 2005). Part of the public funds obtained by these biotechnology lobbyist groups was targeted toward various campaigns designed to alter favourably the Canadian public's perception of the biotechnology industry. As Kuyek (2002, 71) scathingly points out, "by contributing to BIOTECanada, the government is essentially paying the industry to lobby government."

Overall, we have seen that capital, with the aid of the state, is endeavouring to co-opt the emancipatory capacity of science and technology, which is one of the reasons why it is so important to reconstruct science in a manner that admits the interaction between it and technology, capital, and society. Accepting such co-construction legitimizes attempts to inject social values and justice, democracy, and ecology into debates about the development and application of agricultural biotechnology. Situating science and technology in a socio-political context conditioned by both power interests and struggle opens science up to external and political readings that firmly reject the myth that science develops according to its own internal dynamics based on objective truth and reason. That is, it leads the way to an epistemology of science that construes science as a politically contested social construct that can be emancipated from capitalist control and domination to be diverted in service of fulfilling the ethical, social, political, and ecological demands of humanity. As Hardt and Negri (2004, 283) write, "[t]his is not a protest against technology, in other words, but against the political powers that decide without the representation of those primarily affected to privatize the common, enriching the few and exacerbating the misery of the many."

Many of the Canadian activists included in this work ardently believe that science, technology, society, and humans interact and co-evolve in a holistic social context, meaning that science and technology are not neutral but, instead, are influenced in their development and application by particular social interests and biases. In concrete historical terms, the people mobilizing against particular aspects of agricultural biotechnology are struggling to throw open to scrutiny and democratic debate the ways that biotechnology is being successively appropriated by capital in its efforts to expand accumulation and profit – referred to in a previous chapter as contemporary processes of primitive accumulation at the genetic level of existence. Given the far-reaching implications that biotechnology portends for a range of complex organic life forms, including humanity and the environment, a determined group of biopolitical activists are dedicating themselves to ensuring that such resources and their developmental trajectory remain within the BioCommons. While the major battles waged to date have all been protracted, successes have been had against recombinant bovine growth hormone and genetically engineered wheat. At an even more fundamental level, I believe that, by maintaining pressure on corporations and government, the resistance against agricultural biotechnology has thus far achieved an important victory in circumscribing the range of genetically engineered products that industry and government champions first prophesized would be introduced into Canada.

I would like to conclude both this chapter and the book by referencing another admonition articulated by Hardt and Negri (2004, 284-85):

> This series of biopolitical grievances allows us to recognize and engage the ontological conditions on which they are all established, something like what Michel Foucault calls the critical interrogation of the present and ourselves. "The critical ontology of ourselves," Foucault writes, "must be considered not, certainly, as a theory, a doctrine, nor even as a permanent body of knowledge" but rather as "the historical analysis of the limits imposed on us and an experiment with the possibility of going beyond them." The legal, economic, and political protests that we have considered are all posed on this ontological foundation, which is crisscrossed by powerful and bitter conflicts over goals that invest the entire realm of life.

Ideally, the research presented in this volume has illuminated the multiple points of struggle being opened up by an assorted collection of Canadians around the contested nature of agricultural biotechnology in this country. Yet, more importantly, I ardently hope that it has done so in a manner that broadens the terms of the debate surrounding this technoscience by admitting deeper perspectives – perspectives that not only reflect technological and scientific concerns about biotechnology but also question and challenge more fundamental issues about the capitalist control that seeks to structure our lives and capture the developmental trajectory of this science and its technological applications.

Notes

Introduction

1 As defined in Canadian policy documents, biotechnology refers to "the application of science and technology to living organisms, as well as parts, products and models thereof, to alter living or non-living materials for the production of knowledge, goods and services" (Canadian Biotechnology Advisory Committee 2005b, ix).

2 Other countries and regions that have developed national biotechnology strategies include: Australia, the European Union, India, Japan, New Zealand, and Russia, which, in 2006, inaugurated a national biotechnology program that has received 150 billion roubles (US$5.25 billion) in funding until 2015 ("Russia to Spend 150 Billion Roubles" 2006). According to a European Commission document from 1991, "biotechnology is a key technology for the future competitive development of the Community and it will determine the extent to which a large number of industrial activities located within the Community will be leaders in the development of innovatory products and processes" (as cited in McNally and Wheale 1998, 305). Although the United States does not have a national strategy, many individual states have incorporated biotechnology programs into their economic agendas. Similarly, a number of other countries around the world have begun focusing special attention on biotechnology, including Chile, Cuba, Malaysia, Singapore, and South Korea (Canadian Biotechnology Advisory Committee 2006b).

3 All dollar amounts are listed in Canadian dollars unless otherwise indicated.

4 These figures were obtained from the "Plants with Novel Traits" database, which is maintained by the Canadian Food Inspection Agency (http://active.inspection.gc.ca/).

5 The Genome Canada website provides a significant amount of information about the predicted benefits that genetics and proteomics hold for human health and the environment, including a list of all of the research projects currently being funded through this federal organization (http://www.genomecanada.ca/).

6 There have been a number of useful accounts written about the history of genetics, from Gregor Mendel onwards, and the development of biotechnology. See, for example, Freeland Judson (1992); Hindmarsh and Lawrence (2004); and Nelkin and Lindee (2004).

7 Bud (1993); Finn (1989); and Kay (2000) offer helpful volumes that outline the origins of biotechnology.

8 Deoxyribonucleic acid (DNA) is a large molecule (polymer) comprising nucleotides, which themselves are composed of three elements: a nitrogen-containing base, a five-carbon (pentose) sugar, and a phosphate group. Based on James Watson and Francis Crick's work, DNA is typically described as a ladder twisted along its long axis to form a helix. The well-known double helix shape results because there are two paired DNA molecule strands that run in the opposite direction of one another (anti-parallel). The sides of the ladder comprise alternating sugar molecules and phosphate groups, which are often referred to as the sugar-phosphate "backbone." The individual rungs of the ladder are formed by four chemical units called nucleotide bases. The bases are adenine (A), thymine (T), guanine (G), and

cytosine (C). Bases on opposite strands pair specifically; an A always pairs with a T, and a C always pairs with a G. The bases provide the coupling mechanism for the two strands of DNA. As a consequence of these particular pairings, the two halves, or strands, of the DNA molecule are complementary. For example, wherever one strand has an A, the complementary strand will have a T, a G on one strand will be faced with a C on the opposite strand, and so on. The human genome contains approximately three billion base pairs (Hopkins 2006).

9   Parry (2004) also explores this theme in her work.

10   The *Grundrisse* is largely a collection of unedited notes contained in a series of seven notebooks that Marx wrote in 1857-58 mainly for the purposes of self-clarification and preparation for his subsequent books *A Contribution to the Critique of Political Economy* and *Capital*. Foreign language publishers in Moscow first published this work in two volumes in 1939 and 1941, respectively. However, it was not until 1953 that the work became easily available to the Western world.

11   As pointed out in the preface to *Capital*, Marx opens the book with a presentation of capital – in particular, the commodity form – before discussing labour, because this is the way we encounter and experience capitalist society, thus making it a logical starting point. However, Marx also makes it clear that this exposition differs from his research method, which begins with labour, always recognizing its primacy.

12   A notable exception is the collection of essays edited by Schurman and Kelso (2003).

13   As Marx and Engels (1962, 47 [emphasis in original]) point out in the *Manifesto,* "to be a capitalist, is to have not only a purely personal, but a social *status* in production. Capital is a collective product, and only by the united action of many members, nay, in the last resort, only by the united action of all members of society, can it be set in motion. Capital is, therefore, not a personal, it is a social power."

14   My use of the term "discourse" follows that of Ricoeur as being "a shared set of concepts, vocabulary, terms of reference, evaluations, associations, polarities and standards of argument connected to some coherent perspective on the world" (as discussed in Rochon 1998, 9).

Chapter 1: Canadian Biotechnology Policy and Its Critics

1   Abergel and Barrett (2002) offer an historical overview of federal biotechnology policy. Devlin Kuyek (2002) provides a detailed and critical historical account of the development of biotechnology in Canada from 1980 until 2002. His work outlines how the originally publicly funded research and development in biotechnology was successively appropriated by private corporations beginning with the Conservative Mulroney government and continuing under the Liberal Chrétien government. He also examines the increasing infiltration of the federal policy-making phalanx by biotech industry insiders.

2   *Seeds Act,* R.S.C. 1985, c. S-8.

3   The Canadian Food Inspection Agency (CFIA) defines a plant with novel traits in the following manner: "A plant variety possessing a characteristic that is intentionally selected or created through a specific genetic change and is either not previously associated with a distinct and stable population of the plant species in Canada or expressed outside the normal range of a similar existing characteristic in the plant species."

4   Mutagenesis relies on chemicals or irradiation to induce mutation.

5   The CFIA was created in 1997 to consolidate the delivery of federal food inspection and quarantine services as well as plant protection and animal health programs, all of which were previously delivered by the Departments of Agriculture and Agri-Food, Fisheries and Oceans, Health, and Industry.

6   *Canadian Environmental Protection Act,* R.S.C. 1999, c. 33.

7   According to the website of the Canadian Biotechnology Advisory Committee (CBAC) (http://cbac-cccb.ca/), the Harper government's new Science and Technology Strategy announced the government's intention to "create a new Science, Technology and Innovation Council as part of a broader effort to consolidate external advisory committees to strengthen the role of independent expert advisors. The new council will provide the government with policy advice on issues referred to it by the government and will release regular state-of-the-nation reports that track Canada's science and technology performance and progress

against international benchmarks of success." As part of the Harper government's overall new Science and Technology Strategy, which is titled Mobilizing Science and Technology to Canada's Advantage, the Advisory Council on Science and Technology, the Council of Science and Technology Advisors, and the CBAC were disbanded and replaced by a single advisory body – the Science, Technology and Innovation Council – composed of representatives from federal science advisory entities, institutions of higher education, and the private sector. The overall tone of the federal document that lays out the new strategy makes it very clear that Canadian science and technology policy will now be influenced, if not determined, heavily by the private sector, with universities and colleges relegated to the status of incubator for ideas and research, which business is then invited to appropriate and commercialize. The new strategy specifies four key areas where the federal government will concentrate it efforts and resources: environmental science and technologies; natural resources and energy; health and related life sciences and technologies; and information and communication technologies. As we will see more clearly in the pages that follow, biotechnology is implicated in all four of these areas (Government of Canada 2007).

8   These are the latest available data from Statistics Canada as of mid-2012.

9   The notion of the "bioeconomy" is being promoted heavily by the Organisation for Economic Co-operation and Development (OECD), particularly in its latest book *The Bioeconomy to 2030: Designing a Policy Agenda* (2009, para. 1). According to the OECD, "the bioeconomy refers to the set of economic activities relating to the invention, development, production and use of biological products and processes."

10   Members of CBAN include the following: ACT for the Earth (Toronto); Biofreedom (Edmonton); Canadian Organic Growers; Check Your Head; Council of Canadians; Ecological Farmers Association of Ontario; Food Action Committee of Ecology Action Centre (Halifax); GE Free (Yukon); GeneAction (Toronto); Greenpeace Canada; Inter Pares; National Farmers Union; Prince Edward Island Coalition for a GMO-Free Province; Saskatchewan Organic Directorate; Society for a GE Free British Columbia; Union Paysanne; and Unitarian Service Committee of Canada. In addition, CBAN is supported by ETC Group (the Action Group on Erosion, Technology, and Concentration) and Rights and Democracy/Droits et Démocratie.

11   Although they have yet to attract the same degree of attention as agricultural products, the issue of genetically engineered trees is assuming increased prominence among activists opposed to the genetic engineering of living organisms. In April 2008, CBAN drafted an open letter to the Government of Canada, calling on it to immediately halt all current field trials of genetically engineered trees in this country, to not approve any future field trials, to discontinue public funding of field testing, and, at the then upcoming May 2008 Conference of the Parties to the *Convention on Biological Diversity* in Bonn, to support other nations in establishing a moratorium on field testing, planting, and commercial growing of genetically engineered trees. In Canada, the first field trial for 900 square metres of genetically engineered poplar trees that contained a marker gene attached to a wound-inducible promoter received regulatory approval in August 1997. A second field trial involving white spruce, in which an insect-resistant gene *(Bacillus thuringiensis)* was inserted, began in 2000. This later field trial involved poplar trees that contained an enhancer that altered the expression of surrounding genes to ascertain more information about their respective functions. All trials took place in Quebec at the Laurentian Forestry Centre, which is part of the federal Canadian Forest Service. Both trials were terminated in May 2007. In May 2010, the United States Department of Agriculture granted ArborGen approval to plant on 330 acres of land in seven southern American states over a quarter of a million eucalyptus trees genetically engineered to be cold tolerant, produce less lignin, and have altered fertility. This decision was made despite the fact that during the public comment period in early 2009 almost 17,500 comments were sent to the department by people and groups opposed to the planting of genetically engineered trees. Only thirty-nine favourable comments were received. On 1 July 2010, an alliance of various American conservation groups filed a lawsuit against the US Department of Agriculture to stop the release of these genetically engineered eucalyptus trees.

12   As set out in paragraph 2(1)(a) of the *Canadian Environmental Protection Act*, R.S.C. 1999, c. 33, the precautionary principle means that, "where there are threats of serious or

irreversible damage, lack of full scientific certainty shall not be used as a reason for post-poning cost-effective measures to prevent environmental degradation." This is a widely accepted definition that was endorsed at the Earth Summit in Rio de Janeiro in 1992 and written into the *Rio Declaration on Environment and Development,* 13 June 1992, 31 I.L.M. 874 (1992).

13 The *Cartagena Protocol on Biosafety,* which entered into force on 11 September 2003, 39 I.L.M. 1027 (2000), covers the transport across borders of living organisms engineered by biotechnological processes, including the adverse effects such movement could exercise on a signatory country's biodiversity. It is important to recognize that the protocol applies only to living engineered organisms. Lobbyists from the biotech industry, with strong support from the Canadian and American governments, engaged in fervent efforts to weaken the treaty so that signatory countries would not be able to restrict the import of genetically engineered products on the basis of concerns that such products could exercise negative social, economic, health, and environmental impacts on their populations. See, for example, Gwynne Dyer's (1999) commentary in the *Globe and Mail. Convention on Biological Diversity,* 31 I.L.M. 818 (1992).

14 In another stark example of the widespread opposition to genetically engineered foods in the European Union, a million signatures had been gathered by October 2010 in a legal attempt to stop the cultivation of genetically engineered crops in member states. Under the terms of the *Treaty of Lisbon* amending the *Treaty on European Union* and the *Treaty Establishing the European Community,* if a million citizens from a broad swath of member countries sign a petition to change a law, the European Commission must consider the grievance (European Union 2007).

Chapter 2: Enclosure and Resistance on the BioCommons

1 The rights to this invention were licenced to Eli Lilly and Company in 1982 when its human insulin received marketing approval from the US Food and Drug Administration. See Hall (1987) for an account of the race between different companies and research labs to synthesize human insulin.

2 True to the spirit of Wall Street (although, interestingly, Schneider worked out of Washington, DC, rather than New York), Nelson Schneider applied in December 1979 for a trademark on the term "biotechnology" when used "for magazines reporting scientific and financial developments in the field of genetics" (Bud 1993, 256 n 66). US Trademark 1180658, Provisional Registration, US Class 38, 3 December 1979.

3 See Robbins-Roth (2000) for a history, albeit uncritical, of the biotechnology industry up to the late 1990s.

4 Despite an exhaustive search that included, among others, Agriculture and Agri-Food Canada, CropLife, the National Farmers Union, the Canola Council of Canada, the Canadian Seed Growers Association, the Canadian Seed Trade Association, the Canadian Seed Institute, Grain Growers of Ontario, and Grain Growers of Canada, I was unable to find any organization in Canada that tracks the total amount of land devoted to genetically engineered crop production.

5 To calculate this number, I assumed canola is 97 percent genetically engineered (this figure is based on general increases in the amount of genetically engineered canola planted in Canada over the previous six years, as outlined in the next note). Using the Statistics Canada figures for genetically engineered and conventional corn and soybeans in Quebec and Ontario, I was able to calculate what percentage these genetically engineered varieties contributed to overall production of corn and soybeans. The numbers were actually quite similar for both provinces, with genetically engineered soybeans comprising 37 percent of total soybeans planted and genetically engineered corn comprising 42 percent. I then applied the respective percentages to the production of these two crops in other provinces to obtain a rough approximation of genetically engineered hectarage. In this way, I suspect that I have probably overestimated the amount of genetically engineered soybeans and corn in other provinces. In any event, the total production of soybeans and corn outside of Ontario and Quebec amounts to 240,200 hectares and 91,800 hectares, respectively. Although it is doubtful that farmers would spend extra for genetically engineered seeds to

produce fodder corn, I nonetheless included it in my calculations using the same percentages employed to calculate genetically engineered corn in provinces outside of Ontario and Quebec. I also included the full amount of sugarbeets that were genetically engineered since I was unable to obtain an idea of the size of the genetically engineered market for this crop (Statistics Canada 2011).

6   I employed the same methods as used previously to calculate the amount of genetically engineered hectarage, although the proportions of genetically engineered canola were 95 percent, 93 percent, 92 percent, 87 percent, and 84 percent respectively for 2010, 2009, 2008, 2007, and 2006. With the exception of 2010, these figures were provided by Denise Maurice, vice president of crop protection of the Canola Council of Canada. I was unable to confirm percentages of genetically engineered canola for the crop years 2010 and 2011 because Maurice died in late 2011, and no one at the Canola Council of Canada was able to access the information when I inquired in early 2012.

7   E-mail correspondence with Rhodora Aldemita, senior programme officer at the International Service for the Acquisition of Agri-Biotech Applications [on file with the author].

8   *Bacillus thuringiensis* is a naturally occurring soil bacterium genetically engineered into various crop seeds. Known as Bt prototoxin (crops that contain this gene are noted by the prefix "Bt" – for example, Bt-Corn), the bacterium produces a crystal protein that destroys the digestive tract of certain insects when ingested and mixed with stomach acid.

9   Canola is a particular species of the broader rapeseed genus that was developed in Canada in the 1970s. Prior to the development of canola, rapeseed oil was used mainly as an industrial lubricant. Baldur Stephannson and Keith Downey crossbred rapeseed to develop a new variety, canola, which contains important nutritional elements and which has become a major food oil crop. The name canola is actually a neologism for Canadian oil low acid, which is registered as a trademark by the Western Canadian Oilseed Crushers' Association (Busch et al. 1994; Tanaka, Juska, and Busch 1999). Monsanto received regulatory approval from Health Canada to market its first line of glyphosate-tolerant canola MON00073-7 (GT73, RT73) for food use on 21 November 1994. The Canadian Food Inspection Agency (CFIA) granted regulatory approval for environmental release and use in feed in the following year. A relatively up-to-date list of all genetically engineered organisms approved by Health Canada (though in Canadian government parlance the discussion revolves around "novel foods" rather than genetically engineered organisms) can be accessed at the following website: http://www.hc-sc.gc.ca/. The list provides no information about products currently being reviewed by the department. In fact, Health Canada considers such information proprietary and not only refuses to disclose details about a particular application but also rejects access to information requests that seek confirmation about whether or not a company has even submitted an application for regulatory approval of a particular product.

10   Health Canada granted Monsanto regulatory approval to market its first line of glyphosate-resistant soybean seeds MON04032-6 (GTS 40-3-2) for food use on 9 April 1996. Approval for environmental release and use in feed had been granted the year prior by the CFIA.

11   Monsanto's first Roundup Ready corn variety was approved for environmental release and use in food and feed in 1997.

12   "Mode of action" is the classification used to describe the way a pesticide works. Selective herbicides that are tolerated by a particular crop will kill or suppress specific weeds that infest the crop. Since no single selective herbicide is capable of controlling all weed types with equal effectiveness, weed control regimens based on selective herbicides require a combination of herbicides, applied either at the same time or sequentially. Selective herbicides are typically classified according to three criteria: (1) the type of weeds they control (for example, grass or broadleaf); (2) the timing of the application (pre-emergence or post-emergence of the crop); and (3) the duration of time the herbicide controls weeds (residual control). Non-selective herbicides, on the other hand, are active on all vegetation that is present at the time of application, including the crop. Non-selective herbicides such as glyphosate would therefore only be adopted widely once companies were able to genetically engineer seeds to tolerate their application.

13   This price information was provided in May 2012 by a Monsanto representative at the company's CustomCare line in Manitoba at 1-800-667-4944.

14 This information was communicated to the author during a very brief telephone conversation with Douglas Davis on 18 April 2012. And although somewhat enigmatic in his wording (making reference to the black knight in *Monty Python and the Holy Grail*), one wonders whether the West Virginia Office of the Attorney General has grown weary of Monsanto's ability to delay and obfuscate on this issue.

15 *Texas Grain Storage, Inc. d/b/a West Chemical & Fertilizer v Monsanto Company*, United States District Court for the Western District of Texas (2007) [*Texas Grain Storage*].

16 As quoted in ibid., para. 74

17 Ibid., para. 68.

18 *Texas Grain Storage.*

19 Despite an exhaustive search that included Statistics Canada, Agriculture and Agri-Food Canada, various producer organizations, the Canola Council of Canada, the Canadian Seed Growers Association, and the Canadian Seed Trade Association, I must concur with the drafters of the eighth report from the Standing Senate Committee on Agriculture and Forestry (2008) that locating accurate and disaggregated seed input prices in this country is next to impossible.

20 A unit of corn seed contains approximately 80,000 seeds, which, between 2001 and 2009, was enough seed to sow between 2.6 and 2.9 acres. Soybean seeds are typically sold by the bushel, and so a unit contains about sixty pounds (approximately 150,000 seeds), which, between 2001 and 2009, was enough seed to sow between 0.81 and 0.89 of an acre (Benbrook 2009b).

21 The colourful term, biogopolists, was coined by Drahos and Braithwaite (2002).

22 Syngenta was formed in November 2000 through the merger of the agrochemical and seed units of AstraZeneca and Novartis. Novartis emerged from the 1996 merger between Ciba-Geigy and Sandoz.

23 Aventis was created through the 1999 merger between Hoechst and Rhône Poulenc. Bayer acquired Aventis CropScience in 2002 to form Bayer CropScience, the first legally independent subgroup of Bayer.

24 This is, perhaps, changing as new, collaborative biotechnology projects emerge, such as DIYBio, the *Biotech Hobbyist Magazine*, and Personal bio-Computing, which seeks to develop a low-tech DNA computer that employs standard biotechnology techniques such as polymerase chain reaction and gel electrophoresis. The ultimate goal of such hobbyist pursuits is to reappropriate biotechnology and repurpose it for uses beyond those determined solely by capital. Members of the Critical Art Ensemble (2002) also discuss such projects and their potential for subverting capitalist control of biotechnology.

25 The *Convention on Biological Diversity*, 31 I.L.M. 818 (1992) [*CBD*], was opened for signature on 5 June 1992 at the United Nations Conference on Environment and Development (the Rio "Earth Summit"). It remained open for signature until 4 June 1993, by which time it had received 168 signatures. The Convention entered into force on 29 December 1993, which was ninety days after the thirtieth ratification. This pact among the vast majority of the world's governments sets out commitments for maintaining the world's ecological underpinnings in the context of economic development. The convention establishes three main goals: the conservation of biological diversity, the sustainable use of its components, and the fair and equitable sharing of the benefits from the use of genetic resources.

26 Conference of the Parties (COP) to the Convention on Biological Diversity, *Decision V/5 on Agricultural Biological Diversity: Review of Phase I of the Programme of Work and Adoption of a Multi-Year Work Programme*, UN Convention on Biological Diversity Secretariat, 2007, online: http://www.cbd.int/decisions/cop-05.shtml?m=COP-05&id=7147&lg=0 [emphasis added]. The COP is the governing body of the CBD. It is responsible for the implementation of the convention based on the decisions it makes at its periodic meetings. As of mid-2012, the COP had held ten ordinary meetings and one extraordinary meeting (the latter, to adopt the *Cartegena Protocol on Biosafety*, 39 I.L.M. 1027 (2000), was held in two parts). Prior to 2000, the COP held its ordinary meetings annually. Since 2000, when a change in the rules of procedure was agreed upon, these meetings have been held biannually. As of the end of 2010, the COP had taken a total of 299 procedural and substantive decisions. The eleventh COP to the *CBD* was held in Hyderabad, India, on 8-19 October 2012.

27 Shiva (2001b, 49) defines biopiracy as "the use of intellectual property systems to legitimize the exclusive ownership and control over biological resources and biological products and processes that have been used over centuries in non-industrialized cultures. Patent claims over biodiversity and indigenous knowledge that are based on the innovation, creativity and genius of the people of the Third World are acts of 'biopiracy.'" Since 1995, the Coalition against Biopiracy has been holding the Captain Hook Awards for Biopiracy. This annual global awards ceremony recognizes the work of the most courageous cogs (this term comes from the Middle Ages when small ships known as cogs were constructed with high sides to provide some degree of protection from marauding pirates) who are battling against biopiracy, while also bestowing citations of shame on those groups and organizations that have committed the most heinous acts of biopiracy (ETC Group 2006a).

28 The Consultative Group on International Agricultural Research (CGIAR) is the largest consortium of seed banks in the world. There are about 1,400 national and regional seed banks dispersed around the globe. The most recent, and highly publicized, seed bank opened in late February 2008 in Svalbard, Norway. Dubbed the "doomsday vault" by the media, this project will house seeds from almost all of the recognized varieties of 150 food crops grown by humans. The British Millennium Seed Bank Project, which has a target of storing seeds from more than 24,000 wild plant species, is the only other international storage facility (Hopkin 2008).

29 The CGIAR has, however, come under increasing scrutiny and criticism for its close ties to corporate interests, which also have access to the germplasm in the CGIAR vaults, most of which comes from farmers in the developing world. Moreover, according to GRAIN, the apparent benefits of "improved" varieties that are developed as a result of CGIAR seed banks and research fail to consider the higher purchase and growing costs of such varieties and the fact that they tend to replace local varieties (Cummings 2008). According to one prominent Indian activist, the CGIAR has become "an agricultural research outsource for the multinational corporations" (Sharma, as cited in Hisano 2005, 94).

30 In fact, within the scientific literature, some work is beginning to emerge that posits molecular biocontainment as a biosafety solution for genetically engineered crops (Hills et al. 2007).

31 Transcontainer project, online: http://www.transcontainer.wur.nl/.

32 Current gene excision technology, which, similar to other genetic use restriction technologies (GURTs), can be induced by chemical or environmental triggers, leaves behind one of the two excision sequences that becomes part of the plant's chromosome, which could have negative effects on the plant's health. Moreover, the altered chromosome will be inherited by progeny, thus transmitting potential dangers down the reproductive line.

33 "Conditionally lethal" genes would kill a plant only if triggered by a chemical or environmental inducer. The conditionally lethal gene could either code for the expression of a toxin itself or for an enzyme that would convert an applied chemical into a toxin. Dow Agrosciences and the National Research Council of Canada own a patent on such technology. The ETC Group compiles lists of which companies and universities own patents on all of the types of GURTs discussed in this section (ETC Group 2007).

34 *Patent Act*, R.S.C. 1985, c. P-4.

35 Although not without his critics, Braverman (1974) provides a compelling, more contemporary application and elaboration of these themes in his now classic work, *Labor and Monopoly Capital*, which interrogates the continual deskilling of workers and the relationship between technological innovation and social class, including the ways that science and technology are harnessed by capital for both accumulation and control purposes.

36 Depending upon the theorist to whom one refers, the nominal term employed to reflect the phenomenon of primitive accumulation differs. Glassman (2006) discusses "primitive accumulation," "accumulation by dispossession," and "accumulation by extra-economic means," though he seems to favour the original term coined by Marx, "primitive accumulation." McCarthy (2004) speaks of accumulation by "extra-economic means." Bonefeld (2001, 2002) and De Angelis (2001, 2007) remain true to Marx, employing the term "primitive accumulation." Harvey (2003, 2006) prefers to substitute the predicate "accumulation by dispossession" for what he believes is the dated "primitive accumulation."

37   Volume 3 of *Capital*, subtitled *The Process of Capitalist Production as a Whole*, was compiled by Engels based on notes left by Marx. It was originally published in 1894, eleven years after the death of Marx.

38   Marx worked on the three volumes of *Theories of Surplus Value* in the 1860s. Considered by some to be the fourth volume of *Capital*, this work was published posthumously by Karl Kautsky.

39   It is worth noting that Rosa Luxemburg (2004) discussed the historically continuous nature of extra-economic prerequisites to capitalist accumulation. In fleshing out her theory, Luxemburg asserted that the extra-economic prerequisites to capitalist production are mechanisms capital not only must engage continuously but also apply to increasingly larger portions of the globe. That is, extra-economic processes that separate producers from the means of production are continually required as an integral component of ensuring capitalist production, production that must contend with and attempt to overcome the crisis-ridden nature of capitalist accumulation. Although to be fair, Luxemburg did seem to accept the traditional periodizing characterization of primitive accumulation, construing it as marking an historical period of transition to the dominance of the capitalist mode of production.

40   Thirsk (1967) offers an excellent history of British agriculture in the Middle Ages, including a detailed treatment of land enclosures. Turner (1984) and Yelling (1977) both provide engaging accounts of the British enclosure movement.

41   We are now witnessing a resurgence of land enclosures in Africa as foreign countries and transnational corporations seek to buy or lease large tracts of land for agricultural production meant for foreign markets.

42   There is a varying array of literature that elaborates particular examples of capital's enclosure of the commons. See Federici (1992) for an account of how the debt crisis of the 1980s and the successive structural adjustment programs of the 1990s provided a convenient justification for the massive land privatization schemes forced on a great number of African nations. Walton and Seddon (1994) develop a poignant discussion of the effects of, and struggles against, the enclosures imposed on countries and people through structural adjustment policies. See Boyle's (2003) work on intellectual property and enclosure of the human genome. See Shiva (1997, 2001b) on intellectual property rights and the enclosure of indigenous knowledge and common resources such as water. In addition to the work by the World Development Movement (http://www.wdm.org.uk/) and GATSwatch.org (http://www.gatswatch.org/), see Wesselius (2002) for a critique of the way the *General Agreement on Trade in Services* functions as an international agreement designed to both consolidate past and facilitate future corporate enclosures of the commons.

43   *Agreement on Trade-Related Aspects of Intellectual Property Rights*, 15 April 1994, *Marrakesh Agreement Establishing the World Trade Organization*, Annex 1C, contained in the *Legal Texts: The Results of the Uruguay Round of Multilateral Trade Negotiations* 320 (1999), 1869 U.N.T.S. 299, 33 I.L.M. 1197 (1994).

44   Traditionally, capital accumulation is denoted by the following formula: M-C-M', where M denotes an amount of money invested by individual capitalists in the market to buy commodities, given by C in this formula. The transformation of money into commodities, shown as M-C, represents the act of "buying." Individual capitalists, however, purchase such commodities not to satisfy their particular needs but, rather, to generate a profit, which occurs when M' is greater than the amount of money originally invested. In order to realize this potential profit, the commodity C must be placed back on the market to be sold. If buyers are found and the sale is made (C-M') at a price where M' is greater than M, the individual capitalist is able to record a profit. Thus, M' = M + ΔM, where ΔM represents the change in the amount of money in the possession of the individual capitalist after the sale of the commodity. While an individual capitalist might terminate investment at this point, as a system, the "class" of capitalist investors driven by the profit motive will generate a new cycle of accumulation in a process that repeats *ad infinitum*: M'-C'-M''. That is, commodities of a greater value are bought (C-M') and placed back on the market to be sold for a greater amount of money, which provides investors with a new sum of money available for purchase and subsequent sale of commodities in an endless cycle of accumulation.

Bell and Cleaver (2002) provide an excellent and fuller explication that outlines the role of labour in providing surplus value for capital through the labour/manufacturing process.

Chapter 3: Battles to Reclaim and Maintain the BioCommons

1 *The Ram's Horn* is the monthly journal put out by Brewster Kneen and his wife, Cathleen, which treats food and agricultural issues.

2 *The World According to Monsanto* (2008) was written and directed by Marie-Monique Robin and produced by Image and Compagnie.

3 Macdonald (2000) claims that Monsanto did not engage in tactics designed to pressure Canadian regulators. Moreover, his treatment of the recombinant bovine growth hormone (rBGH) debate tends to discount the efforts expended by the opponents of Monsanto's application for regulatory approval. According to Macdonald's account, it was concern about the scientific authority of the Bureau of Veterinary Drugs (the organizational unit within Health Canada responsible for reviewing the scientific data accompanying Monsanto's application) that opened the door to other groups with concerns beyond the science of rBGH, but which, nonetheless, could attack the issue on a scientific basis that couched their underlying socio-economic concerns. Ultimately, he claims that these types of latter concerns, alongside science, will come to assume importance in the regulatory process. It remains unclear whether he perceives this as a good thing or not.

4 No doubt thanks to an investment of at least $1.37 million, Ontario Pork, the producers' association that represents Ontario hog farmers, owns the trademark Enviropig™. As the only private investor in this research by the University of Guelph, Ontario Pork was also given a worldwide licence to distribute the pig to swine breeders and producers (*Who "Created" and Owns "Enviropig™"?* n.d.). According to the Canadian Biotechnology Action Network (CBAN), the Government of Ontario also provided financial support to this research through the Ontario Ministry of Agriculture, Food and Rural Affairs, and the Rural Economic Development Program. The Government of Canada funded the project through Natural Sciences and Engineering Research Council grants and other agreements with Agriculture and Agri-Food Canada.

5 *Canadian Environmental Protection Act,* R.S.C. 1999, c. 33.

6 In 2002, instead of being properly destroyed as biological waste, eleven dead Enviropiglets were inadvertently sent to a rendering plant where they contaminated 675 tonnes of poultry feed, which the Canadian Food Inspection Agency (CFIA) subsequently ordered to be recalled.

7 *United States Endangered Species Act,* 16 U.S.C. s. 1531 et seq. (1973).

8 In its 2007 budget, the Harper government revealed plans to begin transferring federal scientific laboratories to academia and/or the private sector. In early June 2008, the Treasury Board released the report authored by members of the independent panel of experts convened to study the matter. Among the five "early candidate" federal labs recommended for divestiture from the federal government was the Canadian Cereal Research and Innovation Laboratory.

9 In denying this information, the reviewers of the access request invoked paragraphs 20(1)(b) and 20(1)(c) of the *Access to Information Act,* R.S.C. 1985, c. A-1, which set out access exemptions for information given to the Government of Canada in confidence and information that could cause financial loss or gain, respectively.

10 Despite overwhelming opposition from farmers' associations such as the National Farmers Union, the Harper government ended the marketing monopoly of the Canadian Wheat Board, which for decades had been the only producer marketing system in western Canada, and the largest marketer of its kind in the world. Effective 1 August 2012, sales of western wheat and barley to the board became voluntary.

11 In fact, then chairman of the Canadian Wheat Board, Ken Ritter, stated that "our elevator systems are saying they cannot segregate it [genetically engineered and non-genetically engineered wheat] to the level required by our customers. So, if we want to continue to be a major wheat exporter, registering RR [Roundup Ready] wheat in Canada is a mistake" (Sierra Club of Canada n.d.).

12 Direct seeding practices comprise any method of planting and fertilizing that avoids tillage to prepare the soil. In addition to saving on fuel costs, no-tillage direct seeding practices also have the potential to sequester atmospheric carbon dioxide. Also known as minimum soil disturbance seeding, these types of production methods rely on retaining most of the crop residues on the soil surface and extending crop rotations. These practices help to increase soil organic matter, which is vital to feeding the soil microbial activity that helps build soil quality and thus improve crop yields. Surface residues also help to enhance the soil's water infiltration rates and holding capacity as well as reduce erosion potential, thus helping to maintain valuable topsoil for production.

13 *Cartegena Protocol on Biosafety*, 39 I.L.M. 1027 (2000); *Convention on Biological Diversity*, 31 I.L.M. 818 (1992) [*CBD*].

14 "CDC Triffid," a herbicide-tolerant variety developed by Alan McHughen when he worked for the Crop Development Centre at the University of Saskatchewan, received marketing approval by Health Canada and the CFIA in 1998. However, in 2001, the Flax Council of Canada and the Saskatchewan Flax Development Commission successfully pressured the CFIA to rescind variety registration, thus making it illegal in this country to sell "CDC Triffid" seeds. In early September 2009, the European Commission's Rapid Alert System for Food and Feed confirmed that Canadian flax imports were contaminated with genetically engineered flax, although, at the time, it was not clear that the source of the contamination was "CDC Triffid." Nonetheless, these discoveries were particularly problematic because the European Union usually accounts for as much as 70 percent of Canada's total flax export market. Based on rumour of contamination alone, cash bids for flax in Manitoba plummeted by 32 percent. By late October 2009, flax prices were down between $2 and $3 per bushel from around the $11 bushel price prior to reports of contamination. By mid-November 2009, Japan became the thirty-fifth country in which contaminated flax was discovered or where products containing contaminated flax entered the food supply chain (Mittelstaedt 2009b). Laboratories conducting testing on behalf of the Flax Council of Canada revealed in August 2010 that about 10 percent of more than 6,000 samples tested positive for CDC Triffid (Pratt 2010).

15 The global world food crisis was provoked by dramatic increases in food prices that began in 2007 and that have resulted in political and economic instability, including social unrest in a number of developing countries. A number of factors have been attributed to these price spikes, including the skyrocketing price of oil and its consequent impact on agricultural inputs such as fertilizers and food transport. The rise to prominence of biofuels has resulted in large portions of arable land being diverted away from food to fuel crop production. This factor, coupled with dwindling world food stockpiles, exerted substantial upward price pressures on food.

16 The Government of Canada has admitted this impossibility in a public consultation document created in 2011 by the Working Group on Low Level Presence, which is chaired by Agriculture and Agri-Food Canada and includes members from the CFIA, the Canadian Grain Commission, Foreign Affairs and International Trade Canada, Environment Canada, and Health Canada: "Even when best management practices are strictly followed, it is difficult to completely prevent this [contamination of non-genetically engineered crops by genetically engineered crops] from occurring" (Agriculture and Agri-Food Canada 2011, 4). Although not stated explicitly, the coincidence in timing leads to the conclusion that the closure of European markets to Canadian flax delivered crucial impetus to the working group to begin the process of relaxing Canadian regulations in respect of genetic contamination. And as this consultation document makes clear in several passages throughout, an ultimate goal beyond managing the domestic occurrence of genetic contamination is to establish an international model aligned with the Canadian government's championing of agricultural biotechnology that "would provide greater assurances for Canadian exporters who face the risk of trade disruptions related to low-level presence ... Canada intends to engage key international partners and will encourage them to similarly adopt appropriate low-level presence policies" (ibid., 6, 9). Put another way, the social, cultural, political, environmental, and scientific concerns of other sovereign nations cannot be permitted to trump Canadian trade imperatives.

17 Mauro also recounts that once he and his colleagues made public their intentions to conduct the survey, officials from the United States embassy requested a private consultation to discuss the research, contending that it had the potential to affect bilateral trade. In Mauro's own words, "[t]hat what we were doing actually had the ability, in their mind, to potentially unhinge Roundup Ready wheat and Monsanto being a US-based corporation the embassy was concerned. They came and they didn't put pressure on us and they didn't for a second suggest that what we were doing was inappropriate. But simultaneously they conveyed the feeling that the big guy is watching you."

18 The Saskatchewan Organic Directorate is the umbrella organization that unites the province's producers, processors, buyers, traders, certifiers, and consumers of certified organic food and fibre. It is incorporated as a non-profit membership organization.

19 By the second half of the eighteenth century, scientists recognized that legumes not only obtain the bulk of their nitrogen requirements from the air rather than from the soil but also actually enrich the soil's nitrogen levels. Such crops are therefore an essential, natural means to replenish the fertility of soil. According to the Saskatchewan Organic Directorate, "because alfalfa has deep, fibrous roots it improves soil texture by adding organic matter, an important carbon sink. Alfalfa is often used to protect or improve soils on marginal lands and highly erode-able land. Alfalfa can be used to make heavy clay soils more porous, and to make light sandy soils better able to retain moisture. It has a role in making soils more resilient in the face of both drought and flooding, increasing concerns as the climate changes" (Saskatchewan Organic Directorate 2010, 9).

20 *Geertson Seed Farms, et al. v Johanns, et al.,* Case no. C 06-01075 CRB, 2007 WL 1302981 (N.D. Cal., 13 February 2007).

21 *Geertson Seed Farms, et al. v Johanns, et al.,* 541 F.3d 938 (9th Cir., 2 September 2008), para. 21.

22 *Monsanto Co. v Geertson Seed Farms,* 130 S.Ct. 2743 (2010).

23 Ibid.

24 "Feral crop species are those from which individuals escape a managed area to survive, reproduce and establish self-perpetuating populations in either natural or semi-natural habitats" (Bagavathiannan and Van Acker 2009, 70).

25 *Seeds Regulations,* C.R.C., c. 1400.

26 An archived version of the registry can be found at the following website, http://survivingthemiddleclasscrash.wordpress.com/.

27 *National Environmental Protection Act,* 42 U.S.C. s. 4321 et seq. (1969).

28 *Center for Food Safety, et al. v Thomas V. Vilsack, et al.,* Case no. 3:08-cv-00484 (2009), 12.

29 Ibid., 14.

30 A public opinion poll of 2,022 Canadians conducted in September 2001 by Decima Research Incorporated shows that over 90 percent (+/- 2.2 percent 19 times out of 20) of the Canadian public supports mandatory labelling for genetically modified foods (Kuyek 2002). A similar poll of 2,000 Canadians conducted by Decima on behalf of the Consumers Association of Canada in October 2003 revealed almost identical results, with 88 percent of respondents believing that the federal government should compel companies to disclose whether the food products they sell contain genetically engineered ingredients (Chase 2003). The results of the poll also revealed that support for mandatory labelling cuts across income, gender, region, and education differences (Moore 2003). It does bear pointing out that, up until this time, the Consumers Association of Canada was against mandatory labelling. In part, the original stance of this body in favour of a voluntary standard might be explained by its prominent position at the government trough where large amounts of money were handed out to groups supportive of biotechnology, as discussed more fully in Chapter 6.

31 In fact, in 2001, Loblaws sent a letter to all of its health food suppliers to remove genetic engineering-free labelling from their packaging by 1 September of that year or risk losing shelf space at any stores operated by the corporation (Adam 2001).

32 Exact quantification and elaboration of rationales either for or against mandatory labelling is not possible since the CBAC report provides only a very broad summary of the comments received. Nonetheless, according to the information provided, of the 160 respondents who participated overall, 121 offered a response to the labelling question: "The vast majority of respondents from the consumer/interested citizen, academic and non-governmental

organization interests [108] wanted a mandatory labelling system, while the vast majority of industry respondents [eight] supported a voluntary system. Government respondents were essentially split (two for mandatory and three for voluntary)" (Canadian Biotechnology Advisory Committee 2002a, 57).

33   It is precisely this same line of argumentation that successive Canadian governments have advanced at annual meetings of the Codex Committee on Food Labelling, a standing committee of the Codex Alimentarius Commission. The latter was established in 1963 by the UN Food and Agriculture Organization and the World Health Organization. Despite overwhelming domestic public support for mandatory labelling of genetically engineered foods, at the meeting of the Codex Committee on Food Labelling in Calgary from 4-8 May 2009, the Canadian delegation worked in tandem with delegates from the United States, Argentina, and Australia to frustrate efforts by the majority of other committee members to agree on an international standard for labelling genetically engineered foods. Canadian delegates articulated the desire that the free market should be the final arbiter of the labelling issue, completely and conveniently disregarding the fact that neo-classical economics actually assumes full information for markets to function optimally. The American delegates tried unsuccessfully to make the argument that, since no consensus had been reached in the committee for so many years, the issue should be removed from ongoing consideration. While any guidelines that the Codex Alimentarius Commission might agree on do not have the force of law, the organization is recognized by the WTO (World Trade Organization). As a result, it is presumed that any country that adopts a labelling regime sanctioned by the Codex Alimentarius Commission could avoid challenges at the WTO by other countries claiming that labelling requirements represent non-tariff trade barriers. In a surprising shift of position, the US delegation to the 2011 annual Codex summit in Geneva removed its opposition to the genetic modification labelling guidance document, thereby allowing the guidelines to be finalized after twenty years of negotiations.

34   As mentioned previously, the *Cartagena Protocol on Biosafety* to the *CBD, supra* note 13, Chapter 1, which 164 countries had joined as of late 2012, regulates the global trade of genetically engineered organisms but does not include mandatory labelling. The protocol is similarly silent on the issue of accountability of companies in a situation where transgenic plants transmit their genes to other varieties, thus polluting and damaging the ecosystem (de la Perrière and Seuret 2000). Nonetheless, proponents of the biotechnology industry were quick to slam the protocol as a sell-out and source of needless regulation (Miller and Conko 2000).

35   House of Commons, *Hansard* (3 April 2008), 4473 (Robert Thibault).

36   Ibid., 4475 (Nathan Cullen). As this book goes to press in late 2012, a similar battle is being waged in California, where agricultural biotechnology and food and beverage industry giants have contributed more than US$27 million to a campaign to defeat Proposition 37, which asks voters to decide in the November elections whether foods that contain genetically engineered ingredients should be labelled as such. Monsanto, which has contributed over US$7 million to the campaign to defeat the proposition (DuPont has ponied up a little over US$4 million, while Bayer CropScience, BASF, and Dow Agrosciences have each contributed US$2 million), has articulated a public position that completely distorts the rationale behind calls for a mandatory labelling regime. The company goes so far as to assert that proponents of Proposition 37, apparently only special interest groups and individuals opposed to biotechnology, are engaging in deliberate misinformation campaigns designed to ultimately deny consumers the broad food choices they currently enjoy: "Beneath their *right to know* slogan is a deceptive marketing campaign aimed at stigmatizing modern food production." Part of Monsanto's message also speaks of the potential for labels to confuse consumers, adding that "the California proposal would serve the purposes of a few special interest groups at the expense of the majority of consumers ... We ... believe that supporting the NO on 37 coalition is the right thing to do" (Monsanto 2012b). Monsanto's rhetoric hits on the typical talking points it and other uncritical cheerleaders of agricultural biotechnology employ in their attempts to disarm valid critique: agricultural biotechnology equals modern production techniques, which only uninformed, backward luddites would oppose; the public is easily confused (that is, stupid) so better to keep them uninformed;

and there are no safety concerns in respect of genetically engineered products. As the evidence presented in Chapters 5 and 6 will demonstrate, there is a substantial and growing corpus of research that challenges such claims – even if Monsanto and others with a vested interest in the agricultural biotechnology industry seek to will it away.

37 Greenpeace has compiled a list that it makes available on its website outlining how various members of the House of Commons voted on this bill. It can be found at the following website, http://www.greenpeace.org/.

Chapter 4: Intellectual Property Rights

1 The term "patent thickets" refers to multiple and overlapping patent claims that force people to obtain licences from multiple patent holders if they wish to use or commercialize new technology (or increasingly merely to conduct research).

2 *Agreement on Trade-Related Aspects of Intellectual Property Rights,* 15 April 1994, *Marrakesh Agreement Establishing the World Trade Organization,* Annex 1C, contained in the *Legal Texts: The Results of the Uruguay Round of Multilateral Trade Negotiations* 320 (1999), 1869 U.N.T.S. 299, 33 I.L.M. 1197 (1994).

3 *Patent Act,* R.S.C. 1985, c. P-4.

4 Ibid. [emphasis added].

5 The novelty requirement is further reinforced through section 28.2 of the *Patent Act, supra* note 3, which rules that the subject matter of a patent application must not have been previously disclosed.

6 Building on his contention that a person has "a Property in his own Person," Locke (1988, 27) postulated that a person's labour power is also her own. Therefore, "whatsoever then, he removes out of the State that nature hath provided, and ... hath mixed his Labour with, and joyned to it something that is his own, thereby makes it his Property." In the context of intellectual property, the argument invoked by proponents of strict protection is that, once an individual has applied his labour to even intangible things, some form of monopoly protection is warranted – what Gordon (1993) labels a Lockean "I made it-it's mine" justificatory pattern.

7 *Diamond v Chakrabarty,* 447 U.S. 303 (1980). In 1972, Ananda Chakrabarty, a scientist employed by General Electric, filed a patent application for a bacterium he had genetically engineered to break down crude oil. The US patent examiner (his name was Diamond, hence the name of the case), while allowing some of the claims for which Chakrabarty had filed, rejected the claim for the bacterium, arguing that micro-organisms are "products of nature" and that living organisms are not patentable subject matter under Article 35, section 101, of the US *Patent Act,* 35 U.S.C. ss. 1 et seq. Chakrabarty appealed the decision to the US Patent Office Board of Appeals, which upheld the examiner's decision on the second ground. In 1980, the Supreme Court of the United States (US Supreme Court) overturned the decision by the Patent Office, ruling that, although the bacterium was a natural material, it was a non-naturally occurring organism that was a product of human ingenuity. By virtue of combining four plasmids and successfully inserting them into a bacterium, Chakrabarty was deemed to have invented the micro-organism. It should be noted that four of the nine US Supreme Court justices dissented from the majority opinion, arguing that Congress had "chosen carefully limited language granting protection to some kinds of discoveries but explicitly excluding others, including living material." The rationale for such careful language was to prevent anyone from "securing a monopoly on living organisms, no matter how produced or how used" (*Diamond v Chakrabarty,* 447 U.S. 303 (1980), 320). See Kevles (1998) for a good discussion of this case; as well as *Ex parte Hibberd,* 227 USPQ 443 (1985) (PTO Board of Patent Appeals and Interferences), which involved genetically engineered plants; *Ex parte Allen,* 2 USPQ 2d 1425 (1987) (PTO Board of Patent Appeals and Interferences), which dealt with oysters; and Harvard University's Oncomouse, including its political economic implications for gene patenting. Kloppenburg (2004) suggests that the *Hibberd* decision was important for seed companies because it opened the way for patents on component parts of an organism, such as a gene that provides herbicide tolerance, whereas traditionally the protections offered under plant breeders' legislation are limited to one claim for a new plant variety as an inseparable whole. The permissible allocation of multiple

intellectual property claims on one organism vastly expanded the scope of monopoly protection available to capital.

8 Important instances of government policy in Canada that have facilitated this include the extension in 1990 by the federal government of patent-type protection for plant varieties to public and private breeders through the *Plant Breeders' Rights Act,* S.C. 1990, c. 20. Other examples include passage of Bill C-22 and Bill C-91, in 1986 and 1992 respectively, which eliminated compulsory licencing from Canadian patent legislation and then increased the length of patent protection to twenty years from the filing date. *Patent Act, supra* note 3, s. 44.

9 The authors of this 2003 study indicate that, in the United States, private companies hold 74 percent of agricultural biotechnology patents. Moreover, 41 percent of all agricultural biotechnology patents are held by the biggest five companies: Monsanto owns 14 percent, DuPont 13 percent, Syngenta 7 percent, Bayer 4 percent, and Dow 3 percent. The authors caution that these numbers could, in fact, underestimate the level of private intellectual property control of agricultural biotechnology because of the exclusive licencing arrangements often negotiated between public sector researchers and private companies (Graff et al. 2003). Fuglie and Schimmelpfennig (2000) provide an account (albeit somewhat uncritical) of the increasing trends toward agricultural technology transfer from the public to the private sector in the United States and a number of developing countries.

10 This argument is not meant to romanticize some previous era in which sharing was a universal, absolute norm. It is, in fact, the case that information or materials have, in the past, also been withheld from competitors. The point is that, while such practices are not new, they are exacerbated by the burgeoning application of patents to scientific knowledge and innovation.

11 The 1980 *Patent and Trademark Law Amendments Act,* P.L. 96-517, requires federally funded organizations to report any potentially patentable discoveries made as a result of the sponsored research. The institutions are permitted to retain title to their inventions only if they agree to file patent applications and exploit any patent granted. If they fail to do so, the government reserves the right to grant licences to other entities in an attempt to ensure practical application of the invention. Clearly, this law assumes that patents are necessary to facilitate the transfer of technological discoveries from government labs to universities and on to the private sector. This perspective clashes with the traditional justification for patents as a means of spurring innovation in the first place. Instead, it employs patent rights to ensure commercial dissemination of products that embody prior research rather than offering the prospect of future patent rights to stimulate additional research. We do not yet have similar legislation in Canada, though, as pointed out in Chapter 1, federal government policies in respect of science and technology advocate a more rapid and broader degree of commercialization of federally funded research (Government of Canada 1996, 2007).

12 The relevant sections of the *Patent Act, supra* note 3, stipulate as follows:

55.2

(1) It is not an infringement of a patent for any person to make, construct, use or sell the patented invention solely for uses reasonably related to the development and submission of information required under any law of Canada, a province or a country other than Canada that regulates the manufacture, construction, use or sale of any product ...

(6) For greater certainty, subsection (1) does not affect any exception to the exclusive property or privilege granted by a patent that exists at law in respect of acts done privately and on a non-commercial scale or for a non-commercial purpose or in respect of any use, manufacture, construction or sale of the patented invention solely for the purpose of experiments that relate to the subject-matter of the patent.

13 In our neo-liberal context, national patent offices, at least in most Western countries, have typically been transformed to function as special operating agencies that rely on cost recovery to finance their operations. Similar to the confused regulatory situation in this country (which will be explored in greater detail in the following chapter), patent offices are now responsible for reviewing applications from applicants who tend to be construed

as customers. The United States Patent and Trademark Office articulates its mandate in the following terms: "The primary mission of the Patent Business is to help customers get patents" (as cited in May 2004, 133). While the Canadian Intellectual Property Office avoids such blatantly corporate-driven language, its commitment to a barrage of service standards designed to facilitate an efficient and effective review system (two prominent catch words in the neo-liberal inspired New Public Management literature) for the benefit of potential patentees clearly demonstrates that any commitment to the public domain or knowledge commons has long been relegated to secondary status.

14 Ribonucleic acid (RNA) is defined as any of a class of single-stranded molecules transcribed from deoxyribonucleic acid (DNA) in the cell nucleus or in the mitochondrion or chloroplast, containing along the strand a linear sequence of nucleotide bases that is complementary to the DNA strand from which it is transcribed. The composition of the RNA molecule is identical to that of DNA except for the substitution of the sugar ribose for deoxyribose and the substitution of the nucleotide base uracil for thymine.

15 This has negative implications for information theory and its inherent linear model based on source, receiver, encoder, and decoder. If information is found beyond DNA within the entire cell and in the interactions between the cell and its environment, then it is incorrect to conceptualize DNA as the message and the remainder of the cell as the receiver that produces particular proteins based on the code/information contained in the DNA.

16 *President and Fellows of Harvard College v Canada (Commissioner of Patents)*, [1998] 3 F.C. 510 [*Harvard College* FC]. Scassa (2003) develops a detailed analysis of this case. Morrow and Ingram (2005) provide a similarly detailed account that clearly disagrees with the majority position.

17 Patent claims are the definitions included in a patent application that set out the scope of the exclusive right for which the patentee seeks intellectual property protection. Most patent applications contain a number of consecutive claims that define the various aspects of the invention in more detail. Establishing claims is a statutory requirement set out in subsection 27(4) of the *Patent Act, supra* note 3, which states that "the specification must end with a claim or claims defining distinctly and in explicit terms the subject-matter of the invention for which an exclusive privilege or property is claimed." The US patent was issued in April 1988 and covers all non-human "onco-animals." The European Patent Office issued a patent on the genetically modified mouse in 1992, which immediately attracted an oppositional filing. The European Patent Office finally ruled in 2001 that the patent is valid but that it should be restricted to rodents.

18 For purposes of assessing patentable subject matter, the Canadian Patent Office distinguishes between uni-cellular life forms and higher life forms (section 12.04.01). Uni-cellular life forms that are new, useful, and inventive are patentable. In general, a process to produce, or which utilizes these organisms, is patentable. Uni-cellular life forms include: microscopic algae; moulds and yeasts; bacteria; protozoa; viruses; cells in culture; transformed cell lines; and hybridomas. Higher life forms are not patentable subject matter. However, a process for producing a higher life form may be patentable provided the process requires significant technical intervention by man and is not essentially a natural biological process that occurs according to the laws of nature, for example, traditional plant cross-breeding. Higher life forms include animals, plants, seeds, and mushrooms. This is a revised version that relies on the jurisprudential precedents established by the Supreme Court of Canada in the *Harvard College* and *Schmeiser* decisions (Canadian Intellectual Property Office n.d., 12-17). In its previous iteration, the manual defined higher life forms as "multi-cellular differentiated organisms" (Morrow and Ingram 2005, 190).

19 *President and Fellows of Harvard College v Canada (Commissioner of Patents)* [2000] 4 F.C. 528.

20 *President and Fellows of Harvard College v Canada (Commissioner of Patents)*, [2002] 4 S.C.R. 45, 2002 SCC 76, para. B [*Harvard College* SCC]. Determining the boundary between lower and higher life forms for the purposes of patentability has been a continuing debate in Canadian jurisprudence. See, for example, *Re Application of Abitibi Co.* (1982) 62 C.P.R. (2d) 81, and *Pioneer Hi-Bred Ltd. v Canada (Commissioner of Patents)*, [1989] 1 S.C.R. 1623.

21 *Harvard College* SCC, *supra* note 20, para. 166.

22 Ibid., para. 158.

23 Ibid., paras. 160-61.

24 Ibid., paras. 1 and 4.

25 Ibid., para. 13. The American-based Biotechnology Industry Organization placed Canada on its "watch list" in February 2004 because of concerns about the strength of Canada's intellectual property regime, particularly in light of the *Harvard College* decision. Its brief to the Office of the United States Trade Representative claims that "the developments on patent eligibility compound an ongoing problem of erosion in protection of intellectual property in pharmaceutical and medical technology in Canada. For example, the ability of companies to realize the full value of their intellectual property rights is limited by restrictive practices governing the pricing of new, patented pharmaceuticals. In addition, health authorities in Canada interpreted regulations promulgated to implement the NAFTA provision on undisclosed tests and other data in a manner that essentially removes any protection for these data associated with pharmaceutical products and that is inconsistent with that Agreement and the TRIPS Agreement" (Biotechnology Industry Organization 2004, 15).

26 *Monsanto Canada Inc. et al. v Schmeiser et al.* (2001), 12 C.P.R. (4th) 204 (F.C.T.D.), 2001 FCT 256 [*Schmeiser* FC]. The et al. indicates that the plaintiffs in the case were Monsanto Company, the owner of the patent at issue, and Monsanto Canada Inc., its Canadian subsidiary and licencee under the patent. The defendants technically included Percy Schmeiser and his corporation that owns and operates his farming enterprise.

27 This number may not be accurate since the same figure appears on another of the company's webpages (United States), which indicates that this figure of 145 lawsuits against farmers in the United States had been reached in 2010 ("*Why Does Monsanto Sue Farmers Who Save Seeds?*" n.d.).

28 On 23 February 1993, the Canadian Intellectual Property Office issued Monsanto Company Patent no. 1,313,830. Titled glyphosate-resistant plants, the fifty-two claims protected by the patent relate to the invention of genetically engineered genes and cells containing those genes that, when inserted into plants, substantially increase their ability to survive applications of glyphosate-based herbicides. In addition to the genetically engineered genes, patent protection also extends to the process Monsanto developed to create and insert glyphosate-resistant genes into seed, in this case canola.

29 Volunteer canola is considered by farmers to be a weed, and, in fact, herbicide-tolerant volunteer canola is now prevalent across much of the Canadian prairies.

30 Since Monsanto, as the plaintiff in this case, is incorporated pursuant to the *Canada Business Corporations Act*, R.S. 1985, c. C-44, it was permitted to choose which court to use to enforce its patent. Monsanto elected to pursue its claim in Federal Court, which makes strategic sense for a couple of reasons: Federal Court judges have greater expertise in patent law than lower court judges, but more importantly for Monsanto, the Federal Court has no jurisdiction in tort actions between individuals. As a result, Schmeiser was not permitted to launch a counterclaim in Federal Court against Monsanto based on liability for crop contamination. Schmeiser was thus compelled to incur the additional time and expense of filing a separate claim for damages against Monsanto in Saskatchewan's Court of Queen's Bench.

31 *Plant Breeders' Rights Act, supra* note 8. *Schmeiser* FC, *supra note* 26, para. 77.

32 *Schmeiser* FC, *supra* note 26, para. 81.

33 Ibid., para. 88.

34 Ibid., para. 115. Citing *Computalog Ltd. v Comtech Logging Ltd.* (1992), 44 C.P.R. (3d) 77 at 88 (F.C.A.).

35 *Schmeiser* FC, *supra* note 26, para. 120.

36 Ibid., para. 123.

37 Canola, unlike corn and wheat, has not been domesticated and continues to possess a number of wild species traits. Canola seeds can remain dormant for anywhere between six and ten years, and they can germinate at any point in the season, not just in the spring. The physical characteristics of the seed (small, round, and smooth) allow it to be transported easily by wind. Moreover, canola, although mainly a self-pollinating species, is subject to outcrossing (to cross animals or plants by breeding individuals of different strains but usually of the same breed) in the range of 20-30 percent and its pollen can be moved several kilometres by insects. Also, once a field has been planted with Roundup Ready canola, the

soil remains contaminated with shattered seeds from that year's harvest, even if in subsequent years conventional canola seed is planted (Clark 2001). Kloppenburg (1988, 2004), in his study of agricultural biotechnology, outlines a number of cases in the United States in which genetic contamination from pollen flows from field to field has occurred in self-pollinated crops such as soybeans as well as open-pollinated crops such as corn, all of which have proven that the buffer zones between genetically engineered and traditional crops recommended by government are wholly insufficient.

38 Until 2003, the Federal Court of Canada consisted of two divisions: an Appeal and a Trial Division. Following amendments to the *Federal Courts Act*, R.S.C. 1985, c. F-7, these divisions became two separate courts – the Federal Court of Appeal and the Federal Court.

39 In January 2008, the US Supreme Court let stand, without comment, a lower court ruling that awarded Monsanto $375,000 in damages against a farmer the company sued for planting seeds saved from the Roundup Ready crops he had legally acquired the previous year.

40 *Monsanto Canada Inc. v Schmeiser*, [2004] 1 S.C.R. 902, 2004 SCC 34, 5-6 [*Schmeiser* SCC] [emphasis added].

41 Ibid., 3.

42 Ibid., para. 43.

43 Ibid., para. 24.

44 Ibid., para. 78.

45 Although not completely on point, recent judicial decisions handed down in Spain and the United Kingdom have limited Monsanto's control over its genetically engineered seeds to the specific patented function – protecting plants from applications of Roundup herbicide. In these cases, Monsanto tried to block the importation of soy derivatives from Argentina, where Monsanto's Roundup Ready soybeans were not protected by patent protection. Monsanto contended that the use of Argentine Roundup Ready soy for which no royalty payments have been paid should be illegal based on its agreements with importers in Spain, the United Kingdom, the Netherlands, and Denmark (Laursen 2010). Although an out-of-court settlement was eventually negotiated between Monsanto and the Dutch importer Cefetra and Alfred C. Toepfer International, the European Court of Justice did publish its decision, which confirms those made in the national courts in both Spain and the United Kingdom that limits the patent protection to the original claims of the invention. Put more explicitly, the patent can only be enforced when the transgene is performing the function for which it received patent protection (that is, purpose-bound protection).

46 *Schmeiser* SCC, *supra* note 40, para. 160.

47 Ibid., 4-5.

48 Ibid., para. 84.

49 Ibid., para. 165 [emphasis added].

50 To be precise, such a right is not expressed specifically in the *Plant Breeders' Rights Act, supra* note 8. Instead, such a right has been read into the act because it does not expressly prohibit such practices in certain situations.

51 See Right Livelihood Foundation, online: http://www.rightlivelihood.org/.

52 The increasing importance of the concerns raised by this case is reflected in the fact that two issues (Issue nos. 3 and 4 of volume 27) of the *Bulletin of Science, Technology and Society* are devoted to examining the ability of the courts and intellectual property law to address disputes over genetically engineered plants.

53 See Organic Agriculture Protection Fund, online: http://www.saskorganic.com/oapf/.

54 *Class Actions Act*, S.S. 2001, c. C-12.01.

55 *Hoffman and Beaudoin v Monsanto Canada*, 2002 SKQB 67, para. 27 (Statement of Claim) [*Hoffman and Beaudoin* 2002].

56 Ibid., para. 35 (Memorandum of Law in Support of the Certification Application on Behalf of the Plaintiffs, 11 August 2004).

57 See Lee (2003) for a discussion of liability under nuisance in the context of genetically engineered organisms. Other writers provide discussions about liability issues related to the release of genetically engineered organisms, albeit from mainly British and European perspectives (Lee and Burrell 2002; Rodgers 2003).

58   Matthews Glenn (2004) offers a well-researched article that provides a legal analysis of the chances of success of these statements of claim based on existing case law against neighbouring farmers, against the government, and against biotechnology companies.

59   The plaintiffs appear to have pursued the negligence charge from the well-established "products liability" line of argumentation, stating in their Statement of Claim, *Hoffman and Beaudoin* 2002, *supra* note 55, para. 25, that "between 1995 and 2001, with respect to Aventis Canada [now Bayer CropScience] and Liberty Link, and between 1996 and 2001, with respect to Monsanto Canada and Roundup Ready, farmers purchasing either variety were not warned about the potential harm to neighbouring crops caused by GM volunteer canola. In particular, no warnings were given to farmers to keep a buffer zone to minimize the flow of pollen to surrounding crops, to ensure that all farm trucks transporting the seed were properly and securely tarped, to thoroughly clean all farm machinery before leaving a field where the GM crop was being grown, or to warn neighbours that GM volunteers might emanate from the GM crop."

60   Similar to arguments I am endeavouring to advance in this chapter, Judge (2007) develops the argument that such actions, rather than availing themselves of tort remedies, might be better pursued through intellectual property law itself. She considers the role of the exhaustion doctrine in copyright as providing a possible parallel in patent law in respect of genetically engineered seed and the case at bar, ultimately arguing for a Hohfeldian framework that would attach duties to the rights granted under patent legislation.

61   *Environmental Management Protection Act*, S.S. 1983-84, c. E-10.2. This statute was replaced on 1 October 2002 by the similarly sounding *Environmental Management and Protection Act*, S.S. 2002, c. E-10.21.

62   McNaughton (2003) provides a detailed legal analysis of how this act might serve as a statutory mechanism to hold producers of genetically engineered seed liable for genetic contamination of the environment.

63   Ownership would presumably arise from the patent rights owned by the defendants and control would inhere in the reach and effect of the technology use agreements that companies such as Monsanto force farmers to agree to before being able to purchase genetically engineered seed. For a discussion of "owners or persons in control" under the original act, see *Busse Farms Ltd. v Federal Business Development Bank*, (1998) 168 D.L.R. (4th) 27 (Sask. Ct. App.).

64   The original *Environmental Management Protection Act, supra* note 61, sets out in subsection 23(1) that the person alleging loss, damage, or injury "is not required to prove negligence or intention" to inflict it. Subsection 23(2) of this act stipulates further that the burden of proving that the loss, damage, or injury "was not caused by a development is on the person who proceeds with the development." The statement of claim filed with the Saskatchewan Court of Queen's Bench includes discussion of the reversed burden of proof set out in the *Environmental Management Protection Act*.

65   *Environmental Assessment Act*, S.S. 1979-80, c. E-10.1. The Supreme Court of Canada has recognized that "stewardship of the environment" is a "fundamental value in Canadian society" that the courts can protect. *Ontario v Canadian Pacific Ltd.*, [1995] 2 S.C.R. 1031. In fact, the Court has established the precedent that polluters must pay to repair environmental damage for which they are responsible. *Imperial Oil Ltd. v Quebec (Minister of the Environment)*, [2003] S.C.J. No. 59. Moreover, the Court has suggested that class action suits involving environmental cases can be useful given "the rise of mass production ... the advent of the mega-corporation, and the recognition of environmental wrongs." *Hollick v Toronto (City)*, [2001] S.C.J. No. 67) (de Beer and McLeod-Kilmurray 2007).

66   *Hoffman and Beaudoin v Monsanto Canada*, 2005 SKQB 225, para. 64 [*Hoffman and Beaudoin* 2005].

67   Ibid., para. 67.

68   Ibid., para. 114 [emphasis added].

69   *Agricultural Operations Act*, S.S. 1995, c. A-12. Phillipson and Bowden (1999) have penned a deeper discussion of "right to farm" legislation.

70   *Hoffman and Beaudoin* 2005, *supra* note 66, para. 11.

71 *Rylands v Fletcher,* (868) L.R. 3 H.L. 330, was the 1868 English case that set the precedent for the doctrine of strict liability for abnormally dangerous conditions and activities. *Hoffman and Beaudoin* 2005, *supra* note 66, para. 97.

72 *Southport Corporation v Esso Petroleum Co.,* [1954] 2 QB 182. *Hoffman and Beaudoin* 2005, *supra* note 66, para. 133.

73 *Hoffman and Beaudoin* 2005, *supra* note 66, para. 168.

74 *Schmeiser* FC, *supra* note 26, para. 96.

75 As set forth in Article 27 of the *Cartagena Protocol on Biosafety,* 39 I.L.M. 1027 (2000), the Conference of the Parties serving as the Meeting of the Parties (COP-MOP) to the protocol, was required to adopt, at its first meeting, a process for negotiating an international law on liability and redress for damage resulting from transboundary movements of living modified organisms. In response to this requirement, in 2004 the COP-MOP established an Open-Ended Ad Hoc Working Group of Legal and Technical Experts on Liability and Redress in the context of the Protocol (MOP-1 Decision BS-I/8). Following six years of negotiations, delegates to the 2010 COP-MOP-5 in Nagoya, Japan, adopted the *Nagoya–Kuala Lumpur Supplementary Protocol on Liability and Redress to the Cartagena Protocol on Biosafety* (Decision BS-V/11). The new treaty shall be open for signature at the United Nations headquarters in New York from 7 March 2011 to 6 March 2012 and will enter into force ninety days after being ratified by at least forty parties to the *Cartagena Protocol on Biosafety.* Although beyond the scope of the present work, this supplementary protocol and its implementation beg deeper analysis, including an assessment of its potential implications for the agricultural biotechnology industry.

76 *Hoffman and Beaudoin v Monsanto Canada,* 2007 SKCA 47, para. 46.

77 Percy Schmeiser, online: http://percyschmeiser.com/.

78 On 31 January 2012, a US Federal Court judge agreed to hear oral arguments in a lawsuit the Organic Seed Growers and Trade Association and others are trying to bring against Monsanto. Similar to the *Hoffman and Beaudoin* case, the seventy-five plaintiffs involved in this suit were quite concerned about genetic contamination from genetically engineered crops. However, they are employing a different tactic, asking the court to invalidate Monsanto's patents on genetically engineered seeds and to prohibit the company from suing farmers whose crops become genetically contaminated. The case was dismissed by the district court in February 2012, and that dismissal is now pending review by the Court of Appeals (United States District Court for the Southern District of New York in Case no. 11-CV-2163-NRB).

Chapter 5: Regulatory Capture and Its Critics

1 *Food and Drugs Act,* R.S.C. 1985, c. F-27. *Food and Drug Regulations,* C.R.C., c. 870. These regulations were amended in 1999 to add Division 28, SOR/99-392, s. 1, which sets out the regulations in respect of novel foods. Novel foods is the regulatory category to which the federal government assigns genetically engineered crops and other genetically engineered food products.

2 *Seeds Act,* R.S.C. 1985, c. S-8.

3 As defined by the Canadian Food Inspection Agency (CFIA), "variety registration is a pre-market assessment activity performing a "gate keeper" role by allowing CFIA oversight of the varieties available in the marketplace. The purpose of variety registration is to ensure that health and safety requirements are met and that information is available to regulators to prevent fraud, as well as facilitating seed certification, and the international seed trade. To date, pre-registration testing and merit assessment is also required prior to registration to demonstrate that the variety performs equal to or better than reference varieties ... Most agricultural crops are subject to variety registration, the notable exceptions include corn, chickpeas, sorghum, food-type soybeans and turf grasses."

4 *Feeds Act,* R.S.C. 1985, c. F-9. *Health of Animals Act,* S.C. 1990, c. 21.

5 A number of contemporary theorists have developed Marxist critiques of modern scientific theory and practice for being inadequately dialectical. From my perspective, some of the strongest accounts come from Levins and Lewontin (1985, 1994, 1997), Burkett (1996, 1999), and Foster (2000).

6 A plasmid is a small, typically circular unit of deoxyribonucleic acid (DNA) often found in bacteria that can replicate within a cell independently of the chromosomal DNA. Plasmids contain a few genes that usually code for proteins, particularly enzymes, some of which confer resistance to antibiotics.

7 Restriction enzymes are any of several enzymes produced by bacteria as a defence against viral infection that catalyze the cleavage of DNA molecules at specific sites. They are commonly used for gene splicing in recombinant DNA technology and for chromosome mapping. They can also be referred to as restriction endonuclease. Ligation enzymes are employed to connect DNA fragments.

8 A vector is any agent that contains or carries modified genetic material (for example, recombinant DNA) that can be used to introduce exogenous genes into the genome of another organism. The four major types of vectors are plasmids, bacteriophages (any number of viruses that infect bacteria) and other viruses, cosmids (a type of hybrid plasmid suitable for use as a cloning vector), and artificial chromosomes.

9 A phenotype is any observable characteristic or trait of an organism, such as its morphology, development, biochemical or physiological properties, or behaviour. Phenotypes result from the expression of an organism's genes as well as the influence of environmental factors and the interactions between the two.

10 The situation is very similar in the United States where, in January 2009, the Department of Agriculture, the Food and Drug Administration, and the Environmental Protection Agency rebuffed recommendations from the Government Accountability Office (the investigative arm of Congress similar to our Office of the Auditor General) to implement a post-marketing surveillance program for genetically engineered crops. In part, the investigation stemmed from the six (known) instances of genetically engineered crops contaminating the nation's food supply since 2000.

11 The CFIA defines "substantial equivalence" as "the equivalence of a novel trait within a particular plant species, in terms of its specific use and safety to the environment and human health, to those in that same species, that are in use and generally considered as safe in Canada, based on valid scientific rationale" (Canadian Food Inspection Agency 2006, para. 1).

12 In a response to the Royal Society of Canada report drafted by Health Canada, the CFIA, Environment Canada, Agriculture and Agri-Food Canada, and the Department of Fisheries and Oceans, substantial equivalence is similarly defined: "The concept of substantial equivalence is used as a guide in the safety assessment of a GM-food by comparing the novel food to its unmodified counterpart which has a history of safe use ... In the context of environmental safety and feed safety, the concept of substantial equivalence also represents a safety standard approach. Substantial equivalence is not used, and will not be used, as a decision threshold" (Government of Canada 2001, 11).

13 The CFIA defines "familiarity" as "the knowledge of the characteristics of a plant species and experience with the use of that plant species in Canada" (Canadian Food Inspection Agency 2006, para. 1). Directive 94-08 on the Assessment Criteria for Determining Environmental Safety of Plants with Novel Traits, which was issued by the CFIA, states that "substantial equivalence is used in the comparative assessment of a PNT [plant with novel traits] relative to its counterpart to assess its relative and acceptable risk: i) A PNT that is substantially equivalent, in terms of its specific use and safety for the environment, as well as for human and animal health, to plants currently cultivated in Canada, having regards to its potential changes in weediness/invasiveness, gene flow, plant pest properties, impacts on other organisms and impact on biodiversity, should pose no greater risk to the Canadian environment compared with its counterpart. A plant that is substantially equivalent to its counterpart and is derived from seed authorized for unconfined release may be exempted from the notification and authorization requirements under the *Seeds Regulations*." The directive can be accessed on the CFIA website: http://www.inspection.gc.ca/.

14 As set out in section B.28.001. of the *Food and Drug Regulations, supra* note 1, a "novel food" means

   (a) a substance, including a microorganism, that does not have a history of safe use as a food;

(b) a food that has been manufactured, prepared, preserved or packaged by a process that

   (i) has not been previously applied to that food, and

   (ii) causes the food to undergo a major change; and

(c) a food that is derived from a plant, animal or microorganism that has been genetically modified such that

   (i)   the plant, animal or microorganism exhibits characteristics that were not previously observed in that plant, animal or microorganism,

   (ii) the plant, animal or microorganism no longer exhibits characteristics that were previously observed in that plant, animal or microorganism, or

   (iii) one or more characteristics of the plant, animal or microorganism no longer fall within the anticipated range for that plant, animal or microorganism (aliment nouveau)."

This same section defines genetically modify as "chang[ing] the heritable traits of a plant, animal or microorganism by means of intentional manipulation (modifier génétiquement)."

15 According to CFIA Directive 94-08, *supra* note 13, a plant with novel traits "is a plant containing a trait not present in plants of the same species already existing as stable, cultivated populations in Canada, or is present at a level significantly outside the range of that trait in stable, cultivated populations of that plant species in Canada. All PNTs are subject to an environmental safety assessment."

16 Mutagenesis relies on chemicals or irradiation to induce mutation. In fact, BASF, the world's largest chemical company, has been given regulatory approval by the CFIA for its herbicide-tolerant CDC Imagine wheat. This modified wheat is tolerant of BASF's imidazolinone broad-spectrum herbicides (Munro 2005, A8). Rather than insert an altered gene, BASF uses chemicals to alter existing genes within the wheat seeds in order to prevent the herbicide from binding to an enzyme in the wheat. In 2005, more than 200,000 acres of this wheat were grown in Alberta, Saskatchewan, and Manitoba. With virtually no opposition, BASF has almost free reign in this country to market CDC Imagine as "the first and only non-genetically modified" herbicide-tolerant wheat in Canada.

17 Andrée and Sharratt (2004) develop a detailed assessment of the Canadian government's efforts, or lack thereof, to act on the recommendations made by the Royal Society of Canada for the regulation of biotechnology in this country.

18 A committee of the National Academy of Sciences in the United States is similarly critical of the American Department of Agriculture for a lack of transparency. In particular, the committee points to the trend within the department of too often allowing companies to invoke the claim of confidential business information in order to prevent the information submitted as part of regulatory applications from being released publicly for independent verification (Committee on Environmental Impacts Associated with Commercialization of Transgenic Plants and National Research Council 2002).

19 As the Critical Art Ensemble (2002) makes clear, the corporate bias of the regulatory system is just as evident in the United States.

20 In a somewhat similar case, in April 2009, Renessen (a joint venture between Monsanto and Cargill) withdrew its application for regulatory approval of its genetically engineered high lysine corn (LY038) in the European Union (EU). Lysine is an essential amino acid that typically is added as a synthetic supplement to animal feed. The withdrawal, which included a demand satisfied by regulators to return all dossier material to the company, followed on the heels of requests from the EU's regulatory body for genetically engineered crops, the European Food Safety Authority (EFSA), for additional data in support of the application. Specifically, the EFSA wanted the company to repeat animal feeding trials since the first ones were not only methodologically suspect but also had, in fact, demonsrated significant alterations in blood and urine parameters among the rats fed LY038. The EFSA, following concerns articulated by a number of EU member states, wanted additional scientific data about the safety of LY038 when cooked. This genetically engineered corn, which contains high levels of free lysine, has been found to react with sugars when heated to form chemical compounds known as advanced glycation end products, which are known to increase levels of oxidative stress that are associated with increased risks of diabetes,

Alzheimer's disease, and cardiovascular and chronic kidney disease (Vlassara et al. 2009). The EFSA was also critical of the studies conducted by the applicant since Renessen had used another genetically engineered product as the control in its safety studies rather than a near-isogenic line as outlined in the Codex Alimentarius and demanded by EU guidelines. Although Renessen claims "decreased commercial value" as the motivation for its decision to pull its application, this position is at odds with previous claims by Monsanto that the market value for feed-grade lysine corn is worth approximately $US1 billion annually (http://www.slideshare.net/). As might be expected, no public statement was issued by either the company or the EFSA, and there is no mention of this regulatory setback on Renessen, Monsanto, or Cargill websites. Indeed, had it not been for the investigative work of Brian John of GM-Free Cymru, this incident would not have been made public. LY038 was approved in 2006 in Canada for environmental release and use as animal feed by the CFIA and for human consumption by Health Canada.

21 Monsanto has been successful in the United States in portraying biotechnology as an industrial sector that could be threatened by public opposition and overly rigid government regulation, which, according to this message, would ultimately threaten America's international dominance in biotechnology (Kleinman and Kloppenburg 1991). Smith (2003) offers an account of industry influence on regulatory decision making in the United States that has resulted in the approval of a number of genetically engineered foods, despite overwhelming scientific evidence against their safety.

22 An independent report published by the Institute on Governance questions whether the Canadian government possesses the requisite scientific capacity to discharge adequately its regulatory duties. This same report also outlines gaps in the science/policy interface in Canada (Boucher et al. 2002). Indeed, according to Statistics Canada the number of federal government personnel engaged in science and technology activities was significantly circumscribed (approximately 15 percent since 1990-91) during the "terror" of program review that began in the early 1990s (Statistics Canada 2000). Although statistics beyond 2000 are not available, in-house scientific capacity does not appear to have recuperated in the federal government, a most startling situation given the breakneck speed of contemporary scientific innovation. Others have also commented on the science deficit in the federal government and its implications for risk (Doern and Reed 2000). As proof of this lack of capacity, we need only look to the federal government's latest science and technology policy, which is using its personnel and infrastructure weaknesses as part of the justification to transfer selected non-regulatory federal laboratories to universities or the private sector (Government of Canada 2007).

23 The Nature Institute has compiled an extensive, publicly accessible, and fully searchable online database that contains a wealth of information about non-target effects of plant genetic engineering. The entries contained in the database can be searched according to the following categories: type effect (environmental, food and feed quality, physiological, morphological, and scrambled DNA), target effect, manipulated organism, commercialized product, most recently added reports, and all reports. The reports themselves are based on findings from the scientific literature, primarily from peer-reviewed journals, which can be found on the bibliography page of the website. As the authors of the database state, this compilation is not yet exhaustive given the extensive nature of the literature yet to be reviewed. Nature Institute, online: http://natureinstitute.org/. The London-based Institute of Science in Society also offers a wealth of scientific information dedicated to challenging the uncritical and oftentimes unscientific information disseminated by industry and some government regulators in respect of agricultural biotechnology. Institute of Science in Society, online: http://www.i-sis.org.uk/.

24 As stated on the GM Contamination Register's website: "This GM [genetic modification] contamination register is the first of its kind in the world. Although GM crops were grown on over 100 million hectares in 2007, there is no global monitoring system. Because of this failure of national and international agencies, GeneWatch UK and Greenpeace International have launched this joint initiative to record all incidents of contamination arising from the intentional or accidental release of genetically modified (GM) organisms (which are also known as genetically engineered (GE) organisms). It also includes illegal plantings of GM

crops and the negative agricultural side-effects that have been reported. Only those incidents which have been publically documented are recorded here. There may be others that are, as yet, undetected." GM Contamination Register, online: http://www.gmcontamination register.org/.

25 The International Survey of Herbicide Resistant Weeds, a collaborative effort between weed scientists in over eighty countries, maintains a publicly accessible database that monitors the evolution of herbicide-resistant weeds across the globe, including an assessment of their impact. This collaborative effort is supported and funded by the Herbicide Resistance Action Committee, the North American Herbicide Resistance Action Committee, and the Weed Science Society of America. Its aim is to maintain scientific accuracy in the reporting of herbicide resistant weeds globally. As of mid-2012, it had recorded 393 resistant biotypes among 211 species (124 dicots [broad leaf] and 87 monocots [grasses/cereals]) in over 690,000 fields. Weed Science, online: http://www.weedscience.org/.

26 Researchers at the University of Guelph's Ontario Agricultural College made public in early May 2009 their suspicion that a field in Essex County (Ontario) contains Canada's first population of a glyphosate-resistant giant ragweed biotype. They do caution, however, that their results are preliminary and that, to date, the existence of the immune giant ragweed is limited to one small area of a 580-acre field (*U of G Researchers Find Suspected Glyphosate-Resistant Weed* 2009).

27 The database can be accessed online at the following website: http://www.weedscience.org/.

28 In the United States, the problem has become serious enough to warrant review by, and public hearings before, the Domestic Policy Subcommittee of the House Oversight and Government Reform Committee.

29 *Helicoverpa zea* is actually a moth that feeds on many different types of plants during the larval stage. For this reason, it is given various common names. When it feeds on cotton, it is called cotton bollworm and when it consumes corn, it is referred to as corn earworm.

30 It is worth noting that Rosi-Marshall, similar to other scientists who have determined potential ill effects associated with GE crops, has been subject to intense criticism by a small cadre of staunch biotechnology proponents that, as a group, seems to mobilize very rapidly anytime negative research in respect of GE crops is published in peer-reviewed journals.

31 Proteinases belong to a group of enzymes that break the long, chainlike molecules of proteins into shorter fragments (peptides) and eventually into their components, amino acids. Proteolytic enzymes are present in bacteria and plants but are most abundant in animals. In the stomach, protein materials are attacked initially by the gastric enzyme pepsin. When the protein material is passed to the small intestine, proteins, which are only partially digested in the stomach, are further attacked by proteolytic enzymes secreted by the pancreas. Proteinase inhibitors are molecules that inhibit the function of proteinases.

32 The first study to confirm the dangers of hidden toxicity because of genetic engineering involved attempts by Pioneer Hi-Bred International (now wholly owned by DuPont) to augment protein levels in soybeans by injecting a gene from the brazil nut. The University of Nebraska determined that the gene produced proteins that elicited strong, and potentially lethal, allergic reactions in samples from people allergic to the brazil nut. Pioneer eventually discontinued its efforts to market the enhanced soybeans (Ferrara and Dorsey 2001).

33 In late June 1998, Pusztai, with the permission of the director of the Rowett Institute where he conducted his research, was interviewed by journalists producing a British television program about GE food. Immediately following the broadcast of the program on 10 August 1998, Pusztai's boss congratulated him for handling the questions well, and the Rowett Institute put out a press release noting that "a range of carefully controlled studies underlie the basis of Dr. Pusztai's concerns." Yet within a few days, and purportedly on the heels of two telephone calls from the Prime Minister's Office to the Rowett Institute, Pusztai was rewarded for his research efforts with a variety of attacks from the Royal Society, the Blair government, and even the director of the Rowett Research Institute. More directly, Pusztai's research project was almost immediately halted, his team was disbanded, all of his data were confiscated, he eventually would be forced to take retirement, and he was enjoined legally from discussing the case publicly for seven months. Ferrara and Dorsey (2001) offer a deeper discussion of this incident. See also Peekhaus (2010).

34   In May 2009, DuPont/Pioneer Hi-Bred received regulatory approval from Health Canada and the CFIA for its GE high oleic soybean trait.

35   It should be noted that the location of field trials is categorized as proprietary information by the CFIA and therefore kept secret from the public by virtue of an exemption in the federal *Access to Information Act*, R.S.C. 1985, c. A-1. In 2003, the Saskatchewan Association of Rural Municipalities passed almost unanimously a resolution demanding that the locations of test plots for genetically engineered crops be made public and that farmers and municipalities be notified before test plots are established. In a letter from then minister of agriculture and agri-food Canada, Lyle Vanclief, the association was apprised of the federal government's position that "it is in the public interest that the exact location of the trials remains confidential" (Warick 2003, E1). This same position was affirmed to me when my access to information request for details outlining specific geographical co-ordinates of all past and present GE field trials was refused by the CFIA in early 2008. The comment made to me during a telephone discussion with an information officer at the CFIA was that the agency feared that revealing exact locations would make it easier for opponents of agricultural biotechnology to vandalize planted fields in actions similar to those that have occurred in Europe.

36   The GMO Panel of the European Food Safety Authority has rejected the claims advanced by Séralini and his colleagues and thus refused to reconsider its previous safety assessments of these GE seeds. Monsanto has also been quick to challenge the scientific validity of the claims made by these French researchers. When questioned about such dismissals, Séralini contends that

> the CRIIGEN research group has published several papers on the signs of alert of chronic pathologies that could be developed after GMO consumption by mammals ... in international peer reviewed journals. These are counter expertises of Monsanto confidential results previously accepted by authorities as sufficient. By contrast, the public opinions about this paper by "official" committees are not published in scientific journals. We are surprised not to have been contacted by any of those, to ask questions or for counter expertise of the results. Moreover, it is clear that our paper questions the competence or the scientific honesty of committees that have already given positive opinions on the commercialization of these maizes, such as EFSA or Monsanto, or the French committee. In this regard, we are not surprised by their reactions, and we claim that, as they were involved in the process of commercialization accepting previous Monsanto results, they are not independent anymore for the counter expertise of our work.

Séralini's comment can be found at the following website: http://www.testbiotech.org/. In events eerily similar to those in which Pusztai was embroiled over a decade earlier, by mid-2010, the attacks against Séralini had escalated to such an extent that the European Network of Scientists for Social and Environmental Responsibility issued an open letter denouncing what they articulate as a "troubling trend ... where researchers who publish results that suggest unintended and potentially negative effects on health or the environment become the target of aggressive discrediting campaigns from influential members of the scientific community." As of mid-November 2010, 1,100 academics and over 12,000 individuals had signed this letter, available at http://sciencescitoyennes.org/.

37   As if almost on cue, several critics were quick to dismiss this study based on charges of poor methodology and flawed statistical analysis (Butler 2012). For example, the use of the Sprague-Dawley rat strain was critiqued because this breed is prone to tumours. However, this rat strain has been used the most in animal feeding trials to evaluate the safety of genetically engineered crops, including for Monsanto's NK 603 maize and the European toxicity studies of glyphosate. Séralini and his colleagues were thus being consistent in their decision to employ this rat strain in their own study. Of course, one might question whether these same critics would draw this critique to its logical conclusion and dismiss industry studies as being invalid for using this rat breed. In any event, this charge also fails to consider the finding of both quantitative and qualitative differences in the tumours

between the test and control groups. Another charge levied against this study is that the control groups were too small. In addition to being a critique that could just as easily be levelled against industry studies, it obscures the fact that Séralini's study used more rats in more test groups and for a far longer period than any previous investigation completed by Monsanto as part of any of its applications for regulatory approvals.

38 Smith (2007) provides a compilation of reports from academia through media, medicine, and eyewitness accounts that speak to the growing evidence of the negative effects of GE food on human health.

39 Monsanto originally marketed Roundup herbicide as being "biodegradable," claiming that it "leaves the soil clean" and "respects the environment." By the early part of this decade, environmentalists and consumer rights organizations, particularly in France, were challenging such pronouncements. Judges in both France and the United States compelled Monsanto to remove "biodegradable" from its labels on the basis of its own studies demonstrating that twenty-eight days after an initial application of the herbicide only 2 percent of the product had broken down. A court case brought against Monsanto in 2001 in France was finally decided in early 2007 when Monsanto France was found guilty of false advertising in respect of the biodegradable claim and sentenced to a €15,000 fine (*The World According to Monsanto* 2008). Monsanto lost its appeal of the decision at the Lyon Court of Appeal in 2008 as well as at the Supreme Court of France in October 2009.

40 *Auditor General Act,* R.S. 1985, c. A-17.

41 An endotoxin is the toxic protoplasm that forms an integral part of the cell wall of certain bacteria and is only released upon destruction and disintegration of the bacterial cell.

42 *Access to Information Act,* R.S.C. 1985, c. A-1.

43 According to its makers, this stacked seed is genetically engineered to be resistant to both primary and secondary pests. The former include the European corn borer *(Ostrinia nubilalis),* the southwestern corn borer *(Diatraea graandiosella),* northern corn rootworm *(Diabrotica barberi),* and western corn rootworm *(Diabrotica virgifera).* The latter include the western bean cutworm *(Richia albicosta),* black cutworm *(Agrotis ipsilion),* fall armyworm *(Spodoptera frugiperda),* and corn earworm *(Helicoverpa zea).*

44 In the United States, the decision by the Environmental Protection Agency in 2003 to mandate refuge areas of 20 percent for Bt corn was criticized by a number of scientists from its own Scientific Advisory Panel. A majority on the panel recommended a 50 percent refuge area be planted with nontransgenic corn. Monsanto asked for a 20 percent refuge area, contending that anything larger would negate the benefit of insecticide reduction (Powell 2003). Professed concerns about insecticide reductions notwithstanding, an increase in regulatory-mandated refuge areas would have meant reduced sales of Monsanto's Bt corn. On the basis of Compliance Assessment Reports filed with the US Environmental Protection Agency, the Center for Science in the Public Interest, an American non-profit health-advocacy organization, determined that approximately 25 percent of Bt corn farmers in the United States failed to comply with refuge requirements as part of their insect resistance management obligations (*Complacency on the Farm* 2009).

45 CFIA, online: http://www.inspection.gc.ca/.

46 In July 2010, Monsanto agreed to pay a US$2.5 million penalty for distributing its Bollgard and Bollgard II cotton seeds (genetically engineered to produce a Bt insecticide) in ten Texas counties that had been restricted under the US *Federal Insecticide, Fungicide and Rodenticide Act,* (P.L. 80-104) 7 U.S.C. s. 136 et seq. This amount is the largest civil administrative penalty settlement achieved to date under the act. Aside from selling these seeds in areas of the country where such sales were prohibited, Monsanto sold these cotton seeds across the country on at least 1,700 occasions without including the relevant planting restrictions in the grower guides that accompany the seeds.

47 Although Bt corn seed has been available in Canada since 1997, the CFIA only began in 2004 to audit corn producers to assess compliance with refuge requirements as part of insect resistance management programs. Producers of Bt seeds are typically required to develop and implement such programs as a condition for unconfined environmental release of their seeds. The CFIA stepped up its oversight function in response to criticism articulated in a

report by the Office of the Auditor General of Canada, which stated that "due to the limited information on grower compliance obtained by the Agency to date ... its audits of conditions for unconfined release of corn have not yet enabled it to fully verify compliance with conditions imposed to prevent insect resistance from developing." The Office of the Auditor General report also expressed a need for "more assurance of grower compliance with insect resistance conditions for corn" (as cited in Canadian Food Inspection Agency 2009, 4).

48   E-mail correspondence from 30 August 2012 and telephone conversation from 4 September 2012 between the author and Heather Shearer, plant biosafety management analyst at the CFIA's Plant Biosafety Office.

49   E-mail correspondence between author and Margaret Neuspiel, plant biosafety policy specialist at the CFIA's Plant Biosafety Office.

50   As per a telephone conversation between the author and Heather Shearer on 4 September 2012.

51   As of September 2012 the report had yet to be released to the public, pending final approval from senior management at the CFIA.

52   At least one European country agrees with the Canadian Biotechnology Action Network's general position in respect of stacked genetically engineered seeds. The Austrian Federal Ministry for Health contends that "a stacked organism has to be regarded as a new event, even if no new modifications have been introduced. The gene-cassette combination is new and only minor conclusions could be drawn from the assessment of the parental lines, since unexpected effects (e.g. synergistic effects of the newly introduced proteins) cannot automatically be excluded" (Bundesministerium für Gesundheit 2008, 2).

53   The situation is somewhat different in the European Union, where, in 2008, Monsanto and Dow AgroSciences submitted an application for market authorization of SmartStax to the European Food Safety Authority (EFSA), a regulatory body that has been the subject of intense scrutiny and critique over the last few years for the revolving door between it and industry. Testbiotech, a German non-profit organization established to promote independent research and public debate on the impacts of biotechnology, received in 2011 several leaked internal Monsanto and Dow AgroSciences dossiers specific to SmartStax. In order to compare these data with those submitted to the EFSA by these companies in support of their application for regulatory approval, Testbiotech requested and was granted restricted access to EFSA documents specific to this file. On the basis of its analysis of all this information, Testbiotech has leveled the following charges against the scientific evidence marshalled by Monsanto and Dow AgroSciences: the applicants failed to properly assess the potential synergistic and interaction effects between the toxins genetically engineered into the corn seed that could impact animal and human health; they did not adequately assess the further potential synergies between the increased Bt loads in the plants and the herbicide residues if sprayed with glyphosate and/or glufosinate; they neglected to conduct feeding trials that analyzed health impacts rather than only nutritional quality (this is a particularly glaring omission since it remains unclear whether ingesting the multiple Bt toxins combined with residues from sprayed herbicides can impact the composition of intestinal flora); they did not develop or utilize reliable protocols that would facilitate independent control of the toxin load in the plants; they did not investigate the life cycle of the proteins or environmental exposure; the design of several studies was plagued by major defects; and all the studies lacked independent quality controls (Then and Bauer-Panskus 2011). During a three-month consultation period, a number of European Union member states, particularly Austria, Germany, and Belgium, articulated their own concerns about the scientific evidence and its consequent conclusions that were provided by the proponents in support of their application for marketing approval (*Application EFSA-GMO-CZ-2008-62* 2008).

54   *Food and Drug Regulations, supra* note 1.

55   The deficiency of more orthodox renderings of Marx that rely on this passage, in addition to privileging the base over the superstructure as mentioned in the text, is that they tend to invest the unfolding of the capitalist organization of society with a teleological inevitability according to the well-known logic of economic determinism.

56   My use of the term "form" corresponds closely with the conceptualization offered by Blanke, Jürgens, and Kastendiek (1978, 118) in which form "expresses both the basic problem and

the essential characteristic of the historical materialist method: the investigation of the connection between the material process of production and reproduction of the life of socialized people and the relations between these people who constitute themselves in this process of material reproduction."

Chapter 6: Capture and Control of Biotechnology Discourse in Canada

1 Orwell (1987) originally published *Nineteen Eighty-Four* in 1949.
2 As mentioned in the introduction, my use of the term "discourse" follows that of Ricoeur (as discussed in Rochon 1998, 9), as being "a shared set of concepts, vocabulary, terms of reference, evaluations, associations, polarities and standards of argument connected to some coherent perspective on the world."
3 See Kleinman and Kloppenburg (1991) for a detailed and compelling analysis of this discursive strategy.
4 George W. Bush, in a speech at an international biotechnology conference in 2003, invoked the rhetoric of feeding the world in his vitriolic attack on European governments that banned the import of genetically engineered crops: "Acting on unfounded, unscientific fears, many European governments have blocked the import of all new biotech crops ... because of these artificial obstacles, many African nations avoid investing in biotechnology, worrying that their products will be shut out of important European markets ... For the sake of a continent threatened by famine, I urge the European governments to end their opposition to biotechnology" (as cited in Harper 2003, A10).
5 According to its website, BIOTECanada is Canada's voice for biotechnology. In fact it is an industry-funded association that claims to represent the broad spectrum of biotech constituents including emerging, established, and related service companies in the health, agricultural, and industrial sectors. Its mission is the sustainable commercial development of biotechnology in this country, which it seeks to achieve by providing solutions to the challenges its members face. BIOTECanada, online: http://www.biotech.ca/.
6 *Access to Information Act*, R.S.C. 1985, c. A-1.
7 In an effort to mitigate some of the growing bad publicity associated with biotechnology issues, a number of firms created the Council for Biotechnology Information (CBI) in April 2000 as a corporate mouthpiece that extols the virtues of biotechnology. The CBI, which has offices in Canada, the United States, and Mexico, has developed a number of what it calls "advertorials" for television, newspapers, and various magazines such as *National Geographic, Gourmet,* and *Natural History*. The founding members of the CBI include Aventis, BASF, Dow Chemical, DuPont, Monsanto, Novartis, Zeneca, the American Crop Protection Association, and the Biotechnology Industry Organization. In the interim, Bayer CropScience has also joined. A number of their current advertisements can be viewed on their website at http://www.whybiotech.com. However, neither the CBI nor the groups mentioned earlier are alone in their attempts to influence public opinion. For example, the International Life Science Institute, the AgBioWorld Foundation, the International Food Information Council, and the Biotechnology Institute have all been involved in designing media campaigns aimed at various social groups, ranging from journalists to teachers, students, consumers, and farmers.
8 An editorial in the science journal *Nature*, drawing on a Pew Research Center report titled *The State of the News Media 2008*, bemoans the fact that on average five hours of American cable news contain about only one minute of science and technology coverage. Moreover, as newspapers slash budgets in response to falling circulation numbers, science desks are typically among the first casualties ("Critical Journalism" 2008).
9 There is a substantial body of literature that demonstrates that the media tend to be sympathetic to dominant scientific discourses, acting to legitimate current policy flowing from both governments and the corporate sector (Dunwoody 1993; Gandy 1982; Hornig Priest 1995; Nelkin 1987; Ward 1995; White 2001).
10 The *Globe and Mail* did run a story on the rendering of these pigs into the human food chain (Venditti 2004).
11 Greenpeace, online: http://www.greenpeace.org/canada/.
12 Even the International Service for the Acquisition of Agri-Biotech Applications, an unapologetic mouthpiece for the agricultural biotechnology industry that counts Bayer CropScience

and Monsanto among its funders, admits, albeit in subdued tones, that "the impact of GM HT [herbicide-tolerance] traits has, however, contributed to increased reliance on a limited range of herbicides and this poses questions about the possible future increased development of weed resistance to these herbicides. Some degree of reduced effectiveness of glyphosate (and glufosinate) against certain weeds may take place. To the extent to which this may occur, this will increase the necessity to include low dose rate applications of other herbicides in weed control programs (commonly used in conventional production systems) and hence may marginally reduce the level of net environmental and economic gains derived from the current use of the GM technology" (Brookes and Barfoot 2006, xvi). As might be expected from a report issued by an ardent biotechnology industry promoter, this methodologically dubious and superficial work relies almost exclusively on industry literature to downplay the social, economic, and environmental concerns articulated against biotechnology (this despite its title).

13 Pesticide is the broader term used for any of the chemicals applied to crops. It encompasses both insecticides and herbicides.

14 In June 2009, Monsanto provided earnings guidance that forecast gross profits of US$1.9 billion on Roundup products for the fiscal year. The company also predicted an almost 50 percent decline to US$1 billion in the following fiscal year, resulting in projected layoffs of 900 workers (4 percent of its labour force) in the United States (Tomich 2009). By late 2010, Monsanto reduced its forecast range for gross profit from Roundup to between US$250-350 million a year. Moreover, the number of job losses had more than doubled to 2,500 as the company reorganized this business line (Tomich 2010). In fact, in 2010, gross profit from Monsanto's agricultural productivity segment (which produces Roundup brand and other herbicides as well as lawn-and-garden herbicide products for the residential market) plummeted 76 percent to US$529 million from US$2.2 billion in 2009. Although gross profit in this company segment improved in 2011 to reach US$773 million, Monsanto has stated its belief that lower manufacturers' margins are now a permanent structural feature of the market for glyphosate-based herbicides (*"Monsanto's Annual Form 10-K"* 2011).

15 The report argues for the need to distinguish between intrinsic and operational yield. Intrinsic yield is achieved when crops are grown under ideal conditions and thus might also be construed of as potential yield. Operational yield refers to harvest size under actual field conditions when environmental factors such as pests and plant stress result in yields below ideal levels (Gurian-Sherman 2009).

16 StarLink corn was a *Bacillus thuringiensis* (Bt) variety genetically engineered by Aventis Crop Sciences (acquired by Bayer CropScience in 2002) to produce the Cry9C protein. Since humans are unable to digest this protein as quickly as other Cry proteins contained in other Bt varieties, there exists the potential for StarLink corn to induce allergic reactions. As a result, this corn was approved in the United States for use in animal feed only (it was never approved in Canada). Despite this restriction, in September 2000, StarLink corn was detected in human food products, including Kraft and Taco Bell taco shells. Shortly after these very well-publicized events, Aventis voluntarily agreed to cancel its registration with the US Environmental Protection Agency for feed and industrial use, effectively removing this seed variety from the market. The company was also forced to buy back all of the 2000 crop year of StarLink corn from farmers at a 25 percent premium over the corn price on 2 October 2000 of US$1.9925 per bushel. The episode also reverberated in international markets when traces of StarLink corn were found in snack foods and animal feed in Japan, taco shells in South Korea, and food aid destined for various African countries, some of whom incurred the rhetorical wrath of George Bush for refusing acceptance of the contaminated corn.

17 In 2006, two genetically engineered strains of rice, one from the United States (LLRice 601, which contains Bayer CropScience's Liberty Link protein and is designed for herbicide tolerance, was found in southern long grain rice) and one from China (rice designed to resist bacterial blight and insect-resistant rice), neither of which had received regulatory approval, had infiltrated and contaminated both the US domestic and the global rice supply, including Canada's (although it was Greenpeace rather than the Canadian government that made this determination). According to the USA Rice Federation, the national advocate for all segments of the American rice industry, 63 percent of American global exports of rice in

2006 were negatively affected by the contamination (USA Rice Federation 2006). According to attorney Scott Perry, over 500 lawsuits with claims by about 7,000 plaintiffs have been brought in the United States against Bayer CropScience, the company responsible for the American contamination.

The lawsuits, which seek compensatory and punitive awards, allege that farmers suffered monetary damages because of the contamination, which caused foreign countries to limit imports of American rice, subsequently resulting in a decline in the price of rice. According to one complaint filed in federal court in St. Louis, the decline in rice futures that occurred after news of the contamination was released publicly cost American rice growers approximately US$150 million. In the first verdict handed down in these lawsuits, a jury in St. Louis in early December 2009 awarded two Missouri farmers close to US$2 million for the losses they incurred through Bayer CropScience's negligence. In February 2010, the jury in another lawsuit awarded three farmers in Arkansas and Mississippi US$1.5 million for losses incurred as a result of contamination by Bayer CropScience's Liberty Link rice. In a third case heard before a state court in Arkansas, the jury awarded a rice farmer just over US$1 million. This case was also the first time a jury awarded a plaintiff punitive damages (US$500,000) (Sider 2010).

In April 2010, another Arkansas jury awarded US$48 million to a dozen rice farmers in that state. A fifth case decided in July 2010 in St. Louis cost Bayer CropScience another US$500,248. In March 2011, an Arkansas jury meted out the as yet highest award against Bayer CropScience when it awarded US$136.8 million to Riceland Foods, a rice milling and export company. Having suffered these judicial defeats, in July 2011 Bayer CropScience agreed to settle with farmers in Missouri, Arkansas, Texas, Louisiana, and Mississippi and pay up to US$750 million to rice farmers affected by the genetic contamination (Patrick 2011).

18 See flax contamination, *supra* note 14, Chapter 3.

19 Brian Wynne (1995, 377) develops an extended discussion of the social constructionist nature of science and scientific knowledge, including the differences between the restricted scope of issues articulated by scientific discourse and the broader issues considered by social groups. Ultimately, he remains critical of science and scientific discourse for failing to "recognize these social dimensions of its own public forms or the fact that public readiness to 'understand' science is fundamentally affected by whether the public feels able to identify with science's unstated prior framing." Indeed, a major benefit of social constructionist approaches is that they remove the privilege scientific discourse has traditionally assigned itself, leading, in theory at least, to a situation in which all knowledges are granted epistemological equivalence.

20 Though writing about scientific technology in general, John McMurtry (2002, 98) is also critical of the implicit assumption made by proponents of technology that it can function as a *deus ex machina* to resolve disasters unleashed by initial technological applications: "Technology has in this way become to the global market system what divine intervention by Yahweh or Indra was to fundamentalist patriarchies of the past."

21 GE Alert is an independent group of scientists, academics, and agricultural professionals committed to informing Canadians about the implications of agricultural genetic engineering. Members have no ties to the life science industry and are therefore free of potential corporate conflicts of interest. ·

References

People and Organizations Interviewed

Herb Barbolet, founder of Farm Folk/City Folk, associate at the Food Security and Sustainable Community Development, Centre for Sustainable Community Development, Simon Fraser University.

Terry Boehm, vice president, National Farmers Union (Head Office).

Josh Brandon, GE-campaigner, Greenpeace Canada, Vancouver.

Bert Christie, professor emeritus, University of Guelph.

Ann Clark, associate professor, Department of Plant Agriculture, University of Guelph.

Eric Darier, agriculture co-ordinator, Greenpeace Canada, Montreal.

Cathy Holtslander, Beyond Factory Farming and committee member of Organic Agriculture Protection Fund.

Brewster Kneen, founder, The Ram's Horn.

Devlin Kuyek, researcher at GRAIN, Montreal.

Ian Mauro, doctoral candidate, co-director and co-producer, Seeds of Change, Department of Environment and Geography, University of Manitoba.

Pat Mooney, executive director, Action Group on Erosion, Technology, and Concentration.

Lucy Sharratt, co-ordinator, Canadian Biotechnology Action Network, Collaborative Campaigning for Food Sovereignty and Environmental Justice.

Arnold Taylor, chair of the Organic Agriculture Protection Fund and president of the prairie regional chapter of Canadian Organic Growers.

Rene Van Acker, professor and chair, Department of Plant Agriculture, University of Guelph.

Books and Articles

Abergel, Elisabeth, and Katherine Barrett. 2002. "Putting the Cart before the Horse: A Review of Biotechnology Policy in Canada." *Journal of Canadian Studies* 37(3): 135-61.

Abley, Mark. 2000a. "Biotech Lobby Got Millions from Ottawa." *Montreal Gazette,* 28 February, A1, A7.

–. 2000b. "Magazine Insert Leaves a Bad Taste." *Montreal Gazette,* 28 March, A1.

Adam, David. 2008. "Canadian Farmer Forces GM Giant back to Court." *The Guardian,* 22 January, 20.

Adam, Nadège. 2001. *Loblaws Removes GE-Free Labels This Weekend: Memories of Consumer Choice?* Council of Canadians, online: http://www.canadians.org/.

–. 2002. *GM Labelling: Canada Is Crawling before Industry and USA.* Council of Canadians. http://www.canadians.org/.

Agriculture and Agri-Food Canada. 2008. *Growing Forward Framework Agreement: Executive Summary.* Government of Canada. http://www4.agr.gc.ca/.

–. 2011. *Policy Approaches for Managing the Low-Level Presence of Genetically Modified Crops Imported into Canada.* Public Consultation Document no. AAFC AGRIDOC #2671654. Ottawa: Government of Canada.

American Medical Association. 2000. *Interim Meeting of the American Medical Association: Report of the Council on Scientific Affairs.* Chicago: American Medical Association.

Anawalt, Howard C. 2003. "International Intellectual Property, Progress, and the Rule of Law." *Santa Clara Computer and High Technology Law Journal* 19: 383-405.

Andrée, Peter, and Lucy Sharratt. 2004. *Genetically Modified Organisms and Precaution: Is the Canadian Government Implementing the Royal Society of Canada's Recommendations?* Ottawa: Polaris Institute.

Application EFSA-GMO-CZ-2008-62 (MON89034 x 1507 x MON88017 x 59122 maize) – Annex G: Comments and Opinions Submitted by Member States during the Three-Month Consultation Period. 2008. Parma, Italy: European Food Safety Authority.

Arrow, Kenneth J. 1996. "The Economics of Information: An Exposition." *Empirica* 23(23): 119-28.

Aubry, Jack. 2003. "$13M Polishes Biotech Image, Critics Charge: Health Advocates Say Federal Funds Should Go to Testing, Labelling GM Foods." *Ottawa Citizen,* 24 November, A5.

Bagavathiannan, Muthukumar V., and Rene C. Van Acker. 2009. "The Biology and Ecology of Feral Alfalfa (*Medicago sativa L.*) and Its Implications for Novel Trait Confinement in North America." *Critical Reviews in Plant Sciences* 28: 69-87.

Bagla, Pallava. 2010. "Hardy Cotton-Munching Pests Are Latest Blow to GM Crops." *Science* 327(5972): 1439.

Barber, Benjamin R. 2001. *Jihad vs. Mcworld: Terrorism's Challenge to Democracy.* 2nd ed. New York: Ballantine Books.

Bartlett, Donald L., and James B. Steele. 2008. "Monsanto's Harvest of Fear." *Vanity Fair* (May): 156-70.

Belcher, Ken, James Nolan, and Peter W.B. Phillips. 2005. "Genetically Modified Crops and Agricultural Landscapes: Spatial Patterns of Contamination." *Ecological Economics* 53: 387-401.

Bell, Peter, and Harry Cleaver. 2002. "Marx's Theory of Crisis as a Theory of Class Struggle." *The Commoner* (Autumn): 1-61.

Benachour, Nora, and Gilles-Eric Séralini. 2009. "Glyphosate Formulations induce Apoptosis and Necrosis in Human Umbilical, Embryonic, and Placental Cells." *Chemical Research in Toxicology* 22: 97-105.

Benbrook, Charles. 1999. *Evidence of the Magnitude and Consequences of the Roundup Ready Soybean Yield Drag from University-Based Varietal Trials in 1998.* Ag BioTech InfoNet Technical Paper no. 1. http://www.nlpwessex.org/docs/BenbrookRR_yield_drag_98.pdf.

–. 2001. *Troubled Times amid Commercial Success for Roundup Ready Soybeans: Glyphosate Efficacy Is Slipping and Unstable Transgene Expression Erodes Plant Defenses and Yields.* Ag BioTech InfoNet Technical Paper no. 4: http://www.nlpwessex.org/docs/Benbrook troubledtimesfinal-exsum.pdf.

–. 2009a. *Critical Issue Report: Impacts of Genetically Engineered Crops on Pesticide Use: The First Thirteen Years.* Boulder, CO: Organic Center.

–. 2009b. *Critical Issue Report: The Seed Price Premium: The Magnitude and Impacts of the Biotech and Organic Seed Price Premium.* Boulder, CO: Organic Center.

–. 2012a. "Impacts of Genetically Engineered Crops on Pesticide Use in the U.S. – The First Sixteen Years." *Environmental Sciences Europe* 24(24): 1-23.

–. 2012b. Keynote Address. Third Bi-Annual *International Conference on Implications of GM Crop Cultivation at Large Spatial Scales* (GMLS), Bremen, Germany, 14-15 June.

Benjamin, Walter. 1998. *Understanding Brecht.* Translated by A. Bostock. London: Verso.

Benkler, Yochai. 2006. *The Wealth of Networks: How Social Production Transforms Markets and Freedoms.* New Haven, CT: Yale University Press.

Best, Steven, and Douglas Kellner. 2001. *The Postmodern Adventure: Science, Technology, and Cultural Studies at the Third Millennium.* New York: Guilford Press.

–. 2004. "Biotechnology, Democracy, and the Politics of Cloning." In *Biotechnology and Communication: The Meta-Technologies of Information,* edited by S. Braman, 197-226. Mahwah, NJ: Lawrence Erlbaum Associates.

Bettig, Ronald V. 1996. *Copyrighting Culture: The Political Economy of Intellectual Property.* Boulder, CO: Westview Press.

Beurton, Peter J. 2000. "A Unified View of the Gene, or How to Overcome Reductionism." In *The Concept of the Gene in Development and Evolution: Historical and Epistemological Perspectives,* edited by P.J. Beurton, R. Falk and H.-J. Rheinberger, 286-314. Cambridge: Cambridge University Press.

Beurton, P.J., R. Falk, and H.-J. Rheinberger, eds. 2000. *The Concept of the Gene in Development and Evolution: Historical and Epistemological Perspectives.* Cambridge: Cambridge University Press.

Bill to Label Genetically Engineered Foods: Will MPs Vote for Monsanto or Canadians? 2008. Ottawa: Canadian Biotechnology Action Network.

Bieler, Andreas, and Adam David Morton. 2006. "Class Formation, Resistance and the Transnational: Beyond Unthinking Materialism." In *Global Restructuring, State, Capital and Labour: Contesting Neo-Gramscian Perspectives,* edited by A. Bieler, W. Bonefeld, P. Burnham, and A.D. Morton, 196-206. Hampshire, UK: Palgrave MacMillan.

Biotechnology Industry Organization. 2004. *Brief to Sybia Harrison, Special Assistant to the Section 301 Committee, Office of the United States Trade Representative.* Washington, DC: Biotechnology Industry Organization.

Blanke, Bernhard, Ulrich Jürgens, and Hans Kastendiek. 1978. "On the Current Marxist Discussion on the Analysis of Form and Function of the Bourgeois State." In *State and Capital: A Marxist Debate,* edited by J. Holloway and S. Picciotto, 108-47. London: Edward Arnold.

Blumenthal, David, Nancyanne Causino, Eric G. Campbell, and Karen Seashore Louis. 1996. "Relationships between Academic Institutions and Industry in the Life Sciences: An Industry Survey." *New England Journal of Medicine* 334: 368-73.

Bøhn, Thomas, Raul Primicerio, Dag O. Hessen, and Terje Traavik. 2008. "Reduced Fitness of Daphnia Magna Fed a Bt-transgenic Maize Variety." *Archives of Environmental Contamination and Toxicity.* http://www.springerlink.com/content/m55x032626021295/fulltext.pdf.

Bøhn, Thomas, Terje Traavik, and Raul Primicerio. 2010. "Demographic Responses of *Daphnia magna* Fed Transgenic Bt-Maize." *Ecotoxicology* 19: 419-30.

Bollier, David. 2002. *Silent Theft: The Private Plunder of Our Common Wealth.* New York: Routledge.

Bologna, Sergio. 1979. "The Tribe of Moles." In *Working-Class Autonomy and the Crisis: Italian Marxist Texts of the Theory and Practice of a Class Movement, 1964-79,* edited by Red Notes and CSE Books, 67-91. London: Red Notes and CSE Books.

Bonefeld, Werner. 1992. "Social Constitution and the Form of the Capitalist State." In *Open Marxism,* edited by W. Bonefeld, R. Gunn, and K. Psychopedis, 93-132. London: Pluto Press.

–. 2001. "The Permanence of Primitive Accumulation: Commodity Fetishism and Social Constitution." *The Commoner* 2 (September): 1-15.

–. 2002. "History and Social Constitution: Primitive Accumulation Is Not Primitive." *The Commoner* (September): http://www.commoner.org.uk/debbonefeld01.pdf.

Borromeo, Emerlito, and Debal Deb. 2006. *Future of Rice 2006: Examining Long Term, Sustainable Solutions for Rice Production.* Amsterdam: Greenpeace International.

Bott, Sebastian, Tsehaye Tesfamariam, Hande Candan, Ismail Cakmak, Volker Römheld, and Günter Neumann. 2008. "Glyphosate-Induced Impairment of Plant Growth and Micronutrient Status in Glyphosate-Resistant Soybean (*Glycine max* L.)." *Plant and Soil* 312: 185-94.

Boucher, Lesley J., David Cashaback, Tim Plumptre, and Andrea Simpson. 2002. *Linking in, Linking out, Linking up: Exploring the Governance Challenges of Biotechnology.* Ottawa: Institute on Governance.

Bourgaize, David, Thomas R. Jewell, and Rodolfo G. Buiser. 2000. *Biotechnology: Demystifying the Concepts.* San Francisco, CA: Benjamin/Cummings.

Bowker, G. 2000. "Biodiversity Datadiversity." *Social Studies of Science* 30: 643-83.

Boyle, James. 1996. *Shamans, Software, and Spleens: Law and the Construction of the Information Society.* Cambridge, MA: Harvard University Press.

–. 2003. "Enclosing the Genome: What Squabbles over Genetic Patents Could Teach Us." In *Perspectives on Properties of the Human Genome Project,* edited by F.S. Kieff and J.M. Olin, 98-124. Amsterdam: Elsevier.

Bradshaw, Laura D., Stephen R. Padgette, Steven L. Kimball, and Barbara H. Wells. 1997. "Perspectives on Glyphosate Resistance." *Weed Technology* 11: 189-98.

Brandon, Josh. 2007. *Toxic Corn: Summary of New Scientific Evidence on the Health Dangers Posed by Genetically Altered Food.* Vancouver: Greenpeace Canada.

Braverman, Harry. 1974. *Labor and Monopoly Capital: The Degradation of Work in the Twentieth Century.* New York: Monthly Review Press.

British Medical Association. 1999. *The Impact of Genetic Modification on Agriculture, Food and Health: An Interim Statement.* London: British Medical Association.

Broad Alliance Formed in Canada to Stop Genetically Engineered Wheat. 2001. Greenpeace: http://archive.greenpeace.org/.

Brockmann, Miguel d'Escoto. 2008. *Opening Remarks by H.E.M. Miguel d'Escoto Brockmann, President of the General Assembly, at the High-level Event on the Millennium Development Goals.* New York: United Nations.

Brookes, Graham, and Peter Barfoot. 2006. *GM Crops: The First Ten Years – Global Socio-Economic and Environmental Impacts.* Ithaca, NY: International Service for the Acquisition of Agri-Biotech Applications.

Bud, Robert. 1993. *The Uses of Life: A History of Biotechnology.* New York: Cambridge University Press.

Bueckert, Dennis. 2004a. "AgCan Ends Testing of GE Wheat Developed with Monsanto." *Canadian Press,* 9 January.

–. 2004b. "Agriculture Canada Puts Brakes on Roundup Ready Wheat Project." *Globe and Mail,* 10 January, A7.

Bundesministerium für Gesundheit. 2008. *GM Food/Feed Application: Risk Assessment.* Vienna: Government of Austria.

Burk, Dan. 2004. "DNA Rules: Legal and Conceptual Implications of Biological 'Lock-out' Systems." *California Law Review* 92: 1553-87.

Burkett, Paul. 1996. "On Some Common Misconceptions about Nature and Marx's Critique of Political Economy." *Capitalism Nature Socialism* 7(3): 53-80.

–. 1999. *Marx and Nature: A Red and Green Perspective.* New York: St. Martin's Press.

Burnham, Peter. 2006. "Marx, Neo-Gramscianism and Globalization." In *Global Restructuring, State, Capital and Labour: Contesting Neo-Gramscian Perspectives,* edited by A. Bieler, W. Bonefeld, P. Burnham, and A.D. Morton, 187-95. Hampshire, UK: Palgrave MacMillan.

Busch, Lawrence, V. Gunter, T. Mentele, M. Tachikawa, and Keiko Tanaka. 1994. "Socializing Nature: Technoscience and the Transformation of Rapeseed into Canola." *Crop Science* 34: 607-14.

Butler, Declan. 2012. "Rat Study Sparks GM Furore: Cancer Claims Put Herbicide-Resistant Transgenic Maize in the Spotlight." *Nature* 489: 484.

Cakmak, Ismail, Atilla Yazici, Yusuf Tutus, and Levent Ozturk. 2009. "Glyphosate Reduced Seed and Leaf Concentrations of Calcium, Manganese, Magnesium, and Iron in Non-Glyphosate Resistant Soybean." *European Journal of Agronomy* 31: 114-19.

Canadian Biotechnology Action Network. 2009a. *Genetically Engineered Wheat Rejected Globally, Groups Remind Monsanto: Tri-National Statement Responds to Industry Pledge to Commercialize GE Wheat.* Press Release. Canadian Biotechnology Action Network: http://www.cban.ca/.

–. 2009b. *No Safety Assessment of GE Corn by Health Canada: Canada Ignores International Food Safety Guidelines.* Press Release. Canadian Biotechnology Action Network. http://www.cban.ca/.

Canadian Biotechnology Advisory Committee. 2002a. *Improving the Regulation of Genetically Modified Foods and Other Novel Foods in Canada: Report to the Government of Canada Biotechnology Ministerial Coordinating Committee.* Ottawa: Government of Canada.

–. 2002b. *Patenting of Higher Life Forms and Related Issues.* Ottawa: Government of Canada.

–. 2004a. *Biotechnology and the Health of Canadians.* Ottawa: Government of Canada.

–. 2004b. *Protecting Privacy in the Age of Genetic Information.* Ottawa: Government of Canada.

–. 2005a. *About Us.* Canadian Biotechnology Advisory Committee. http://www.cbac -cccb.ca/.

–. 2005b. *Canadian Trends in Biotechnology.* 2nd ed. Ottawa: Government of Canada.

–. 2006a. *Human Genetic Materials, Intellectual Property and the Health Sector.* Ottawa: Government of Canada.

–. 2006b. *Toward a Canadian Action Agenda for Biotechnology: A Report from the Canadian Biotechnology Advisory Committee.* Ottawa: Government of Canada.

Canadian Food Inspection Agency. 1993. *Federal Government Agrees on New Regulatory Framework for Biotechnology.* Canadian Food Inspection Agency. http://people.ucalgary.ca.

–. 2004. *Insect Resistance Management Inspection.* Ottawa: Government of Canada.

–. 2006. *The Biology of Triticum turgidum ssp. durum (Durum Wheat)*, Biology Document no. BIO2006-07. Canadian Food Inspection Agency: http://www.inspection.gc.ca/.

–. 2009. *Insect Resistance Management Audit of Canadian Bt Corn Producers.* Ottawa: Government of Canada.

Canadian Intellectual Property Office. n.d. *Manual of Patent Office Practice.* Government of Canada: http://www.ic.gc.ca/eic/site/cipointernet-internetopic.nsf/vwapj/rpbb-mopop -eng.pdf/$file/rpbb-mopop-eng.pdf.

Caruso, Denise. 2007. "A Challenge to Gene Theory: A Tougher Look at Biotech." *New York Times,* 1 July, 33.

Castells, Manuel. 1989. *The Informational City: Information Technology, Economic Restructuring, and the Urban-Regional Process.* Oxford, UK: Basil Blackwell.

–. 2000. *The Rise of the Network Society.* Vol. 1, *The Information Age: Economy, Society and Culture.* 2nd ed. Malden, MA: Blackwell Publishers.

Catangui, Michael A., and Robert K. Berg. 2009. "Western Bean Cutworm, *Striacosta albicosta* (Smith) (Lepidoptera: Noctuidae), as a Potential Pest of Transgenic Cry1Ab *Bacillus thuringiensis* Corn Hybrids in South Dakota." *Environmental Entomology* 35: 1439-52.

Caulfield, Timothy, and Colin Feasby. 1998. "The Commercialization of Human Genetics in Canada: An Overview of Policy and Legal Issues." In *Socio-Ethical Issues in Human Genetics,* edited by B. M. Knoppers, 339-401. Cowansville, QC: Les Éditions Yvon Blais.

Center for Food Safety. 2004. *Monsanto vs. U.S. Farmers.* Washington, DC: Center for Food Safety.

–. 2010. *FDA Committee (VMAC) Split on Recommendations about Controversial Genetically Engineered Salmon.* Press Release. Washington, DC: Center for Food Safety.

Chase, Steven. 2003. "Canadians Want GM Foods Labelled, Poll Finds." *Globe and Mail,* 4 December, A8.

Cheeke, Tanya E., Todd N. Rosenstiel, and Mitchell B. Cruzan. 2012. "Evidence of Reduced Arbuscular Mycorrhizal Fungal Colonization in Multiple Lines of Bt Maize." *American Journal of Botany* 99: 700-07.

Clark, E. Ann. 2001. "On the Implications of the Schmeiser Decision: The Crime of Percy Schmeiser." *Genetics Society of Canada Bulletin* 32: 19-22.

–. 2002. "Government and GM ... for Whom ... by Whom?" Paper presented to Association Canadienne Francaise pour l'Avancement des Sciences, Laval, QC, 15 May.

–. 2003. "Genetically Engineered Crops: Myths and Realities." Paper presented to Yale School of Forestry and Environmental Studies, New Haven, CT, February.

Clarke, Simon. 1983. "State, Class Struggle and the Reproduction of Capital." *Kapitalistate* 10(11): 113-30.

Cleaver, Harry. 1992. *Marxian Categories, the Crisis of Capital and the Constitution of Social Subjectivity Today.* http://libcom.org/.

Coad, Lindsey. 2007. "Farmer Loses Case over Licenced Soybeans." *Sarnia Observer,* 26 June, A1.

Codex Alimentarius Commission. 2003. *Guidelines for the Conduct of Food Safety Assessment of Foods Derived from Recombinant-DNA Plants*, Doc. CAC/GL 45-2003. Rome: Codex Alimentarius.

Cohen, Jon. 1995. "Research Materials: Share and Share Alike Isn't Always the Rule in Science." *Science* 268: 1715-18.

Committee on Environmental Impacts Associated with Commercialization of Transgenic Plants, and National Research Council. 2002. *Environmental Effects of Transgenic Plants: The Scope and Adequacy of Regulation.* Washington, DC: National Academy Press.

Complacency on the Farm. 2009. Washington, DC: Center for Science in the Public Interest.

Cook, Guy. 2005. *Genetically Modified Language: The Discourse of Arguments for GM Crops and Food.* New York: Routledge.

Cozzens, Susan E, and Edward J. Woodhouse. 1995. "Science, Government, and the Politics of Knowledge." In *Handbook of Science and Technology Studies,* edited by S. Jasanoff, G.E. Markle, J.C. Petersen, and T. Pinch, 533-53. Thousand Oaks, CA: Sage Publications.

Crick, Francis. 1962a. "The Genetic Code." *Scientific American* 207: 66-75.

–. 1962b. "Towards the Genetic Code." *Discovery* 23(3): 8-16.

–. 1963. "The Recent Excitement in the Coding Problem." *Progress in Nucleic Acids Research* 1: 163-217.

Critical Art Ensemble. 2002. *The Molecular Invasion.* Brooklyn, NY: Autonomedia.

"Critical Journalism." 2008. *Nature* 452: 387-88.

Cummings, Claire Hope. 2008. *Uncertain Peril: Genetic Engineering and the Future of Seeds.* Boston: Beacon Press.

Davies, Philip. 2004. "Gene Flow and Genetically Engineered Crops." In *Recoding Nature: Critical Perspectives on Genetic Engineering,* edited by R. Hindmarsh and G. Lawrence, 71-81. Sydney, Australia: University of New South Wales Press.

De Angelis, Massimo. 2001. "Marx and Primitive Accumulation: The Continuous Character of Capital's 'Enclosures.'" *The Commoner* 2 (September): 1-22.

–. 2007. *The Beginning of History: Value Struggles and Global Capital.* London: Pluto.

de Beer, Jeremy, and Heather McLeod-Kilmurray. 2007. "Commentary: The SCC Should Step up to the Environmental Plate." *Lawyers Weekly* 27: 7.

de la Perrière, Robert Ali Brac, and Franck Seuret. 2000. *Brave New Seeds: The Threat of GM Crops to Farmers.* Translated by M. Sovani and V. Rao. London: Zed Books.

de Landa, Manuel. 1991. *War in the Age of Intelligent Machines.* New York: Zone.

De Marcellus, Olivier. 2003. "Commons, Communities and Movements: Inside, Outside and against Capital." *The Commoner* 6 (Winter). http://www.commoner.org.uk/demarcellus06.pdf.

De Schrijver, A., Y. Devos, M. Van den Bulcke, P. Cadot, M. De Loose, D. Reheul, and M. Sneyers. 2007. "Risk Assessment of GM Stacked Events Obtained from Crosses between GM Events." *Trends in Food Science and Technology* 18: 101-09.

de Vendômois, Joël Spiroux, François Roullier, Dominique Cellier, and Gilles-Eric Séralini. 2009. "A Comparison of the Effects of Three GM Corn Varieties on Mammalian Health." *International Journal of Biological Sciences* 5: 706-26.

Dean, Amy, and Jennifer Armstrong. 2009. "Genetically Modified Foods Position Paper: The American Academy of Environmental Medicine." American Academy of Environmental Medicine, 8 May. http://www.aaemonline.org/gmopost.html.

Decima Research. 2004. *Public Opinion Research on Biotechnology: Canada-U.S. Tracking Survey.* Ottawa: Decima Research.

Definitive Global Rejection of Genetically Engineered Wheat: Statement of Australian, Canadian and US Farmer, Environmental and Consumer Organizations. 2009. Canadian Biotechnology Action Network. http://www.cban.ca/.

Desser, Chris. 2000. "Unnatural Selection or Bad Choice." In *Made Not Born: The Troubling World of Biotechnology,* edited by C. Walker, 16-26. San Francisco, CA: Sierra Club Books.

D'Hertefeldt, Tina, Rikke B. Jørgensen, and Lars B. Pettersson. 2008. *Long-Term Persistence of GM Oilseed Rape in the Seedbank.* Biology Letters. http://www.metapress.com/content/g623581733561560/fulltext.pdf.

Dhurua, Sanyasi, and Govind T. Gujar. 2011. "Field-Evolved Resistance to Bt toxin Cry1Ac in the Pink Bollworm, *Pectinophora gossypiella* (Saunders) (Lepidoptera: Gelechiidae), from India." *Pest Management Science* 67: 898-903.

Diamond, Norman. 1981. "The Politics of Scientific Conceptualization." In *Science, Technology and the Labour Process: Marxist Studies,* edited by L. Levidow and B. Young, 32-45. Atlantic Highlands, NJ: Humanities Press.

Diels, Johan, Mario Cunha, Célia Manaia, Bernardo Sabugosa-Madeira, and Margarida Silva. 2011. "Association of Financial or Professional Conflict of Interest to Research Outcomes on Health Risks or Nutritional Assessment Studies of Genetically Modified Products." *Food Policy* 36: 197-203.

Doern, G. Bruce, and Ted Reed. 2000. "Canada's Changing Science-Based Policy and Regulatory Regime: Issues and Framework." In *Risky Business: Canada's Changing Science-Based Policy and Regulatory Regime,* edited by G.B. Doern and T. Reed, 3-28. Toronto: University of Toronto Press.

Doern, G. Bruce, and Markus Sharaput. 2000. *Canadian Intellectual Property: The Politics of Innovating Institutions and Interests.* Toronto: University of Toronto Press.

Domingo, José L. 2007. "Toxicity Studies of Genetically Modified Plants: A Review of the Published Literature." *Critical Reviews in Food Science and Nutrition* 47: 721-33.

Domingo, José L., and Jordi Giné Bordonaba. 2011. "A Literature Review on the Safety Assessment of Genetically Modified Plants." *Environment International* 37: 734-42.

Domingo-Roig, José L., and Mercedes Gómez Arnáiz. 2000. "Riesgos Sobre la Salud de los Alimentos Modificados Genéticamente: Una Revision Bibliografica." *Revista Española de Salud Pública* 74: 255-61.

Dona, Artemis, and Ioannis S. Arvanitoyannis. 2009. "Health Risks of Genetically Modified Foods." *Critical Reviews in Food Science and Nutrition* 49: 164-75.

Dorhout, David L., and Marlin E. Rice. 2010. "Intraguild Competition and Enhanced Survival of Western Bean Cutworm (Lepidoptera: Noctuidae) on Transgenic Cry1Ab (MON810) *Bacillus thuringiensis* Corn." *Journal of Economic Entomology* 103: 54-62.

Downes, Sharon, Tracey Parker, and Rod Mahon. 2010. "Incipient Resistance of *Helicoverpa punctigera* to the Cry2Ab Bt Toxin in Bollgard II® Cotton." *PLoS ONE* 5(9): e12567.

Doyle, Jack. 1985. *Altered Harvest: Agriculture, Genetics, and the Fate of the World's Food Supply.* New York: Viking.

Drahos, Peter, and John Braithwaite. 2002. *Information Feudalism: Who Owns the Knowledge Economy?* London: Earthscan.

Duke, Stephen O., and Stephen B. Powles. 2008. "Glyphosate: A Once-in-a-Century Herbicide." *Pest Management Science* 64: 319-25.

Dunlop, Greg. 2010. *Bt Corn IRM Compliance in Canada: Canadian Corn Pest Coalition Report.* Cambridge, ON: AgData (Canada) Limited.

Dunwoody, Sharon. 1993. *Reconstructing Science for Public Consumption: Journalism as Science Education.* Victoria, Australia: Deakin University Press.

Dyer, Gwynne. 1999. "Frankenstein Foods: Eight Days Ago, Twenty Scientists from Thirteen Countries Demanded the Reinstatement of a British Researcher Fired for Warning That Genetically Modified Foods Could Pose a Danger to Human Health. That's When All Hell Broke Loose." *Globe and Mail,* 20 February, D1.

Dyer-Witheford, Nick. 1999. *Cyber-Marx: Cycles and Circuits of Struggle in High Technology Capitalism.* Urbana, IL: University of Illinois Press.

The Ecologist. 1993. *Whose Common Future? Reclaiming the Commons.* Philadelphia, PA: New Society Publishers.

Eisenberg, Rebecca S. 1992. "Patent Rights in the Human Genome Project." In *Gene Mapping: Using Law and Ethics as Guides,* edited by G.J. Annas and S. Elias, 226-45. New York: Oxford University Press.

–. 1997a. "Commentary: The Move toward the Privatization of Biomedical Research." In *The Future of Biomedical Research,* edited by C.E. Barfield and B.L.R. Smith, 121-24. Washington, DC: American Enterprise Institute and the Brookings Institution.

–. 1997b. "Genomic Patents and Product Incentives." In *Human DNA: Law and Policy: International and Comparative Perspectives,* edited by B.M. Knoppers, C.M. Laberge and M. Hirtle, 373-78. Cambridge, MA: Kluwer Law International.

Eker, Selim, Levent Ozturk, Atilla Yazici, Bulent Erenoglu, Volker Romheld, and Ismail Cakmak. 2006. "Foliar-Applied Glyphosate Substantially Reduced Uptake and Transport of Iron and Manganese in Sunflower (*Helianthus annuus* L.) Plants." *Journal of Agricultural and Food Chemistry* 54: 10019-25.

Elmore, Roger W., Fred W. Roeth, Lenis A. Nelson, Charles A. Shapiro, Robert N. Klein, Stevan Z. Knezevic, and Alex Martin. 2001. "Glyphosate-Resistant Soybean Cultivar Yields Compared with Sister Lines." *Agronomy Journal* 93: 408-12.

Engels, Friedrich. 1940 [1898]. *Dialectics of Nature*. Translated by C. Dutt. New York: International Publishers.

Enviropig™: Environmental Benefits. n.d. University of Guelph. http://www.uoguelph.ca/enviropig/environmental_benefits.shtml.

Epstein, Samuel S. 1990a. "Potential Public Health Hazards of Biosynthetic Milk Hormones." *International Journal of Health Services* 20: 73-84.

–. 1990b. "Questions and Answers on Synthetic Bovine Growth Hormones." *International Journal of Health Services* 20: 573-81.

–. 1996. "Unlabeled Milk from Cows Treated with Biosynthetic Growth Hormones: A Case of Regulatory Abdication." *International Journal of Health Services* 26: 173-85.

ETC Group. 2006a. *Captain Hook Awards for Biopiracy*. Communiqué no. 922006. Ottawa: ETC Group.

–. 2006b. *Terminator Rejected! A Victory for the People*. Ottawa: ETC Group.

–. 2006c. *Terminator Seed Battle Begins: Farmers Face Billions of Dollars in Potential Costs*. Ottawa: ETC Group.

–. 2007. *Terminator: The Sequel*. Communiqué no. 95. Ottawa: ETC Group.

–. 2008. *Who Owns Nature? Corporate Power and the Final Frontier in the Commodification of Life*. Communiqué no. 100. Ottawa: ETC Group.

–. 2009. *Who Will Feed Us? Questions for the Food and Climate Crisis*. Communiqué no. 102. Ottawa: ETC Group.

European Union. 2007. *Treaty of Lisbon Amending the Treaty on European Union and the Treaty Establishing the European Community*. Doc. 2007/C 306/01. Brussels, Belgium: European Commission.

Ewen, Stanley W.B., and Arpad Pusztai. 1999. "Effect of Diets Containing Genetically Modified Potatoes Expressing *Galanthus nivalis* Lectin on Rat Small Intestine." *Lancet* 354: 1353-54.

Expert Working Party on Biotechnology and Sustainable Development. 2006. *Biopromise? Biotechnology, Sustainable Development and Canada's Future Economy: Executive Report to the Canadian Biotechnology Advisory Committee from the BSDE Expert Working Party*. Ottawa: Government of Canada.

Expert Working Party on Human Genetic Materials Intellectual Property and the Health Sector. 2005. *Human Genetic Materials: Making Canada's Intellectual Property Regime Work for the Health of Canadians*. Ottawa: Government of Canada.

Falk, Raphael. 2000. "The Gene: A Concept in Tension." In *The Concept of the Gene in Development and Evolution: Historical and Epistemological Perspectives*, edited by P.J. Beurton, R. Falk, and H.-J. Rheinberger, 317-48. Cambridge: Cambridge University Press.

"Farmers, Greenpeace Launch Anti-GM Wheat Ads: *Western Producer* Refuses to Run Ad." 2004. *Canada News Wire*. http://www.organicconsumers.org/ge/wheat032304.cfm#farmer.

Faure, N., H. Serieys, and A. Berville. 2002. "Potential Gene Flow from Cultivated Sunflower to Volunteer, Wild *Helianthus* Species in Europe." *Agriculture Ecosystems and Environment* 89: 183-90.

Federici, Silvia. 1992. "The Debt Crisis, Africa and the New Enclosures." In *Midnight Oil: Work, Energy, War, 1973-1992*, edited by Midnight Notes Collective, 303-16. Jamaica Plains, MA: Autonomedia.

Feenberg, Andrew. 1999. *Questioning Technology*. New York: Routledge.

–. 2002. *Transforming Technology: A Critical Theory Revisited*. 2nd ed. New York: Oxford University Press.

Fernandez, M.R., F. Selles, D. Gehl, R.M. DePauw, and R.P. Zentner. 2005. "Crop Production Factors Associated with Fusarium Head Blight in Spring Wheat in Eastern Saskatchewan." *Crop Science* 45: 1908-16.

Fernandez, M.R., R.P. Zentner, P. Basnyat, D. Gehl, F. Selles, and D. Huber. 2009. "Glyphosate Associations with Cereal Diseases Caused by *Fusarium spp.* in the Canadian Prairies." *European Journal of Agronomy* 31: 133-43.

Ferrara, Jennifer, and Michael K. Dorsey. 2001. "Genetically Engineered Foods: A Minefield of Safety Hazards." In *Redesigning Life? The Worldwide Challenge to Genetic Engineering*, edited by B. Tokar, 51-66. Montreal and Kingston: McGill-Queen's University Press.

Finamore, Alberto, Marianna Roselli, Serena Britti, Giovanni Monastra, Roberto Ambra, Aida Turrini, and Elena Mengheri. 2008. "Intestinal and Peripheral Immune Response to MON810 Maize Ingestion in Weaning and Old Mice." *Journal of Agricultural and Food Chemistry* 56: 11533-39.

Finn, R.K. 1989. "Some Origins of Biotechology." *Swiss Review for Biotechnology* 7: 15-17.

Fitzgerald, Alison. 2010. "Monsanto Seven-State Probe Threatens Profit from 93 percent Soybean Share." *BusinessWeek,* 10 March. http://www.businessweek.com/.

Foster, John Bellamy. 2000. *Marx's Ecology: Materialism and Nature.* New York: Monthly Review Press.

Fox Keller, Evelyn. 2000. *The Century of the Gene.* Cambridge, MA: Harvard University Press.

Freeland Judson, Horace. 1992. "A History of the Science and Technology behind Gene Mapping and Sequencing." In *The Code of Codes: Scientific and Social Issues in the Human Genome Project,* edited by D.J. Kevles and L. Hood, 37-80. Cambridge, MA: Harvard University Press.

Freeman, Aaron. 2001. "Federal Government's Pro-Biotech Bias Is Most Evident at CFIA." *The Hill Times,* 19 November, B4.

–. 2003. "Feds, Consumers' Association Help Market GM Foods: Federal Government Funnels Money to Consumers Association of Canada." *The Hill Times,* 19 November, B4.

Freeze, Colin. 2002. "Ottawa Promoting Safety of GMOs: Critics Say Money Spent on Food Pamphlets Shows Watchdog Too Close to Biotech Firms." *Globe and Mail,* 5 February, A9.

Friesen, Lyle F., Alison G. Nelson, and Rene C. Van Acker. 2003. "Evidence of Contamination of Pedigreed Canola *(Brassica napus)* Seedlots in Western Canada with Genetically Engineered Herbicide Resistance Traits." *Agronomy Journal* 95: 1342-47.

Fuglie, Keith Owen, and David Schimmelpfennig. 2000. *Public-Private Collaboration in Agricultural Research: New Institutional Arrangements and Economic Implications.* Ames, IA: Iowa State University Press.

Gaines, Todd A., Wenli Zhang, Dafu Wang, Bekir Bukun, Stephen T. Chisholm, Dale L. Shaner, Scott J. Nissen, William L. Patzoldt, Patrick J. Tranel, A. Stanley Culpepper, Timothy L. Grey, Theodore M. Webster, William K. Vencill, R. Douglas Sammons, Jiming Jiang, Christopher Preston, Jan E. Leach, and Philip Westra. 2010. "Gene Amplification Confers Glyphosate Resistance in Amaranthus palmeri." *Proceedings of the National Academy of Sciences* 107(3): 1029-34.

Gandy, Oscar H, Jr. 1982. *Beyond Agenda Setting: Information Subsidies and Public Policy.* Norwood, NJ: Ablex.

Gasnier, Céline, Coralie Dumont, Nora Benachour, Emilie Clair, Marie-Christine Chagnon, and Gilles-Eric Séralini. 2009. "Glyphosate-Based Herbicides Are Toxic and Endocrine Disruptors in Human Cell Lines." *Toxicology* 262: 184-91.

Gassmann, Aaron J., Jenifer L. Petzold-Maxwell, Ryan S. Keweshan, and Mike W. Dunbar. 2011. "Field-Evolved Resistance to Bt Maize by Western Corn Rootworm." *PLoS ONE* 6(7): 1-7.

Gilbert, Natasha. 2010. "GM Crop Escapes into the American Wild." *Nature.* http://www.nature.com/news/2010/100806/full/news.2010.393.html.

Glassman, Jim. 2006. "Primitive Accumulation, Accumulation by Dispossession, Accumulation by 'Extra-Economic' Means." *Progress in Human Geography* 30: 608-25.

GM Freeze. 2009. *Undoing the ISAAA Myths on GM Crops.* Barnsley, UK: GM Freeze.

"GM Sugar Beets Found in Soil Mix Sold to Gardeners." 2009. *Organic and Non-GMO Report* (July/August). http://www.non-gmoreport.com/articles/july09/gm_sugar_beets_in_gardeners_soil.php.

Gold, E. Richard, and Wendy A. Adams. 2001. "The Monsanto Decision: The Edge or the Wedge." *Nature Biotechnology* 19: 587.

Gordon, Barney. 2007. "Manganese Nutrition of Glyphosate-Resistant and Conventional Soybeans." *Better Crops with Plant Food* 91(4): 12-13.

Gordon, Wendy J. 1993. "A Property Right in Self-Expression: Equality and Individualism in the Natural Law of Intellectual Property." *Yale Law Journal* 102: 1533-609.

Gorz, André. 1976. "On the Class Character of Science and Scientists." In *The Political Economy of Science: Ideology of/in the Natural Sciences*, edited by H. Rose and S. Rose, 59-71. London: Macmillan.

Government of Canada. 1996. *Science and Technology for the New Century: A Federal Strategy*. Ottawa: Government of Canada.

–. 2001. *Action Plan of the Government of Canada in Response to the Royal Society of Canada Expert Panel Report Elements of Precaution: Recommendations for the Regulation of Food Biotechnology in Canada*. Edited by Health Canada, Canadian Food Inspection Agency, Environment Canada, Agriculture and Agri-Food Canada and Department of Fisheries and Oceans. Ottawa: Government of Canada.

–. 2004. *Biotechnology Transforming Society: Creating an Innovative Economy and a Higher Quality of Life: Report on Biotechnology (1998-2003)*. Ottawa: Government of Canada.

–. 2007. *Mobilizing Science and Technology to Canada's Advantage*. Ottawa: Government of Canada.

–. 2009. *Mobilizing Science and Technology to Canada's Advantage: Progress Report*. Ottawa: Government of Canada.

Government of Canada Invests $51.8 Million in R&D to Drive Growth in Atlantic Canada, $10.2 Million in Prince Edward Island. 2009. Government of Canada. http://www.acoa-apeca.gc.ca/eng/Agency/mediaroom/NewsReleases/Pages/2504.aspx.

Graff, Gregory D., Susan E. Cullen, Kent J. Bradford, David Zilberman, and Alan B. Bennett. 2003. "The Public-Private Structure of Intellectual Property Ownership in Agricultural Biotechnology." *Nature Biotechnology* 21: 989-95.

Gray, Mike. 2011a. "Additional Reports of Severe Root Damage to Bt Corn Received: Questions and Answers." *The Bulletin* (22). http://bulletin.ipm.illinois.edu/article.php?id=1569.

–. 2011b. "Severe Root Damage to Bt Corn Observed in Northwestern Illinois." *The Bulletin* (20). http://bulletin.ipm.illinois.edu/article.php?id=1555.

Greenpeace Canada. 2003a. *Greenpeace Quarantines Genetically Engineered Wheat in Canada*. http://www.greenpeace.org/international/press/releases/greenpeace-quarantines-genetic.

–. 2003b. *How to Avoid Genetically Engineered Food: A Greenpeace Shoppers Guide*. 2nd ed. Toronto: Greenpeace Canada.

–. 2004. *Tell Canada to Keep Monsanto out of Your Bread*. http://weblog.greenpeace.org/ge/archives/001275.html.

Greenpeace International. 2008. *GM Contamination Register Report 2007: Annual Review of Cases of Contamination, Illegal Planting and Negative Side Effects of Genetically Modified Organisms*. Amsterdam: Greenpeace International.

Guattari, Félix. 1992. "Regimes, Pathways, Subjects." In *Incorporations*, edited by J. Crary and S. Kwinter, 16-35. New York: Zone.

Gurian-Sherman, Doug. 2009. *Failure to Yield: Evaluating the Performance of Genetically Engineered Crops*. Cambridge, MA: Union of Concerned Scientists.

Hall, Angela. 2004. "Monsanto Withdraws Bids: GM Wheat Project 'Essentially Over.'" *Star-Phoenix*, 22 June, D1.

Hall, Linda, Keith Topinka, John Huffman, Lesley Davis, and Allen Good. 2000. "Pollen Flow between Herbicide-Resistant *Brassica napus* Is the Cause of Multiple-Resistant *B. napus* Volunteers." *Weed Science* 48: 688-94.

Hall, Stephen. 1987. *Invisible Frontiers: The Race to Synthesise a Human Gene*. London: Sidgwick and Jackson.

Hansen, Michael. 2010. Veterinary Medicine Advisory Committee, *Comments of Consumers Union on Genetically Engineered Salmon, Food and Drug Administration Docket no. FDA-201034-N-0001*, 16 September.

Haraway, Donna J. 1997. *Modest Witness@Second Millennium: FemaleMan© Meets OncoMouse™: Feminism and Technoscience*. New York: Routledge.

Hardell, Lennart, and Mikael Eriksson. 1999. "A Case-Control Study of Non-Hodgkin Lymphoma and Exposure to Pesticides." *Cancer* 85: 1353-60.

Hardt, Michael, and Antonio Negri. 2004. *Multitude: War and Democracy in the Age of Empire.* New York: Penguin Press.

Harper, Tim. 2003. "Bush Links EU Ban to Famine in Africa; Cites Resistance to Biotech Crops; Calls on Europe to Accept Imports." *Toronto Star,* 24 June, A10.

Hart, Miranda M., Jeff R. Powell, Robert H. Gulden, David J. Levy-Booth, Kari E. Dunfield, K. Peter Pauls, Clarence J. Swanton, John N. Klironomos, and Jack T. Trevors. 2009. "Detection of Transgenic *cp4 epsps* Genes in the Soil Food Web." *Agronomy for Sustainable Development* 29: 497-501.

Harvey, David. 1976. "The Marxian Theory of the State." *Antipode* 8(2): 80-89.

–. 2000. *Spaces of Hope.* Edinburgh: Edinburgh University Press.

–. 2003. *The New Imperialism.* Oxford: Oxford University Press.

–. 2006. *Spaces of Global Capitalism.* New York: Verso.

Hauter, Wenonah. 2010. "Letter from Food and Water Watch to Commissioner Margaret Hamburg, U.S. Food and Drug Administration re. Genetically Engineered Salmon." Washington, DC, 15 November. http://documents.foodandwaterwatch.org/doc/FOIA -FWW-GEsalmon-letter-FDA.pdf.

Heinemann, Jack A. 2009. *Report on Animals Exposed to GM Ingredients in Animal Feed.* Auckland, New Zealand: Commerce Commission of New Zealand.

Heller, Chaia. 2001. "McDonald's, MTV and Monsanto: Resisting Biotechnology in the Age of Informational Capital." In *Redesigning Life? The Worldwide Challenge to Genetic Engineering,* edited by B. Tokar, 405-19. Montreal and Kingston: McGill-Queen's University Press.

Heller, Michael A., and Rebecca S. Eisenberg. 1998. "Can Patents Deter Innovation? The Anticommons in Biomedical Research." *Science* 280: 698-701.

Hess, Charlotte, and Elinor Ostrom, eds. 2007. *Understanding Knowledge as a Commons: From Theory to Practice.* Cambridge, MA: MIT Press.

Hills, Melissa J., Linda Hall, Paul G. Arnison, and Allen G. Good. 2007. "Genetic Use Restriction Technologies (GURTs): Strategies to Impede Transgene Movement." *Trends in Plant Science* 12: 177-83.

Hindmarsh, Richard. 2001. "Constructing Bio-Utopia: Laying Foundations amidst Dissent." In *Altered Genes II: The Future?* edited by R. Hindmarsh and G. Lawrence, 36-52. Melbourne, Australia: Scribe Publications.

Hindmarsh, Richard, and Geoffrey Lawrence. 2004. "Recoding Nature: Deciphering the Script." In *Recoding Nature: Critical Perspectives on Genetic Engineering,* edited by R. Hindmarsh and G. Lawrence, 23-40. Sydney, Australia: University of New South Wales Press.

Hirsch, Joachim. 1978. "The State Apparatus and Social Reproduction: Elements of a Theory of the Bourgeois State." In *State and Capital: A Marxist Debate,* edited by J. Holloway and S. Picciotto, 57-107. London: Edward Arnold.

Hisano, Shuji. 2005. "A Critical Observation on the Mainstream Discourse of Biotechnology for the Poor." *Tailoring Biotechnologies* 1(2): 81-105.

Ho, Mae-Wan. 1999. *Genetic Engineering: Dream or Nightmare? Turning the Tide on the Brave New World of Bad Science and Big Business.* 2nd ed. Dublin, Ireland: Gateway.

Ho, Mae-Wan, Angela Ryan, and Joe Cummins. 1999. "Cauliflower Mosaic Viral Promoter: A Recipe for Disaster?" *Microbial Ecology in Health and Disease* 11: 194-97.

Holloway, John. 1992. "Crisis, Fetishism, Class Composition." In *Open Marxism,* edited by W. Bonefeld, R. Gunn, and K. Psychopedis, 145-69. London: Pluto Press.

Holloway, John, and Sol Picciotto. 1978. "Introduction: Towards a Materialist Theory of the State." In *State and Capital: A Marxist Debate,* edited by J. Holloway and S. Picciotto, 1-31. London: Edward Arnold.

Hopkin, Michael. 2008. "Frozen Futures." *Nature* 452: 404-5.

Hopkins, William G. 2006. *Plant Biotechnology.* New York: Chelsea House.

Hornig Priest, Susanna. 1995. "Information Equity, Public Understanding of Science, and the Biotechnology Debate." *Journal of Communication* 45: 39-54.

–. 2006. "The Public Opinion Climate for Gene Technologies in Canada and the United States: Competing Voices, Contrasting Frames." *Public Understanding of Science* 15: 55-71.

Hornig Priest, Susanna, and Toby Ten Eyck. 2004. "Transborder Information, Local Resistance and the Spiral of Silence: Biotechnology and Public Opinion in the United States." In *Biotechnology and Communication: The Meta-Technologies of Information*, edited by S. Braman, 175-96. Mahwah, NJ: Lawrence Erlbaum Associates.

Hubbard, Kristina. 2009. *Out of Hand: Farmers Face the Consequences of a Consolidated Seed Industry*. Stoughton, WI: Farmer to Farmer Campaign on Genetic Engineering/National Family Farm Coalition.

Hubbard, Ruth, and Elijah Wald. 1993. *Exploding the Gene Myth: How Genetic Information Is Produced and Manipulated by Scientists, Physicians, Employers, Insurance Companies, Educators, and Law Enforcers*. Boston: Beacon Press.

Individual Action Not the Way to Go: Organic Farmers. 2008. Organic Agriculture Protection Fund. http://www.saskorganic.com/oapf/.

Industry Canada. 1996. *Science and Technology for the New Century: A Federal Strategy*. Ottawa: Government of Canada.

–. 1998. *The 1998 Canadian Biotechnology Strategy: An Ongoing Renewal Process*. Ottawa: Government of Canada.

–. 2000. *Pathways to Growth: Opportunities in Biotechnology*. Ottawa: Government of Canada.

Is "Enviropig™" Safe? n.d. Canadian Biotechnology Action Network. http://www.cban.ca/Resources/Topics/Enviropig/Is-Enviropig-Safe.

James, Clive. 2006. *Global Status of Commercialized Biotech/GM Crops: 2006*. Ithaca, NY: The International Service for the Acquisition of Agri-Biotech Applications.

–. 2007. *Global Status of Commercialized Biotech/GM Crops: 2007*. Ithaca, NY: The International Service for the Acquisition of Agri-Biotech Applications.

–. 2008. *Global Status of Commercialized Biotech/GM Crops: 2008*. Ithaca, NY: The International Service for the Acquisition of Agri-Biotech Applications.

–. 2011. *Global Status of Commercialized Biotech/GM Crops: 2011*. Ithaca, NY: International Service for the Acquisition of Agri-Biotech Applications.

Jarvis, Bill. 2000. "A Question of Balance: New Approaches for Science-Based Regulation." In *Risky Business: Canada's Changing Science-Based Policy and Regulatory Regime*, edited by G.B. Doern and T. Reed, 307-33. Toronto: University of Toronto Press.

Jenkins, Matt. 2007. "Brave New Hay." *High Country News*. http://www.hcn.org/servlets/hcn.Article?article_id=17054.

Jiao, Zhe, Xiao-Xi Si, Gong-Ke Li, Zhuo-Min Zhang, and Xin-Ping Xu. 2010. "Unintended Compositional Changes in Transgenic Rice Seeds (*Oryza sativa L.*) Studied by Spectral and Chromatographic Analysis Coupled with Chemometrics Methods." *Journal of Agricultural and Food Chemistry* 58: 1746-54.

Johal, G.S., and D.M. Huber. 2009. "Glyphosate Effects on Diseases of Plants." *European Journal of Agronomy* 31: 144-52.

Johnson, William G., Vince M. Davis, Greg R. Kruger, and Stephen C. Weller. 2009. "Influence of Glyphosate-Resistant Cropping Systems on Weed Species Shifts and Glyphosate-Resistant Weed Populations." *European Journal of Agronomy* 31: 162-72.

Jordan, Trish, and Brenda Harris. 2009. "Stacked Traits Good for Growers." *Western Producer*, 5 November. http://www.producer.com/2009/11/stacked-traits-good-for-growers/.

Jost, P., D. Shurley, S. Culpepper, P. Roberts, R. Nichols, J. Reeves, and S. Anthony. 2008. "Economic Comparison of Transgenic and Nontransgenic Cotton Production Systems in Georgia." *Agronomy Journal* 100: 42-51.

Judge, Elizabeth F. 2007. "Intellectual Property Law as an Internal Limit on Intellectual Property Rights and Autonomous Source of Liability for Intellectual Property Owners." *Bulletin of Science, Technology and Society* 27: 301-13.

Kaskey, Jack. 2009. "Monsanto Facing 'Distrust' As It Seeks to Stop DuPont." *Bloomberg.com*, 10 November. http://www.bloomberg.com/.

Katz, Cindi. 1998. "Whose Nature, Whose Culture? Private Productions of Space and the 'Preservation' of Nature." In *Remaking Reality: Nature at the Millennium*, edited by B. Braun and N. Castree, 46-63. London: Routledge.

Kautsky, Karl. 1988 [1899]. *The Agrarian Question*. Translated by P. Burgess. Vol. 1. London: Zwan Publications.

Kay, Lily E. 2000. *Who Wrote the Book of Life? A History of the Genetic Code*. Stanford, CA: Stanford University Press.

Kenney, Martin. 1986. *Biotechnology: The University-Industrial Complex*. New Haven, CT: Yale University Press.

Kevles, Daniel J. 1998. "*Diamond v. Chakrabarty* and Beyond: The Political Economy of Patenting Life." In *Private Science: Biotechnology and the Rise of Molecular Sciences*, edited by A. Thackray, 65-79. Philadelphia, PA: University of Pennsylvania Press.

King, Jonathan. 1997. "The Biotechnology Revolution: Self-Replicating Factories and the Ownership of Life Forms." In *Cutting Edge: Technology, Information Capitalism and Social Revolution*, edited by J. Davis, T.A. Hirschl, and M. Stack, 145-56. London: Verso.

King, Jonathan, and Doreen Stabinsky. 1999. "Biotechnology under Globalisation: The Corporate Expropriation of Plant, Animal and Microbial Species." *Race and Class* 40(2-3): 73-89.

Klein, Roger D. 2007. "Gene Patents and Genetic Testing in the United States." *Nature Biotechnology* 25: 989-90.

Kleinman, Daniel Lee, and Jack Kloppenburg, Jr. 1991. "Aiming for the Discursive High Ground: Monsanto and the Biotechnology Controversy." *Sociological Forum* 6: 427-47.

Kloppenburg, Jack Ralph, Jr. 1988. *First the Seed: The Political Economy of Plant Biotechnology, 1492-2000*. New York: Cambridge University Press.

–. 2004. *First the Seed: The Political Economy of Plant Biotechnology*. 2nd ed. Madison, WI: University of Wisconsin Press.

Kneen, Brewster. 2000. *A Common Script: The Sham of Public Consultation, Analysis of Federal Government's Strategy to Push Biotech*. The Ram's Horn. http://ramshorn.ca/issue-179.

Knezevic, Irena. 2005. "How Breakfast Is Served: Globalization, the Press in Canada and Genetically Modified Food." Master's thesis, University of Windsor.

Knispel, Alexis L., Stéphane M. McLachlan, Rene C. Van Acker, and Lyle F. Friesen. 2009. "Gene Flow and Multiple Herbicide Resistance in Escaped Canola Populations." *Weed Science* 56: 72-80.

Kremer, Robert J., and Nathan E. Means. 2009. "Glyphosate and Glyphosate-Resistant Crop Interactions with Rhizosphere Microorganisms." *European Journal of Agronomy* 31: 153-61.

Kremer, Robert J., Nathan E. Means, and Sujung Kim. 2005. "Glyphosate Affects Soybean Root Exudation and Rhizosphere Micro-Organisms." *International Journal of Environmental Analytical Chemistry* 85: 1165-74.

Krimsky, Sheldon. 1991. *Biotechnics and Society: The Rise of Industrial Genetics*. New York: Praeger.

Kuruganti, Kavitha. 2009. "Bt Cotton and the Myth of Enhanced Yields." *Economic and Political Weekly* 44(22): 29-33.

Kuyek, Devlin. 2002. *The Real Board of Directors: The Construction of Biotechnology Policy in Canada, 1980-2002*. Sorrento, BC: The Ram's Horn.

Laidlaw, Stuart. 2001. "Genetically Modified Spuds Cleared: Inspectors Had Blasted 'Extremely Poor' Field Trials." *Toronto Star*, 23 January, B1.

Lalonde, Michelle. 2007. "Cost to Label Genetic Food Is Overblown." *The Gazette*, 19 March, A3.

Latham, Jonathan R., Allison K. Wilson, and Ricarda A. Steinbrecher. 2006. "The Mutational Consequences of Plant Transformation." *Journal of Biomedicine and Biotechnology* (2006). http://downloads.hindawi.com/journals/jbb/2006/025376.pdf.

Laursen, Lucas. 2010. "How Green Biotech Turned White and Blue." *Nature Biotechnology* 28: 393-95.

Lavigne, C, E. Klein, and D. Couvet. 2002. "Using Seed Purity Data to Estimate an Average Pollen Mediated Gene Flow from Crops to Wild Relatives." *Theoretical and Applied Genetics* 104: 139-45.

Leahy, Stephen. 2004. "Activists Wary as Monsanto Withdraws GE Wheat." *Inter Press Service*, 11 May. http://www.ipsnews.net/.

Lean, Geoffrey. 2005. "Revealed: Health Fears over Secret Study into GM Food." *The Independent*, 22 May. http://www.independent.co.uk/.

Lee, Maria. 2003. "What Is Nuisance?" *Law Quarterly Review* 119: 298-325.

Lee, Maria, and Robert Burrell. 2002. "Liability for the Escape of GM Seeds: Pursuing the 'Victim'?" *Modern Law Review* 65: 517-37.

Lei, Zhen, Rakhi Juneja, and Brian D. Wright. 2009. "Patents versus Patenting: Implications of Intellectual Property Protection for Biological Research." *Nature* 27: 36-40.

Leiss, William. 2000. "Between Expertise and Bureaucracy: Risk Management Trapped at the Science-Policy Interface." In *Risky Business: Canada's Changing Science-Based Policy and Regulatory Regime,* edited by G.B. Doern and T. Reed, 49-74. Toronto: University of Toronto Press.

–. 2001. *In the Chamber of Risks: Understanding Risk Controversies.* Montreal and Kingston: McGill-Queen's University Press.

Leonard, Christopher. 2009. "AP Investigation: Monsanto Seed Business Role Revealed." *Seattle Post-Intelligencer,* 13 December. http://www.seattlepi.com/.

Lessig, Lawrence. 1999. *Code and Other Laws of Cyberspace.* New York: Basic Books.

Let Nature's Harvest Continue: Statement from All the African Delegates (Except South Africa) to FAO Negotiations on the International Undertaking for Plant Genetic Resources. 1998. http://www.docstoc.com/docs/22622699/LET-NATURES-HARVEST-CONTINUE.

Levuax, Ari. 2012. "The Genetically Engineered Salmon That Could Soon Run Wild. *Outside,* 6 June: http://www.outsideonline.com/.

Levidow, Les. 1995. "Whose Ethics for Agricultural Biotechnology?" In *Biopolitics: A Feminist and Ecological Reader on Biotechnology,* edited by V. Shiva and I. Moser, 175-90. London: Zed Books.

–. 1999. "Democratizing Technology – or Technologizing Democracy? Regulating Agricultural Biotechnology in Europe." In *Democratising Technology: Theory and Practice of Deliberative Technology Policy,* edited by R. von Schomberg, 51-69. Hengelo, Netherlands: International Centre for Human and Public Affairs.

Levidow, Les, and Joyce Tait. 1995. "The Greening of Biotechnology: GMOs as Environment-Friendly Products." In *Biopolitics: A Feminist and Ecological Reader on Biotechnology,* edited by V. Shiva and I. Moser, 121-38. London: Zed Books.

Levins, Richard, and Richard C. Lewontin. 1985. *The Dialectical Biologist.* Cambridge, MA: Harvard University Press.

–. 1994. "Holism and Reductionism in Ecology." *Capitalism Nature Socialism* 5(4): 33-40.

–. 1997. "The Biological and the Social." *Capitalism Nature Socialism* 8(3): 89-92.

Lewontin, Richard C. 1982. "Organism and Environment." In *Learning, Development, and Culture: Essays in Evolutionary Epistemology,* edited by H.C. Plotkin, 151-68. Toronto: J. Wiley.

–. 2000a. "The Maturing of Capitalist Agriculture: Farmer as Proletarian." In *Hungry for Profit: The Agribusiness Threat to Farmers, Food, and the Environment,* edited by F. Magdoff, J.B. Foster, and F.H. Buttel, 93-106. New York: Monthly Review Press.

–. 2000b. *The Triple Helix: Gene, Organism, and Environment.* Cambridge, MA: Harvard University Press.

Lindner, Roland. 2009. "Die Bauern und die Detektive." *Frankfurter Allgemeine Zeitung,* 13 June, 3.

Little, Matthew. 2009. "Agency's Decision Lacks Scientific Support, Allege Researchers." *Epoch Times Online,* 12 August. http://www.theepochtimes.com/.

Locke, John. 1988 [1690]. *Two Treatises of Government.* Edited by P. Laslett. Cambridge, UK: Cambridge University Press.

Loeppky, Rodney. 2005. *Encoding Capital: The Political Economy of the Human Genome Project.* New York: Routledge.

Long, Clarisa, and Richard A. Johnson. 1997. "Intellectual Property and Open Communication in Biomedical Research." In *The Future of Biomedical Research,* edited by C.E. Barfield and B.L.R. Smith, 105-20. Washington, DC: American Enterprise Institute and the Brookings Institution.

Lotter, Don. 2009. "The Genetic Engineering of Food and the Failure of Science – Part 1: The Development of a Flawed Enterprise." *International Journal of the Sociology of Agriculture and Food* 16(1): 31-49.

Lövei, Gabor L., David A. Andow, and Salvatore Arpaia. 2009. "Transgenic Insecticidal Crops and Natural Enemies: A Detailed Review of Laboratory Studies." *Environmental Entomology* 38: 293-306.

Lövei, Gabor L., and Salvatore Arpaia. 2005. "The Impact of Transgenic Plants on Natural Enemies: A Critical Review of Laboratory Studies." *Entomologia Experimentalis et Applicata* 114: 1-14.

Lu, Yanhui, Kongming Wu, Yuying Jiang, Bing Xia, Ping Li, Hongqiang Feng, Kris A.G. Wyckhuys, and Yuyuan Guo. 2010. "Mirid Bug Outbreaks in Multiple Crops Correlated with Wide-Scale Adoption of Bt Cotton in China." *Science* 328(5982): 1151-54.

Luxemburg, Rosa. 2004. "The Historical Conditions of Accumulation, from *The Accumulation of Capital.*" In *The Rosa Luxemburg Reader,* edited by P. Hudis and K.B. Anderson, 32-70. New York: Monthly Review Press.

Macdonald, Mark R. 2000. "Socioeconomic versus Science-Based Regulation: Informal Influences on the Formal Regulation of rBST in Canada." In *Risky Business: Canada's Changing Science-Based Policy and Regulatory Regime,* edited by G.B. Doern and T. Reed, 156-81. Toronto: University of Toronto Press.

MacRae, Rod, Holly Penfound, and Charles Margulis. 2002. *Against the Grain: The Threat of Genetically Engineered Wheat.* Toronto: Greenpeace.

Magat, Wesley A., and W. Kip Viscusi. 1992. *Informational Approaches to Regulation.* Cambridge, MA: MIT Press.

Magdoff, Fred, John Bellamy Foster, and Frederick H. Buttel. 2000a. "An Overview." In *Hungry for Profit: The Agribusiness Threat to Farmers, Food, and the Environment,* edited by F. Magdoff, J.B. Foster, and F.H. Buttel, 7-21. New York: Monthly Review Press.

–, eds. 2000b. *Hungry for Profit: The Agribusiness Threat to Farmers, Food, and the Environment.* New York: Monthly Review Press.

Magnan, André. 2006. "Refeudalizing the Public Sphere: 'Manipulated Publicity' in the Canadian Debate on GM Foods." *Canadian Journal of Sociology* 31: 25-53.

Malatesta, M., M. Biggiogera, E. Manuali, M.B.L. Rocchi, B. Baldelli, and G. Gazzanelli. 2003. "Fine Structural Analyses of Pancreatic Acinar Cell Nuclei from Mice Fed on Genetically Modified Soybean." *European Journal of Histochemistry* 47: 385-88.

Malatesta, Manuela, Federica Boraldi, Giulia Annovi, Beatrice Baldelli, Serafina Battistelli, Marco Biggiogera, and Daniela Quaglino. 2008. "A Long-Term Study on Female Mice Fed on a Genetically Modified Soybean: Effects on Liver Ageing." *Histochemistry and Cell Biology* 130: 967-77.

Mandel, Ernest. 1975. *Late Capitalism.* Translated by J. De Bres. London: NLB.

Mansour, Salah. 2009. *USDA Foreign Agricultural Service GAIN Report: Egypt Biotechnology.* Report no. EG9012. Washington, DC: United States Department of Agriculture.

Marc, Julie, Magali Le Breton, Patrick Cormier, Julia Morales, Robert Bellé, and Odile Mulner-Lorillon. 2005. "A Glyphosate-Based Pesticide Impinges on Transcription." *Toxicology and Applied Pharmacology* 203: 1-8.

Marcuse, Herbert. 1968a. *Negations: Essays in Critical Theory.* Translated by J.J. Shapiro. London: Penguin Press.

–. 1968b. *One Dimensional Man: Studies in the Ideology of Advanced Industrial Society.* Boston: Beacon Press.

Martinez, Jeannette. 2011. *Memorandum to Mike Mendelsohn, Senior Regulatory Action Leader, United States Environmental Protection Agency, Microbial Pesticides Branch, Biopesticides and Pollution Prevention Division.* Washington, DC: United States Environmental Protection Agency.

Marvier, Michelle, Yves Carrière, Norman Ellstrand, Paul Gepts, Peter Kareiva, Emma Rosi-Marshall, Bruce E. Tabashnik, and L. LaReesa Wolfenbarger. 2008. "Harvesting Data from Genetically Engineered Crops." *Science* 320: 452-53.

Marvier, Michelle, and Rene C. Van Acker. 2005. "Can Crop Transgenes Be Kept on a Leash?" *Frontiers in Ecology and the Environment* 3: 93-100.

Marx, Karl. 1962a. "The Eighteenth Brumaire of Louis Bonaparte." In *Karl Marx and Friedrich Engels: Selected Works in Two Volumes,* 243-344. Moscow: Foreign Languages Publishing House.

–. 1962b. "Preface to a Contribution to the Critique of Political Economy." In *Karl Marx and Friedrich Engels: Selected Works in Two Volumes*, 361-65. Moscow: Foreign Languages Publishing House.

–. 1963. *Theories of Surplus Value: Part I*. Moscow: Progress Publishers.

–. 1967. *Capital: A Critique of Political Economy: The Process of Capitalist Production as a Whole*. Edited by F. Engels. Vol. 3. New York: International Publishers.

–. 1972. *Theories of Surplus Value: Part III*. Translated by J. Cohen. London: Lawrence and Wishart.

–. 1975a [1844]. "Economic and Philosophical Manuscripts." In *Early Writings*, 279-400. New York: Vintage Books.

–. 1975b [1844]. "Letter from Marx to P.V. Annenkov (28 December, 1846)." In *Early Writings*, 441-52. New York: Vintage Books.

–. 1992. *Capital: A Critique of Political Economy*. Translated by B. Fowkes. Vol. 1. London: Penguin Books.

–. 1993. *Grundrisse: Foundations of the Critique of Political Economy*. Translated by M. Nicolaus. London: Penguin Books.

–. 1994. "Economic Manuscript of 1861-63 (Conclusion)." In *Karl Marx, Frederick Engels: Collected Works*, 7-336. New York: International Publishers.

Marx, Karl, and Friedrich Engels. 1962. "Manifesto of the Communist Party." In *Karl Marx and Friedrich Engels: Selected Works in Two Volumes*, 33-65. Moscow: Foreign Languages Publishing House.

Maskus, Keith, and Jerome Reichman. 2004. "The Globalization of Private Knowledge Goods and the Privatization of Global Public Goods." *Journal of International Economic Law* 7: 279-320.

Matthaei, J. Heinrich, Oliver W. Jones, Robert G. Martin, and Marshall W. Nirenberg. 1962. "Characteristics and Composition of RNA Coding Units." *Proceedings of the National Academy of Sciences* 48: 666-77.

Matthews Glenn, Jane. 2004. "Footloose: Civil Responsibility for GMO Gene Wandering in Canada." *Washburn Law Journal* 43: 547-73.

Maurin, Jost. 2008. "Transgene Pflanzen in der Landwirtschaft: Gentechnik-Vorteile nicht bewiesen." *Die tageszeitung*, 29 April. http://www.taz.de/!16530/.

Mauro, Ian J., and Stéphane M. McLachlan. 2008. "Farmer Knowledge and Risk Analysis: Postrelease Evaluation of Herbicide-Tolerant Canola in Western Canada." *Risk Analysis* 28: 463-76.

Mauro, Ian J., Stéphane M. McLachlan, and Rene C. Van Acker. 2009. "Farmer Knowledge and a Priori Risk Analysis: Pre-Release Evaluation of Genetically Modified Roundup Ready Wheat across the Canadian Prairies." *Environmental Science and Pollution Research International* 16: 689-701.

May, Christopher. 2004. "Justifying Enclosure? Intellectual Property and Meta-Technologies." In *Biotechnology and Communication: The Meta-Technologies of Information*, edited by S. Braman, 119-43. Mahwah, NJ: Lawrence Erlbaum Associates.

May, Christopher, and Susan K Sell. 2006. *Intellectual Property Rights: A Critical History*. Boulder, CO: Lynne Rienner Publishers.

Mayer, Sue. 2002. *Genetically Engineered Wheat – Changing Our Daily Bread*. Berlin: Greenpeace International.

McAfee, Kathleen. 2003. "Neoliberalism on the Molecular Scale: Economic and Genetic Reductionism in Biotechnology Battles." *Geoforum* 34: 203-19.

McCarthy, James. 2004. "Privatizing Conditions of Production: Trade Agreements as Neoliberal Environmental Governance." *Geoforum* 35: 327-41.

McGiffen, Steven P. 2005. *Biotechnology: Corporate Power versus the Public Interest*. London: Pluto Press.

McIntyre, Beverly D., Hans R. Herren, Judi Wakhungu, and Robert T. Watson, eds. 2009. *Synthesis Report: A Synthesis of the Global and Sub-Global IAASTD Reports*. Washington, DC: Island Press.

McKie, David. 2000. *Food Agency Accused of Funding Propaganda*. Radio broadcast. Ottawa: CBC-Radio.

McLeod-Kilmurray, Heather. 2007. "*Hoffman v. Monsanto:* Courts, Class Actions, and Perceptions of the Problem of GM Drift." *Bulletin of Science, Technology and Society* 27: 188-201.

McMurtry, John. 1998. *Unequal Freedoms: The Global Market as an Ethical System.* Toronto: Garamond and Kumarian Press.

–. 2002. *Value Wars: The Global Market versus the Life Economy.* London: Pluto Press.

McNally, Ruth, and Peter Wheale. 1998. "The Consequences of Modern Genetic Engineering: Patents, 'Nomads' and the 'Bio-industrial Complex.'" In *The Social Management of Genetic Engineering,* edited by P. Wheale, R. von Schomberg, and P. Glasner, 303-30. Aldershot, UK: Ashgate.

McNaughton, Jodi. 2003. "GMO Contamination: Are GMOs Pollutants under *the Environmental Management and Protection Act?" Saskatchewan Law Review* 66: 183-216.

Mészáros, István. 2008. *The Challenge and Burden of Historical Time: Socialism in the Twenty-First Century.* New York: Monthly Review Press.

Mgbeoji, Ikechi. 2007. "Adventitious Presence of Patented Genetically Modified Organisms on Private Premises: Is Intent Necessary for Actions in Infringement against the Property Owner?" *Bulletin of Science, Technology and Society* 27: 314-21.

Midnight Notes Collective. 1992. "The New Enclosures." In *Midnight Oil: Work, Energy, War, 1973-1992,* edited by Midnight Notes Collective, 317-33. Jamaica Plains, MA: Autonomedia.

Mikkelsen, T., B. Anderson, and R. Jørgensen. 1996. "The Risk of Crop Transgene Spread." *Nature* 380: 31.

Miller, Henry I., and Gregory Conko. 2000. "The Great Biotech Sell-Out: Capitulation on UN's Biosafety Protocol Leads to Needless Regulation." *National Post,* 29 February, C7.

Millstone, Erik, Eric Brunner, and Sue Mayer. 1999. "Beyond 'Substantial Equivalence.'" *Nature* 401: 525-26.

Ministry of State for Science and Technology. 1980. *Biotechnology in Canada.* Ottawa: Ministry of State for Science and Technology.

Mitchell, W.J. Thomas. 2003. "The Work of Art in the Age of Biocybernetic Reproduction." *Modernism/Modernity* 10: 481-500.

Mittelstaedt, Martin. 2009a. "Modified Corn Seeds Sow Doubts." *Globe and Mail,* 4 August, A6.

–. 2009b. "Who Contaminated Canada's Crops? Prairie Whodunit Has Flax Farmers Baffled." *Globe and Mail,* 28 October, A1.

Monsanto. 2012a. "The Myth about Accidental Pollination." *Beyond the Rows* [weblog], 14 August. http://monsantoblog.com/.

–. 2012b. "Taking a Stand: Proposition 37, the California Labeling Proposal." *Beyond the Rows* [weblog], 14 August: http://monsantoblog.com/.

Monsanto and ISAAA's Hype Exposed, Part I. 2006. GM Watch. http://www.gmwatch.org/.

Monsanto Canada. n.d. *Seed the Techology, Harvest the Rewards: Technology Use Agreement Terms and Conditions.* Winnipeg, MB: Monsanto Canada.

Monsanto Canada: Technology Use Guide 2010. 2010. Winnipeg, MB: Monsanto Canada.

"Monsanto's Annual Form 10-K". 2011. *Annual Report to the United States Securities and Exchange Commission.* St. Louis, MO: Monsanto Company.

Mooney, Pat Roy. 1983. "The Law of the Seed: Another Development in Plant Genetic Resources." *Development Dialogue* 12: 1-176.

Moore, Oliver. 2003. "Poll Shows Huge Support for GMO Labelling." *Globe and Mail,* 3 December. http://www.theglobeandmail.com/.

Morrow, A. David, and Colin B. Ingram. 2005. "Of Transgenic Mice and Roundup Ready Canola: The Decisions of the Supreme Court of Canada in *Harvard College v. Canada* and *Monsanto v. Schmeiser." University of British Columbia Law Review* 38: 189-222.

Mortensen, David A., J. Franklin Egan, Bruce D. Maxwell, Matthew R. Ryan, and Richard G. Smith. 2012. "Navigating a Critical Juncture for Sustainable Weed Management." *BioScience* 62: 75-84.

Moser, Ingunn. 1995. "Introduction: Mobilizing Critical Communities and Discourses on Modern Biotechnology." In *Biopolitics: A Feminist and Ecological Reader on Biotechnology,* edited by V. Shiva and I. Moser, 1-24. London: Zed Books.

Mowery, David C., Richard R. Nelson, Bhaven N. Sampat, and Arvids A. Ziedonis. 1999. "The Effects of the Bayh-Dole Act on U.S. University Research and Technology Transfer." In *Industrializing Knowledge: University-Industry Linkages in Japan and the United States,* edited by L.M. Branscomb, F. Kodama, and R. Florida, 269-306. Cambridge, MA: MIT Press.

Munn-Venn, Trevor, and Paul Mitchell. 2005. *Biotechnology in Canada: A Technology Platform for Growth.* Ottawa: Conference Board of Canada.

Munro, Margaret. 2005. "Modified Wheat Takes Root with Little Opposition: Keeps Growing When Sprayed with Herbicides." *National Post,* 29 December, A8.

Murdock, Graham. 2004. "Popular Representation and Postnormal Science: The Struggle over Genetically Modified Foods." In *Biotechnology and Communication: The Meta-Technologies of Information,* edited by S. Braman, 227-59. Mahwah, NJ: Lawrence Erlbaum Associates.

National Agricultural Statistics Service. 2012. *Quick Stats.* United States Department of Agriculture. http://quickstats.nass.usda.gov/.

National Farmers Union. 2003. *Ten Reasons Why We Don't Want GM Wheat.* Saskatoon, SK: National Farmers Union.

–. 2009. *NFU Briefs Federal Agriculture Committee: Offers Chilling Analysis of Farm Crisis and Promising Solutions.* Press release. Saskatoon, SK: National Farmers Union.

–. n.d. *NFU Policy on Genetically Modified (GM) Foods.* National Farmers Union. http://www.nfu.ca/.

Navdanya. 2009. *Effect on Soil Biological Activities Due to Cultivation of Bt. Cotton.* New Delhi, India: Navdanya.

Negri, Antonio. 2005. *The Politics of Subversion: A Manifesto for the Twenty-First Century.* Translated by J. Newell. 2nd ed. Cambridge: Polity Press.

Neill, Monty, George Caffentzis, and Johnny Machete. n.d. *Toward the New Commons: Working Class Strategies and the Zapatistas.* http://archive.org/.

Nelkin, Dorothy. 1987. *Selling Science: How the Press Covers Science and Technology.* New York: W.H. Freeman.

Nelkin, Dorothy, and M. Susan Lindee. 2004. *The DNA Mystique: The Gene as a Cultural Icon.* 2nd ed. Ann Arbor, MI: University of Michigan Press.

Nestle, Marion. 2003. *Safe Food: Bacteria, Biotechnology, and Bioterrorism.* Berkeley, CA: University of California Press.

Neumann, G., S. Kohls, E. Landsberg, K. Stock-Oliveira Souza, T. Yamada, and V. Römheld. 2006. "Relevance of Glyphosate Transfer to Non-Target Plants via the Rhizosphere." *Journal of Plant Diseases and Protection* 20 (Special Issue): 963-69.

NGO Letter to Chrétien on GE Wheat. 2001. Council of Canadians. http://www.canadians.org/.

Nirenberg, Marshall. 1963. "The Genetic Code II." *Scientific American* 208: 80-94.

North, Douglass. 1990. *Institutions, Institutional Change, and Economic Performance.* Cambridge: Cambridge University Press.

Nottingham, Stephen. 2002. *Genescapes: The Ecology of Genetic Engineering.* London: Zed Books.

Nunn, Jim. 2002. "CAC and Industry Group Develop Biotech Booklet." As broadcast on *Marketplace,* 6 March. Toronto: Canadian Broadcasting Corporation.

Office of the Auditor General of Canada. 2000a. *Petition no. 23 under the Auditor General Act.* Ottawa: Government of Canada.

–. 2000b. *Response of the Federal Departments and Agencies to the Petition Filed by the Sierra Legal Defence Fund under the Auditor General Act.* Ottawa: Government of Canada.

–. 2002. *Response of the Federal Departments and Agencies to the Petition Filed January 18, 2002, by Greenpeace Canada under the Auditor General Act.* Ottawa: Government of Canada.

"'Opposition to GE Wheat Growing,' Says Canadian Wheat Board." 2004. *CropChoice News/ Canadian Newswire,* 19 March. http://www.cropchoice.com/.

Organic Agriculture Protection Fund. 2004. *Canada's Organic Farmers Tell Monsanto to Drop GE Wheat Project.* Canwood, SK: Organic Agriculture Protection Fund.

–. n.d. *Lobby to Stop GE Wheat.* Organic Agriculture Protection Fund. http://www.saskorganic.com/oapf/lobby.html.

Organisation for Economic Co-operation and Development. 2009. *The Bioeconomy to 2030: Designing a Policy Agenda.* Organisation for Economic Co-operation and Development. http://www.oecd.org/.

Orwell, George. 1987. *Nineteen Eighty-Four.* London: Secker and Warburg.

Ostlie, Ken, and Bruce Potter. 2012. "Performance Problems Surface Again with Bt Corn Rootworm Traits." *Minnesota Crop News,* University of Minnesota Extension Office.

Packer, Kathryn, and Andrew Webster. 1996. "Patenting Culture in Science: Reinventing the Scientific Wheel of Credibility." *Science, Technology, and Human Values* 21: 427-53.

Paganelli, Alejandra, Victoria Gnazzo, Helena Acosta, Silvia L. López, and Andrés E. Carrasco. 2010. "Glyphosate-Based Herbicides Produce Teratogenic Effects on Vertebrates by Impairing Retinoic Acid Signaling." *Chemical Research in Toxicology* 23: 1586-95.

Parry, Bronwyn. 2004. *Trading the Genome: Investigating the Commodification of Bio-Information.* New York: Columbia University Press.

Patrick, Robert. 2011. "Genetic Rice Lawsuit in St. Louis Settled for $750 Million." *St. Louis Post-Dispatch,* 2 July: http://www.stltoday.com/.

Patterson, Kelly. 2007. "Genetically Modified 'Zombie Seeds' Raise Environmental Concerns." *Ottawa Citizen,* 13 June, A5.

Peekhaus, Wilhelm. 2010. "The Neo-Liberal University and Agricultural Biotechnology: Reports from the Field." *Bulletin of Science, Technology and Society* 30: 415-29.

Peixoto, Francisco. 2005. "Comparative Effects of the Roundup and Glyphosate on Mitochondrial Oxidative Phosphorylation." *Chemosphere* 61: 1115-22.

Peters, Ted. 2003. *Playing God? Genetic Determinism and Human Freedom.* 2nd ed. New York: Routledge.

Phillipson, Martin. 2005. "Giving Away the Farm? The Rights and Obligations of Biotechnology Multinationals: Canadian Developments." *King's Law Journal* 16: 362-72.

Phillipson, Martin, and Marie-Anne Bowden. 1999. "Environmental Assessment and Agriculture: An Ounce of Prevention Is Worth a Pound of Manure." *Saskatchewan Law Review* 62: 415-35.

Pollack, Andrew. 2007. "U.S. Agency Violated Law in Seed Case, Judge Rules." *New York Times,* 14 February. http://www.nytimes.com/.

–. 2010. "Monsanto's Fortunes Turn Sour." *New York Times,* 5 October, B1.

Pollara Earnscliffe. 2003. *Public Opinion Research into Biotechnology Issues in the United States and Canada: Eighth Wave.* Ottawa: Pollara Research and Earnscliffe Research and Communications.

Powell, Kendall. 2003. "Concerns over Refuge Size for US EPA-approved Bt Corn." *Nature Biotechnology* 21: 467-68.

Powles, Stephen B. 2008. "Evolved Glyphosate-Resistant Weeds around the World: Lessons to Be Learnt." *Pest Management Science* 64: 360-65.

–. 2010. "Gene Amplification Delivers Glyphosate-Resistant Weed Evolution." *Proceedings of the National Academy of Sciences* 107(3): 955-56.

Powles, Stephen B., and Christopher Preston. 2006. "Evolved Glyphosate Resistance in Plants: Biochemical and Genetic Basis of Resistance." *Weed Technology* 20: 282-89.

Pratt, Sean. 2010. "Flax Tests Show GM Contamination Widespread." *Western Producer,* 19 August. http://www.producer.com/.

Prudham, Scott. 2007. "The Fictions of Autonomous Invention: Accumulation by Dispossession, Commodification and Life Patents in Canada." *Antipode* 39: 406-29.

Pryme, Ian F., and Rolf Lembcke. 2003. "*In vivo* Studies on Possible Health Consequences of Genetically Modified Food and Feed – with Particular Regard to Ingredients Consisting of Genetically Modified Plant Materials." *Nutrition and Health* 17: 1-8.

Reidenberg, Joel R. 1998. "Lex informatica: The Formulation of Information Policy Rules through Technology." *Texas Law Review* 76: 553-93.

Relyea, Rick A. 2005. "The Lethal Impact of Roundup on Aquatic and Terrestrial Amphibians." *Ecological Applications* 15: 1118-24.

Richard, Sophie, Safa Moslemi, Herbert Sipahutar, Nora Benachour, and Gilles-Eric Séralini. 2005. "Differential Effects of Glyphosate and Roundup on Human Placental Cells and Aromatase." *Environmental Health Perspectives* 113: 716-20.

Rifkin, Jeremy. 1998. *The Biotech Century: Harnessing the Gene and Remaking the World*. New York: Jeremy P. Tarcher/Putnam.

Riley, Susan. 1999. "Seeds of Discontent Blowing Past Cabinet." *Ottawa Citizen,* 27 October, A14.

Robbins-Roth, Cynthia. 2000. *From Alchemy to IPO: The Business of Biotechnology.* Cambridge, MA: Perseus.

Robins, Kevin, and Frank Webster. 1999. *Times of the Technoculture: From the Information Society to the Virtual Life*. London: Routledge.

Rochon, Thomas R. 1998. *Culture Moves: Ideas, Activism, and Changing Values*. Princeton, NJ: Princeton University Press.

Rodgers, Christopher P. 2003. "Liability for the Release of GMOs into the Environment: Exploring the Boundaries of Nuisance." *Cambridge Law Journal* 62: 371-402.

Romahn, Jim. 2007. "Grower Fined in RR Patent Case." *Ontario Farmer,* 18 December, B18.

Rose, Steven. 1997. *Lifelines: Biology beyond Determinism*. Oxford: Oxford University Press.

Roseboro, Ken. 2010. "USDA Stance on GM Alfalfa Threatens 'Fabric of Organic Industry.'" *Organic and Non-GMO Report,* February. http://www.non-gmoreport.com/.

Rosi-Marshall, E.J., J.L. Tank, T.V. Royer, M.R. Whiles, M. Evans-White, C. Chambers, N.A. Griffiths, J. Pokelsek, and M.L. Stephen. 2007. "Toxins in Transgenic Crop Byproducts May Affect Headwater Stream Ecosystems." *Proceedings of the National Academy of Sciences* 104: 16204-08.

Rossiter, Ned. 2006. *Organized Networks: Media Theory, Creative Labour, New Institutions*. Rotterdam, The Netherlands: NAI Publishers.

Routledge, Paul. 2004. "Convergence of Commons: Process Geographies of People's Global Action." *The Commoner* 8 (Autumn-Winter). http://www.commoner.org.uk/08routledge.pdf.

Royal Society of Canada. 2001. *Elements of Precaution: Recommendations for the Regulation of Food Biotechnology in Canada*. Ottawa: Royal Society of Canada.

Ruivenkamp, Guido. 2005. "Tailor-Made Biotechnologies: Between Bio-Power and Sub-Politics." *Tailoring Biotechnologies* 1(1): 11-33.

"Russia to Spend 150 Billion Roubles ($5.25 Billion) on Biotechnology Next Decade." 2006. *Biotechnology Journal* 1: 23.

Sagstad, A., M. Sanden, Ø. Haugland, A.-C. Hansen, P.A. Olsvik, and G.-I. Hemre. 2007. "Evaluation of Stress- and Immune-Response Biomarkers in Atlantic Salmon, Salmo salar L., Fed Different Levels of Genetically Modified Maize (Bt Maize), Compared with Its Near-Isogenic Parental Line and a Commercial Suprex Maize." *Journal of Fish Diseases* 30: 201-12.

Salter, Liora. 1993. "Capture or Co-Management: Democracy and Accountability in Regulatory Agencies." In *A Different Kind of State? Popular Power and Democratic Administration,* edited by G. Albo, D. Lagille and L. Panich, 87-100. Toronto: Oxford University Press.

Sammons, R. Douglas, David C. Heering, Natalie Dinicola, Harvey Glick, and Greg A. Elmore. 2009. "Sustainability and Stewardship of Glyphosate and Glyphosate-Resistant Crops." *Weed Technology* 21: 347-54.

Saskatchewan Organic Directorate. 2006. *Position Paper on the Introduction of Genetically Modified Alfalfa*. Saskatchewan Organic Directorate. http://www.saskorganic.com/oapf/pdf/SOD_GMO_Alfalfa_Position_Paper.pdf.

–. 2010. *Comments Submitted to the USDA APHIS re: Glyphosate-tolerant Alfalfa Events J101 and J163: Request for NonRegulated Status, Draft Environmental Impact Statement – November 2009*. Saskatchewan Organic Directorate.

Sasu, Miruna A., Matthew J. Ferrari, Daolin Du, James A. Winsor, and Andrew G. Stephenson. 2009. "Indirect Costs of a Nontarget Pathogen Mitigate the Direct Benefits of a Virus-Resistant Transgene in Wild *Cucurbita*." *Proceedings of the National Academy of Sciences* 106: 19067-71.

Scassa, Teresa. 2003. "A Mouse Is a Mouse Is a Mouse: A Comment on the Supreme Court of Canada's Decision on the Harvard Mouse Patent." *Oxford University Commonwealth Law Journal* 3: 105-17.

Schiller, Dan. 2007. *How to Think about Information*. Chicago: University of Illinois Press.

Schimmelpfennig, David E., Carl E. Pray, and Margaret F. Brennan. 2004. "The Impact of Seed Industry Concentration on Innovation: A Study of US Biotech Market Leaders." *Agricultural Economics* 30: 157-67.

Schmidt, Sarah. 2010. "Developer of Genetically Engineered Salmon Eyes Canadian Regulators." *Vancouver Sun,* 28 August, B2.

–. 2011. "GE Fish May Pose Risks to Wild Stock: Documents." *Vancouver Sun,* 17 October, B2.

–. 2012a. "Plans for First GE Pig for Human Consumption Now in Limbo." *Vancouver Sun,* 2 April: http://www.vancouversun.com/.

–. 2012b. "U.S. OK Could See GE Fish Enter Canada." *Leader-Post,* 5 January, C7.

Schmitz, Sonja A. 2001. "Cloning Profits: The Revolution in Agricultural Biotechnology." In *Redesigning Life? The Worldwide Challenge to Genetic Engineering,* edited by B. Tokar, 44-50. Montreal and Kingston: McGill-Queen's University Press.

Schubert, David R. 2008. "The Problem with Nutritionally Enhanced Plants." *Journal of Medicinal Food* 11: 601-05.

Schurman, Rachel A., and Dennis Doyle Takahashi Kelso, eds. 2003. *Engineering Trouble: Biotechnology and Its Discontents.* Berkeley, CA: University of California Press.

"A Seedy Practice." 2009. *Scientific American* 301(2): 28.

Séralini, Gilles-Eric, Dominique Cellier, and Joël Spiroux de Vendômois. 2007a. "New Analysis of a Rat Feeding Study with a Genetically Modified Maize Reveals Signs of Hepatorenal Toxicity." *Archives of Environmental Contamination and Toxicity* 52: 596-602.

–. 2007b. *Report on NK 603 GM Maize Produced by Monsanto Company: Controversial Effects on Health Reported after Subchronic Toxicity Test: 90-Day Study Feeding Rats.* Paris, France: Committee for Independent Research and Information on Genetic Engineering.

Séralini, Gilles-Eric, Emilie Clair, Robin Mesnage, Steeve Gress, Nicolas Defarge, Manuela Malatesta, Didier Hennequin, and Joël Spiroux de Vendômois. 2012. "Long-Term Toxicity of a Roundup Herbicide and a Roundup-Tolerant Genetically Modified Maize." *Food and Chemical Toxicology* 50: 4221-31.

Séralini, Gilles-Eric, Joël Spiroux de Vendômois, Dominique Cellier, Charles Sultan, Marcello Buiatti, Lou Gallagher, Michael Antoniou, and Krishna R. Dronamraju. 2009. "How Subchronic and Chronic Health Effects Can Be Neglected for GMOs, Pesticides or Chemicals." *International Journal of Biological Sciences* 5: 438-43.

Shand, Hope. 2005. "New Enclosures: Why Civil Society and Governments Should Look beyond Life Patents." In *Rights and Liberties in the Biotech Age: Why We Need a Genetic Bill of Rights,* edited by S. Krimsky and P. Shorett, 40-48. Lanham, MD: Rowman and Littlefield.

Shannon, Grover. 2008. *Conventional Soybeans Offer High Yields at Lower Cost.* University of Missouri. http://agebb.missouri.edu/news/.

Shapiro, Robert B. 1999. *Open Letter from Monsanto CEO Robert B. Shapiro to Rockefeller Foundation President Gordon Conway.* http://www.monsanto.com/.

Sharratt, Lucy. 2001. "No to Bovine Growth Hormone: Ten Years of Resistance in Canada." In *Redesigning Life? The Worldwide Challenge to Genetic Engineering,* edited by B. Tokar, 385-96. Montreal and Kingston: McGill-Queen's University Press.

–. 2002. *Regulating Genetic Engineering for Profit: A Guide to Corporate Power and Canada's Regulation of Genetically Engineered Foods.* Ottawa: Polaris Institute.

–. 2010. "Bill C-474: Concrete Change to Challenge Monsanto's GM Crops." *Peace and Environment News* (September-October): http://www.cban.ca/.

Shiva, Vandana. 1997. *Biopiracy: The Plunder of Nature and Knowledge.* Boston: South End Press.

–. 2000. "The World on the Edge." In *Global Capitalism,* edited by W. Hutton and A. Giddens, 112-29. New York: New Press.

–. 2001a. "Biopiracy: The Theft of Knowledge and Resources." In *Redesigning Life? The Worldwide Challenge to Genetic Engineering,* edited by B. Tokar, 283-89. Montreal and Kingston: McGill-Queen's University Press.

–. 2001b. *Protect or Plunder? Understanding Intellectual Property Rights.* New York: Zed Books.

Sider, Alison. 2010. "Bayer Ordered to Pay Farmer: $1 Million Is Tab for Modified Rice." *Arkansas Democratic Gazette,* 10 March, 29.

Sierra Club of Canada. n.d. *Genetically Engineered Wheat Fact Sheet.* Ottawa: Sierra Club Canada.

Simpson, Greg, and Kees de Lange. 2004. *Nutritional Strategies to Decrease Nutrients in Swine Manure.* Toronto: Ontario Ministry of Agriculture, Food and Rural Affairs.

Smith, Jeffrey M. 2003. *Seeds of Deception: Exposing Industry and Government Lies about the Safety of the Genetically Engineered Foods You're Eating.* Fairfield, IA: Yes! Books.

–. 2007. *Genetic Roulette: The Documented Health Risks of Genetically Engineered Foods.* Fairfield, IA: Yes! Books.

Smyth, Stuart, George G. Khachatourians, and Peter W.B. Phillips. 2002. "Liabilities and Economics of Transgenic Crops." *Nature Biotechnology* 20: 537-41.

Snow, A.A., D.A. Andow, P. Gepts, E.M. Hallerman, A. Power, J.M. Tiedje, and L.L. Wolfenbarger. 2005. "Genetically Engineered Organisms and the Environment: Current Status and Recommendations." *Ecological Applications* 15: 377-404.

Spears, John. 2001. "Monsanto Pulls Altered Potatoes in Wake of Consumer Resistance." *Toronto Star,* 23 March, B1.

Standing Senate Committee on Agriculture and Forestry. 2008. *"Growing" Costs for Canadian Farmers.* Ottawa: Senate of Canada.

State of Market Acceptance and Non-Acceptance of Genetically Engineered Wheat. 2004. Council of Canadians. http://www.canadians.org/.

Statistics Canada. 1998. "Biotechnology Scientific Activities in Selected Federal Government Departments and Agencies, 1997-98." *Science Statistics* 22(4). http://www5.statcan.gc.ca/.

–. 2000. *Federal Government Personnel Engaged in Scientific and Technological (S&T) Activities, 1990-1991 to 1999-2000.* Ottawa: Government of Canada.

–. 2001. "Biotechnology Scientific Activities in Selected Federal Government Departments and Agencies, 1999-2000." *Science Statistics* 25(3). http://www5.statcan.gc.ca/.

–. 2002. "Biotechnology Scientific Activities in Selected Federal Government Departments, and Agencies, 2000-2001." *Science Statistics* 26(2). http://www5.statcan.gc.ca/.

–. 2003. "Biotechnology Scientific Activities in Selected Federal Government Departments, and Agencies, 2001-2002." *Science Statistics* 27(1). http://www5.statcan.gc.ca/.

–. 2004. "Biotechnology Scientific Activities in Selected Federal Government Departments, and Agencies, 2002-2003." *Science Statistics* 28(7). http://www5.statcan.gc.ca/.

–. 2005. "Biotechnology Scientific Activities in Selected Federal Government Departments, and Agencies, 2003-2004." *Science Statistics* 29(3). http://www5.statcan.gc.ca/.

–. 2010. "Biotechnology Scientific Activities in Selected Federal Government Departments, and Agencies, 2008-2009." *Science Statistics* 34(2): http://www5.statcan.gc.ca/.

–. 2011. *Field Crop Reporting Series: November Estimate of Production of Principal Field Crops.* Ottawa: Government of Canada.

Steinbrecher, Ricarda A. 2001. "Ecological Consequences of Genetic Engineering." In *Redesigning Life? The Worldwide Challenge to Genetic Engineering,* edited by B. Tokar, 75-102. Montreal and Kingston: McGill-Queens's University Press.

Stewart, Lyle. 2000. "Food Agency Accused of Funding Propaganda." *Montreal Gazette,* 2 April, B3.

–. 2001. "Who Can You Trust? There's a Too-Cozy Relationship between Food Regulators and Producers." *Montreal Gazette,* 20 July, B3.

–. 2002a. "Canada's GM Food Fight." *Montreal Gazette,* 29 March, B3.

–. 2002b. "GM Food Meeting Left Bad Taste." *Montreal Gazette,* 1 March, B3.

–. 2002c. "Good PR Is Growing." *THIS Magazine,* May-June. http://healthcoalition.ca/wp-content/uploads/2010/02/This-Magazine-GM-Food.pdf.

–. 2002d. "A Question of Credibility." *Montreal Gazette,* 24 May, B3.

Storer, Nicholas P., Jonathan M. Babcock, Michele Schlenz, Thomas Meade, Gary D. Thompson, James W. Bing, and Randy M. Huckaba. 2010. "Discovery and Characterization of Field Resistance to Bt Maize: *Spodoptera frugiperda* (Lepidoptera: Noctuidae) in Puerto Rico." *Journal of Economic Entomology* 103: 1031-38.

Sunder Rajan, Kaushik. 2006. *Biocapital: The Constitution of Postgenomic Life.* Durham, NC: Duke University Press.

Szarek, J., A. Siwicki, A. Andrzejewska, E. Terech-Majewska, and T. Banaszkiewicz. 2000. "Effects of the Herbicide Roundup™ on the Ultrastructural Pattern of Hepatocytes in Carp (*Cyprinus carpio*)." *Marine Environmental Research* 50: 263-66.

Tabashnik, Bruce E., Aaron J. Gassmann, David W. Crowder, and Yves Carriére. 2008. "Insect Resistance to Bt Crops: Evidence versus Theory." *Nature Biotechnology* 26: 199-202.

Tabashnik, Bruce E., Gopalan C. Unnithan, Luke Masson, David W. Crowder, Xianchun Li, and Yves Carrière. 2009. "Asymmetrical Cross-Resistance between *Bacillus thuringiensis* Toxins Cry1Ac and Cry2Ab in Pink Bollworm." *Proceedings of the National Academy of Sciences*, 6 July. http://www.pnas.org/.

Tabashnik, Bruce E., J.B.J. Van Rensburg, and Yves Carrière. 2009. "Field-Evolved Insect Resistance to Bt Crops: Definition, Theory, and Data." *Journal of Economic Entomology* 106: 11889-94.

Tam, Pauline. 1999. "Government Fast-Tracked Monsanto's GM Potatoes: Private Deal Struck Quietly to Speed up Regulatory System." *Ottawa Citizen*, 30 November, A1.

Tanaka, Keiko, Arunas Juska, and Lawrence Busch. 1999. "Globalization of Agricultural Production and Research: The Case of the Rapeseed Subsector." *Sociologia Ruralis* 39: 54-77.

Tank, Jennifer L., Emma J. Rosi-Marshall, Todd V. Royer, Matt R. Whiles, Natalie A. Griffiths, Therese C. Frauendorf, and David J. Treering. 2010. "Occurrence of Maize Detritus and a Transgenic Insecticidal Protein (Cry1Ab) within the Stream Network of an Agricultural Landscape." *Proceedings of the National Academy of Sciences* 107: 17645-50.

Tansey, James. 2003. *The Prospects for Governing Biotechnology in Canada*. Vancouver: W. Maurice Young Centre for Applied Ethics, University of British Columbia.

Task Force on Biotechnology. 1981. *Biotechnology: A Development Plan for Canada*. Report of the Task Force on Biotechnology to the Minister of State for Science and Technology. Ottawa: Task Force on Biotechnology.

Tesfamariam, Tsehaye, S. Bott, I. Cakmak, V. Römheld, and G. Neumann. 2009. "Glyphosate in the Rhizosphere: Role of Waiting Times and Different Glyphosate Binding Forms in Soils for Phytotoxicity to Non-Target Plants." *European Journal of Agronomy* 31: 126-32.

Thacker, Eugene. 2005. *The Global Genome: Biotechnology, Politics, and Culture*. Cambridge, MA: MIT Press.

Then, Christoph. 2010. *Agro-Biotechnology: New Plant Pest Caused by Genetically Engineered Corn*. Munich: Testbiotech e.V.

Then, Christoph, and Andreas Bauer-Panskus. 2011. *Industrie und Europäische Lebensmittelbehörde EFSA untergraben Risikoabschätzung beim gentechnisch veränderten Mais SmartStax*. Munich, Germany: Testbiotech.

Then, Christoph, and Antje Lorch. 2009. *Schadensbericht Gentechnik*. Berlin, Germany: Bund Ökologische Lebensmittelwirtschaft.

Thirsk, Joan. 1967. "Enclosing and Engrossing." In *The Agrarian History of England and Wales*, edited by J. Thirsk, 200-55. Cambridge: Cambridge University Press.

Thurow, Lester C. 1997. "Needed: A New System of Intellectual Property Rights." *Harvard Business Review* 75(5): 95-105.

Tomich, Jeffrey. 2009. "Monsanto Trims 900 Jobs as Roundup Sales Plunge." *St. Louis Post-Dispatch*, 25 June, A9.

–. 2010. "Roundup Woes Spur More Monsanto Cuts." *St. Louis Post-Dispatch*, 1 September, A10.

Treasury Board of Canada Secretariat. 2006. *Canadian Biotechnology Strategy (CBS)*. Treasury Board of Canada Secretariat. http://www.tbs-sct.gc.ca/.

Tronti, Mario. 1979. "The Strategy of the Refusal." In *Working Class Autonomy and the Crisis: Italian Marxist Texts of the Theory and Practice of a Class Movement: 1964-79*, edited by Red Notes and CSE Books, 7-21. London: Red Notes and CSE Books.

Trosow, Samuel E. 2003. "The Illusive Search for Justificatory Theories: Copyright, Commodification and Capital." *Canadian Journal of Law and Jurisprudence* 16: 217-41.

Turner, Michael. 1984. *Enclosures in Britain, 1750-1830*. London: Macmillan.

"U of G Researchers Find Suspected Glyphosate-Resistant Weed." 2009. University of Guelph. http://www.uoguelph.ca/news/.

US Food and Drug Administration. 2010. *Environmental Assessment for AquAdvantage®* *Salmon.* Washington, DC: Government of the United States.

USA Rice Federation. 2006. *US Rice Export Markets Impacted by the Presence of LLrice601.* USA Rice Federation. http://www.usarice.com/industry/communication/exportimpact.pdf.

Van Acker, R.C., A.L. Brûlé-Babel, and L.F. Friesen. 2003. *An Environmental Safety Assessment of Roundup Ready® Wheat: Risks for Direct Seeding Systems in Western Canada.* Winnipeg: Canadian Wheat Board.

van Rensburg, J.B.J. 2007. "First Report of Field Resistance by the Stem Borer, *Busseola fusca* (Fuller) to Bt-transgenic Maize." *South African Journal of Plant Soil* 24: 147-51.

van Wijk, Jeroen. 1995. "Broad Biotechnology Patents Hamper Innovation." *Biotechnology and Development Monitor* 25: 15-17.

Vaver, David. 1991. "Some Agnostic Observations on Intellectual Property." *Intellectual Property Journal* 6: 125-53.

Venditti, Pat. 2004. "Three Dangerous Little Pigs." *Globe and Mail,* 20 February, A21.

Verzola, Roberto. 2000. "Cyberlords: The Rentier Class of the Information Sector." In *Readme! Filtered by Nettime: ASCII Culture and the Revenge of Knowledge,* edited by J. Bosma, 91-97. New York: Autonomedia.

Vlassara, Helen, Weijing Cai, Susan Goodman, Renata Pyzik, Angie Yong, Xue Chen, Li Zhu, Tina Neade, Michal Beeri, Jeremy M. Silverman, Luigi Ferrucci, Laurie Tansman, Gary E. Striker, and Jaime Uribarri. 2009. "Protection against Loss of Innate Defenses in Adulthood by Low Advanced Glycation End Products (AGE) Intake: Role of the Antiinflammatory AGE Receptor-1." *Journal of Clinical Endocrinology and Metabolism* 94: 4483-91.

Voigt, Christopher A. 2008. "Life from Information." *Nature Methods* 5: 27-28.

Volkmann, Kelsey. 2010. "West Virginia Sues Monsanto over Soybean Prices." *St. Louis Business Journal,* 25 October. http://www.bizjournals.com/.

von Schomberg, René. 1998. "Democratising the Policy Process for the Environmental Release of Genetically Engineered Organisms." In *The Social Management of Genetic Engineering,* edited by P. Wheale, R. von Schomberg, and P. Glasner, 237-48. Aldershot, UK: Ashgate.

Walsh, Lance P., Chad McCormick, Clyde Martin, and Douglas M. Stocco. 2000. "Roundup Inhibits Steroidogenesis by Disrupting Steroidogenic Acute Regulatory (StAR) Protein Expression." *Environmental Health Perspectives* 108: 769-76.

Walton, John, and David Seddon. 1994. *Free Markets and Food Riots: The Politics of Global Adjustment.* Cambridge, MA: Blackwell.

Wang, Shenghui, David Just, and Per Pinstrup-Andersen. 2006. "Tarnishing Silver Bullets: Bt Technology Adoption, Bounded Rationality and the Outbreak of Secondary Pest Infestations in China." Paper presented at the Annual Conference of the American Agricultural Economics Association, "Envisioning the Future," Long Beach, California, 23-23 July. http://ageconsearch.umn.edu/bitstream/21230/1/sp06wa07.pdf.

Ward, Ian. 1995. *Politics of the Media.* South Melbourne, Australia: Macmillan.

Warick, Jason. 2003. "Lining up against GM Wheat." *Star Phoenix,* 9 August, E1.

Warwick, Suzanne I., Anne Légère, Marie-Josée Simard, and Tracey James. 2008. "Do Escaped Transgenes Persist in Nature? The Case of an Herbicide Resistance Transgene in a Weedy *Brassica rapa* Population." *Molecular Ecology* 17: 1387-95.

Wesselius, Erik. 2002. *Behind GATS 2000: Corporate Power at Work.* Amsterdam, The Netherlands: Transnational Institute.

Wheale, Peter, and Ruth McNally. 1998. "The Social Management of Genetic Engineering: An Introduction." In *The Social Management of Genetic Engineering,* edited by P. Wheale, R. von Schomberg, and P. Glasner, 1-27. Aldershot, UK: Ashgate.

Wheat Biotechnology Commercialization: Statement of Canadian, American and Australian Wheat Organizations. 2009. US Wheat. http://www.uswheat.org/.

White, Tiffany. 2001. "'Get out of My Lab, Lois!' In Search of the Media Gene." In *Altered Genes II: The Future?* edited by R. Hindmarsh and G. Lawrence, 69-82. Melbourne, Australia: Scribe Publications.

Who "Created" and Owns "Enviropig™"? n.d. Canadian Biotechnology Action Network. http://www.cban.ca/.

"Why Does Monsanto Sue Farmers Who Save Seeds?" n.d. Monsanto. http://www.monsanto. com/.

Willison, Donald J., and Stuart M. MacLeod. 2002. "Patenting of Genetic Material: Are the Benefits to Society Being Realized?" *Canadian Medical Association Journal* 167: 259-62.

Wills, Peter R. 2001. "Disrupting Evolution: Biotechnology's Real Result." In *Altered Genes II: The Future?* edited by R. Hindmarsh and G. Lawrence, 53-68. Melbourne, Australia: Scribe Publications.

Wilson, Allison K., Jonathan R. Latham, and Ricarda A. Steinbrecher. 2006. "Transformation-Induced Mutations in Transgenic Plants: Analysis and Biosafety Implications." *Biotechnology and Genetic Engineering Reviews* 23: 209-34.

Wilson, Barry. 2002. "Canada Afraid to Upset U.S. with GM Labels." *Western Producer*, 21 November. http://www.producer.com/.

–. 2003. "Market 'Risk' Once Part of Process." *Western Producer*, 10 April, A1.

The World According to Monsanto (DVD). 2008. Directed by Marie-Monique Robin. Paris: Image and Compagnie.

Wynne, Brian. 1995. "Public Understanding of Science." In *Handbook of Science and Technology Studies,* edited by S. Jasanoff, G.E. Markle, J.C. Petersen, and T. Pinch, 361-88. Thousand Oaks, CA: Sage Publications.

Yamada, Tsuioshi, Robert J. Kremer, Paulo Roberto de Camargo e Castro, and Bruce W. Wood. 2009. "Glyphosate Interactions with Physiology, Nutrition, and Diseases of Plants: Threat to Agricultural Sustainability?" *European Journal of Agronomy* 31: 111-13.

Yelling, J.A. 1977. *Common Field and Enclosure in England, 1450-1850.* Hamden, CT: Archon Books.

Young, Bob. 1985. "Is Nature a Labour Process?" In *Science, Technology and the Labour Process: Marxist Studies,* edited by L. Levidow and B. Young, 206-32. Atlantic Highlands, NJ: Humanities Press.

Yoxen, Edward. 1981. "Life as a Productive Force: Capitalising the Science and Technology of Molecular Biology." In *Science, Technology and the Labour Process: Marxist Studies,* edited by L. Levidow and B. Young, 66-122. Atlantic Highlands, NJ: Humanities Press.

–. 1983. *The Gene Business: Who Should Control Biotechnology?* New York: Harper and Row.

Zakreski, Terry. 2007. Memorandum of Argument, *Supreme Court Act,* s. 40(1).

Zeller, Simon L., Olena Kalinina, Susanne Brunner, Beat Keller, and Bernhard Schmid. 2010. "Transgene x Environment Interactions in Genetically Modified Wheat." *PLoS ONE* 5(7): e11405.

Zobiole, Luiz H. S., Rubem S. Oliveira, Jesui V. Visentainer, Robert J. Kremer, Nacer Bellaloui, and Tsuioshi Yamada. 2010. "Glyphosate Affects Seed Composition in Glyphosate-Resistant Soybean." *Journal of Agricultural and Food Chemistry* 58: 4517-22.

Zolla, Lello, Sara Rinalducci, Paolo Antonioli, and Pier Giorgio Righetti. 2008. "Proteomics as a Complementary Tool for Identifying Unintended Side Effects Occurring in Transgenic Maize Seeds as a Result of Genetic Modifications." *Journal of Proteome Research* 7: 1850-61.

Zweiger, Gary. 2001. *Transducing the Genome: Information, Anarchy, and Revolution in the Biomedical Sciences.* New York: McGraw-Hill.

Index

Note: "(t)" after a page number indicates an illustration.

Printed and bound in Canada by Friesens

Set in Stone by Artegraphica Design Co. Ltd.

Cartographer: Eric Leinberger

Copy editor: Stacy Belden

Proofreader: Lana Okerlund